MULTILEVEL ANALYSIS

MULTILEVEL ANALYSIS

An introduction to
basic and advanced multilevel modeling

Tom A. B. Snijders and Roel J. Bosker

SAGE Publications
London • Thousand Oaks • New Delhi

First published 1999

SAGE Publications Ltd
6 Bonhill Street
London EC2A 4PU

SAGE Publications Inc
2455 Teller Road
Thousand Oaks, California 91320

SAGE Publications India Pvt Ltd
32, M-Block Market
Greater Kailash - I
New Delhi 110 048

British Library Cataloguing in Publication data

A catalogue record for this book is
available from the British Library

ISBN 0-7619-5889-4
ISBN 0-7619-5890-8 (pbk)

Library of Congress catalog record available

Printed in Great Britain by The Cromwell Press Ltd, Trowbridge, Wiltshire

Contents

Preface

This book grew out of our teaching and consultation activities in the domain of multilevel analysis. It is intended as well for the absolute beginner in this field as for those who have already mastered the fundamentals and are now entering more complicated areas of application. The reader is referred to Section 1.2 for an overview of this book and for some reading guidelines.

We are grateful to various people from whom we got reactions on earlier parts of this manuscript and also to the students who were exposed to it and helped us realize what was unclear. We received useful comments and benefited from discussions about parts of the manuscript with, among others, Joerg Blasius, Marijtje van Duijn, Wolfgang Langer, Ralf Maslowski, and Ian Plewis. Moreover we would like to thank Hennie Brandsma, Mieke Brekelmans, Jan van Damme, Hetty Dekkers, Miranda Lubbers, Lyset Rekers-Mombarg and Jan Maarten Wit, Carolina de Weerth, Beate Völker, Ger van der Werf, and the Zentral Archiv (Cologne) who kindly permitted us to use data from their respective research projects as illustrative material for this book. We would also like to thank Annelies Verstappen-Remmers for her unfailing secretarial assistance.

Tom Snijders
Roel Bosker
June, 1999

1 Introduction

1.1 Multilevel analysis

Multilevel analysis is a methodology for the analysis of data with complex patterns of variability, with a focus on nested sources of variability: e.g., pupils in classes, employees in firms, suspects tried by judges in courts, animals in litters, longitudinal measurements of subjects, etc. In the analysis of such data, it usually is illuminating to take account of the variability associated with each level of nesting. There is variability, e.g., between pupils but also between classes, and one may draw wrong conclusions if either of these sources of variability is ignored. Multilevel analysis is an approach to the analysis of such data including the statistical techniques as well as the methodology of how to use these. The name of multilevel analysis is used mainly in the social sciences (in the wide sense: sociology, education, psychology, economics, criminology, etc.), but also in other fields such as the bio-medical sciences. Our focus will be on the social sciences.

In its present form, multilevel analysis is a stream which has two tributaries: contextual analysis and mixed effects models.

Contextual analysis is a development in the social sciences which has focused on the effects of the social context on individual behavior. Some landmarks before 1980 were the paper by Robinson (1950) who discussed the ecological fallacy (which refers to confusion between aggregate and individual effects), the paper by Davis, Spaeth, and Huson (1961) about the distinction between within-group and between-group regression, the volume edited by Dogan and Rokkan (1969), and the paper by Burstein, Linn, and Capell (1978) about treating regression intercepts and slopes on one level as outcomes on the higher level.

Mixed effects models are statistical models in the analysis of variance and in regression analysis where it is assumed that some of the coefficients are fixed and others are random. This subject is too vast even to mention some landmarks. The standard reference book on random effects models and mixed effects models is Searle, Casella, and McCulloch (1992), who give an extensive historical overview in their Chapter 2. The name 'mixed model' seems to have been used first by Eisenhart (1947).

Contextual modeling until about 1980 focused on the definition of appropriate variables to be used in ordinary least squares regression analysis. The main focus in the development of statistical procedures for mixed models was until the 1980s on random effects (i.e., random differences between classes in some classification system) more than on random coefficients (i.e., random effects of numerical variables). Multilevel analysis as we now know it was formed by these two streams coming together. It was realized that in contextual modeling, the individual and the context are distinct sources of variability, which should both be modeled as random influences. On the other hand, statistical methods and algorithms were developed that allowed the practical use of regression-type models with nested random coefficients. There was a cascade of statistical papers: Aitkin, Anderson, and Hinde (1981); Laird and Ware (1982); Mason, Wong, and Entwisle (1983); Goldstein (1986); Aitkin and Longford (1986); Raudenbush and Bryk (1986); De Leeuw and Kreft (1986), and Longford (1987) proposed and developed techniques for calculating estimates for mixed models with nested coefficients. These techniques, together with the programs implementing them which were developed by a number of these researchers or under their supervision, allowed the practical use of models of which until that moment only special cases were accessible for practical use. By 1986 the basis of multilevel analysis was established, many further elaborations have been developed since then, and the methodology has proved to be quite fruitful for applications. On the organizational side, the 'Multilevel Models Project' in London stimulates developments by its Newsletter and its web site http://www.ioe.ac.uk/multilevel/ with the mirror web sites http://www.medent.umontreal.ca/multilevel/ and also http://www.edfac.unimelb.edu.au/multilevel/.

In the biomedical sciences mixed models were proposed especially for longitudinal data; in economics mainly for panel data (Swamy, 1971), the most common longitudinal data in economics. One of the issues treated in the economic literature was the pooling of cross-sectional and time series data (e.g., Madalla, 1971 and Hausman and Taylor, 1981), which is closely related to the difference between within-group and between-group regressions. Overviews are given by Chow (1984) and Baltagi (1995).

A more elaborate history of multilevel analysis is presented in the bibliographical sections of Longford (1993a) and in Kreft and de Leeuw (1998). For an extensive bibliography, see Hüttner and van den Eeden (1995).

1.1.1 Probability models

The main statistical model of multilevel analysis is the hierarchical linear model, an extension of the multiple linear regression model to a model that includes nested random coefficients. This model is explained in Chapter 5 and forms the basis of most of this book.

There are several ways to argue why it makes sense to use a probability model for data analysis. In sampling theory a distinction is made between *design-based inference* and *model-based inference* (see, e.g., Särndal, Swensson, and Wretman, 1991). The former means that the researcher draws a probability sample from some finite population, and wishes to make inferences from the sample to this finite population. The probability model then follows from how the sample is drawn by the researcher. Model-based inference means that the researcher postulates a probability model, usually aiming at inference to some large and sometimes hypothetical population like all English primary school pupils in the 1990s or all human adults living in a present-day industrialized culture. If the probability model is adequate then so are the inferences based on it, but checking this adequacy is possible only to a limited extent.

It is possible to apply model-based inference to data collected by investigating some entire research population, like all twelve-year-old pupils in Amsterdam at a given moment. Sometimes the question is posed why one should use a probability model if no sample is drawn but an entire population is observed. Using a probability model that assumes statistical variability, even though an entire research population was investigated, can be justified by realizing that conclusions are sought which apply not only to the investigated research population but to a wider population. The investigated research population is supposed to be representative for this wider population – for pupils also in earlier or later years, in other towns, maybe in other countries. Applicability to such a wider population is not automatic, but has to be carefully argued by considering whether indeed the research population may be considered to be representative for the larger (often vaguely outlined) population. The inference then is not primarily about a given delimited set of individuals but about social, behavioral, biological, etc., mechanisms and processes. The random effects, or residuals, playing a role in such probability models can be regarded as the resultants of the factors that are not included in the explanatory variables used. They reflect the approximating nature of the model used. The model-based inference will be adequate to the extent that the assumptions of the probability model are an adequate reflection of the effects that are not explicitly included by means of observed variables.

As we shall see in Chapters 3, 4, and 5, the basic idea of multilevel analysis is that data sets with a nesting structure that includes unexplained variability at each level of nesting, such as pupils in classes or employees in firms, are usually not adequately represented by the probability model of multiple linear regression analysis, but are often adequately represented by the hierarchical linear model. Thus, the use of the hierarchical linear model in multilevel analysis is in the tradition of model-based inference.

1.2 This book

This book is meant as an introductory textbook and as a reference book for practical users of multilevel analysis. We have tried to include all the main points that come up when applying multilevel analysis. Some of the data sets used in the examples, and corresponding commands to run the examples in the computer packages MLn/MLwiN and HLM (see Chapter 15), are available at the web site http://stat.gamma.rug.nl/snijders/multilevel.htm.

After this introductory chapter, the book proceeds with a conceptual chapter about multilevel questions and a chapter about ways for treating multilevel data that are not based on the hierarchical linear model. Chapters 4 to 6 treat the basic conceptual ideas of the hierarchical linear model, and how to work with it in practice. Chapter 4 introduces the random intercept model as the primary example of the hierarchical linear model. This is extended in Chapter 5 to random slope models. Chapters 4 and 5 focus on understanding the hierarchical linear model and its parameters, paying only very limited attention to procedures and algorithms for parameter estimation (estimation being work that most researchers delegate to the computer). Testing parameters and specifying a multilevel model is the topic of Chapter 6.

An introductory course on multilevel analysis could cover Chapters 1 to 6 and Section 7.1, with selected material from other chapters. A minimal course would focus on Chapters 4 to 6. The later chapters are about topics which are more specialized or more advanced, but important in the practice of multilevel analysis.

The text of this book is not based on a particular computer program for multilevel analysis. The last chapter, 15, gives a brief review of programs that can be used for multilevel analysis and makes the link (to the extent that this is still necessary) between the terminology used in these programs and the terminology of the book.

Chapters 7 (about the explanatory power of the model) and 9 (about model assumptions) are important for the interpretation of results of statistical analyses using the hierarchical linear model. Chapter 10 helps the researcher in setting up a multilevel study, and in choosing sample sizes at the various levels.

Chapters 8, and 11 to 14, treat various extensions of the basic hierarchical linear model that are useful in practical research. The topic of Chapter 8, heteroscedasticity (non-constant residual variances), may seem rather specialized. Modeling heteroscedasticity, however, can be very useful. It also allows model checks and model modifications that are used in Chapter 9. Chapter 11 treats crossed random coefficients, a model ingredient which strictly speaking is outside the domain of multilevel models, but which is practically important and can be implemented in currently available multilevel software. Chapter 12 is about longitudinal data, with a fixed occasion design (i.e., repeated measures data) as well as those with a variable occasion design. This chapter indicates how the flexibility of the multilevel model gives important opportunities for data analysis (e.g., for incomplete multivariate or longitudinal data) that were unavailable earlier. Chapter 13 is about multilevel analysis for multivariate dependent variables. Chapter 14 describes possibilities of multilevel modeling for dichotomous, ordinal, and frequency data.

If additional textbooks are sought, one could consider Hox (1994) and Kreft and de Leeuw (1998), good introductions; Bryk and Raudenbush (1992), an elaborate treatment of the hierarchical linear model; and Longford (1993a) and Goldstein (1995) for more of the mathematical background.

1.2.1 Prerequisites

For reading this textbook, it is required that you have a good working knowledge of statistics. It is assumed that you know the concepts of probability, random variable, probability distribution, population, sample, statistical independence, expectation (= population mean), variance, covariance, correlation, standard deviation, and standard error. Further it is assumed that you know the basics of hypothesis testing and multiple regression analysis, and that you can understand formulae of the kind that occur in the explanation of regression analysis.

Matrix notation is used only in a few more advanced sections. These sections can be skipped without loss of understanding of other parts of the book.

1.2.2 Notation

The main notational conventions are as follows.

Abstract variables and random variables are denoted by italicized capital letters, like X or Y. Outcomes of random variables and other fixed values are denoted by italicized small letters, like x or z. Thus we speak about the variable X, but in formulae where the value of this variable is considered as a fixed, non-random value, it will be denoted x. There are some exceptions to this, e.g., in Chapter 2, and the use of the letter N for the number of groups ('level-two units') in the data.

The letter $\mathcal{E}$ is used to denote the *expected value*, or population average, of a random variable. Thus, $\mathcal{E}Y$ and $\mathcal{E}(Y)$ denote the expected value of Y. For example, if P_n is the fraction of tails obtained in n coin flips, and the coin is fair, then the expected value is $\mathcal{E}P_n = \frac{1}{2}$.

Statistical parameters are indicated by Greek letters. Examples are μ, σ^2, and β. The following Greek letters are used.

α	alpha
β	beta
γ	gamma
δ	delta
η	eta
θ	theta
λ	lambda
μ	mu
π	pi
ρ	rho
σ	sigma
τ	tau
φ	phi
χ	chi
ω	omega
Σ	capital sigma
T	capital tau

2 Multilevel Theories, Multi-stage Sampling, and Multilevel Models

In many cases simple random sampling is not a very cost-efficient strategy, and multi-stage samples may be more efficient instead. In that case the clustering of the data is, in the phase of data analysis, a nuisance which should be taken into consideration. In many other situations, however, multi-stage samples are employed because one is interested in relations between variables at different layers in a hierarchical system. In this case the dependency of observations within groups is of focal interest, because it reflects that groups differ in certain respects. In either case, the use of single-level statistical models is no longer valid. The fallacies to which their use can lead are described in the next chapter.

2.1 Dependence as a nuisance

From textbooks on statistics it is learned that, as a standard situation, observations should be sampled *independently* from each other. The standard sampling design to which statistical models are linked accordingly is simple random sampling with replacement from an infinite population: the result of one selection is independent of the result of any other selection, and the chances of selecting a certain single unit are constant (and known) across all units in the population. Textbooks on sampling, however, make clear that there are more cost-efficient sampling designs, based on the idea that probabilities of selection should be known but do not have to be constant. One of those cost-efficient sampling designs is the *multi-stage sample*: the population of interest consists of subpopulations, and selection takes place via those subpopulations. If there is only one subpopulation level, the design is a *two-stage sample*. Pupils, for instance, are grouped in schools, so the population of pupils consists of subpopulations of schools that contain pupils. Other examples are: families in neighborhoods, teeth in jawbones, animals in litters, employees in firms, children in families, etc. In a random two-stage sample, a random sample of the primary units (schools, neighborhoods, jawbones, litters, firms, families) is taken in the first stage, and then the secondary units (pupils, families, teeth, animals, employees, children) are sampled at random from the selected primary units in the second stage. A common mistake in research is to ignore the fact that the sampling scheme was a two-stage one, and to pretend that the secondary units were selected independently. The mistake in this case would be, that the researcher overlooks the fact that the secondary units were not sampled independently from each other: having selected a primary unit (a school, for example) increases the chances of selection of secondary units (pupils, for example) from that primary unit. Stated otherwise: the multi-stage sampling design leads to *dependent* observations. The multi-stage sampling design can be graphically depicted as in Figure 2.1. In this figure we see a population that

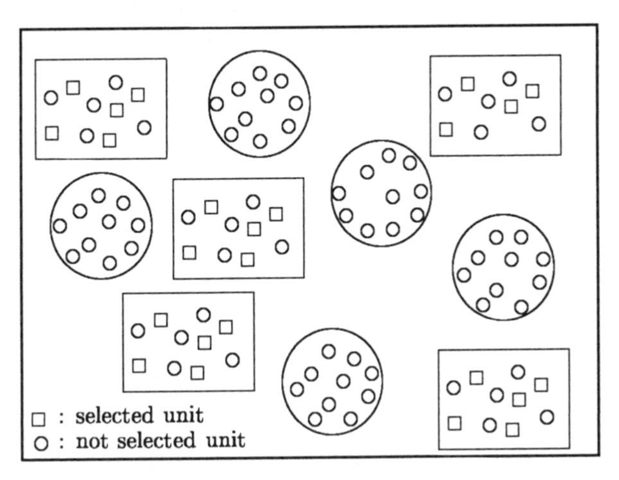

Figure 2.1 Multi-stage sampling.

consists of 10 subpopulations, each containing 10 micro-units. A sample of 25 percent is taken by randomly selecting 5 out of 10 subpopulations and within these – again at random of course – 5 out of 10 micro-units.

Multi-stage samples are preferred in practice, because the costs of interviewing or testing persons are reduced enormously if these persons are geographically or organizationally grouped. It is cheaper to travel to 100 neighborhoods and interview ten persons per neighborhood on their political preferences than to travel to 1,000 neighborhoods and interview one person per neighborhood. In the next chapters we will see how we can make adjustments to deal with these dependencies.

2.2 Dependence as an interesting phenomenon

The previous section implies that, if we want to make inferences on, e.g., the earnings of employees in the profit sector, it is cost-efficient to use a multi-stage sampling design, where employees are selected via the firms in which they work. A common feature in social research, however, is that in many cases we want to make inferences on the firms as well as on the employees.

Questions that we seek to answer may be: do employees in multinationals earn more than employees in other firms? Or: is there a relation between the performance of pupils and the experience of their teacher? Or: is the sentence differential between black and white suspects different between judges, and if so, can we find characteristics of judges to which this sentence differential is related? In this case a variable is defined at the primary unit level (firms, teachers, judges) as well as on the secondary unit level (employees, pupils, cases). From here on we will refer to primary units as macro-level units (or macro-units for short) and to the secondary units as micro-level units (or micro-units for short). Moreover, for the time being, we will restrict ourselves to the two-level case, and thus to two-stage samples only. In Table 2.1 a summary of the terminology is given. Examples of macro-units and the micro-units nested within them are presented in Table 2.2.

Table 2.1 Summary of terms to describe units at either level in the two-level case.

macro-level units	micro-level units
macro-units	micro-units
primary units	secondary units
clusters	elementary units
level-2 units	level-1 units

Table 2.2 Some examples of units at the macro and micro level.

Macro-level	Micro-level
schools	teachers
classes	pupils
neighborhoods	families
firms	employees
jawbones	teeth
families	children
litters	animals
doctors	patients
subjects	measurements
interviewers	respondents
judges	suspects

Most of the examples presented in the table have been dealt with in the text already. It is important to note that what is defined as a macro-unit and a micro-unit, respectively, depends on the theory at hand. Teachers are nested within schools, if we study organizational effects on teacher burn-out: then teachers are the micro-units and schools the macro-units. But when studying teacher effects on student achievement, teachers are the macro-units and students the micro-units. The same goes, *mutatis mu-*

tandis, for neighborhoods and families (e.g., when studying the effects of housing conditions on marital problems), and for families and children (e.g., when studying effects of income on educational performance of siblings).

In all these instances the dependency of the observations on the micro-units within the macro-units is of focal interest. If we stick to the example of schools and pupils, then the dependency (e.g., in mathematics achievement of pupils within a school) may stem from:

1. pupils within a school sharing the same school environment;
2. pupils within a school sharing the same teachers;
3. pupils within a school affecting each other by direct communication or shared group norms;
4. pupils within a school coming from the same neighborhood.

The more the achievement levels of pupils within a school are alike (as compared to pupils from other schools), the more likely it is that causes for the achievement have to do with the organizational unit (in this case: the school). Absence of dependency in this case implies absence of institutional effects on individual performance.

A special kind of nesting is defined by longitudinal data, represented in Table 2.2 as 'measurements within subjects'. The measurement occasions here are the micro-units and the subjects the macro-units. The dependence of the different measurements for a given subject is of primary importance in longitudinal data, but the following section about relations between variables defined at either level is not directly intended for the nesting structure defined by longitudinal data. Because of the special nature of this nesting structure, a separate chapter (Chapter 12) is devoted to it.

2.3 Macro-level, micro-level, and cross-level relations

For the study of hierarchical, or multilevel systems, having two distinct layers, Tacq (1986) distinguished between three kinds of propositions: about micro-units (e.g., 'employees have on average 4 effective working hours per day'; 'boys lag behind girls in reading comprehension'), about macro-units (e.g., 'schools have on average a budget of $20,000 to spend on resources'; 'in neighborhoods with bad housing conditions crime rates are above average'), or about macro–micro relations (e.g., 'if firms have a salary bonus system, employees will have increased productivity'; 'a child suffering from a broken family situation will affect classroom climate').

Multilevel statistical models are always needed if a multi-stage sampling design has been employed. The use of such a sampling design is quite obvious if we are interested in macro–micro relations, less obvious – but often necessary from a cost-effectiveness point of view – if micro-level propositions are our primary concern, and hardly obvious – but sometimes still applicable – if macro-level propositions are what we are focusing on. These three instances will be dealt with in the sequel of this chapter. To facilitate comprehension, following Tacq (1986) we use figures with the following conventions:

> a dotted line indicates that there are two levels;
> below the line is the micro-level;
> above the line is the macro-level;
> macro-level variables are denoted by capitals;
> micro-level variables are denoted by lower case letters;
> arrows denote presumed causal relations.

Multilevel propositions
Multilevel propositions can be represented as in Figure 2.2.

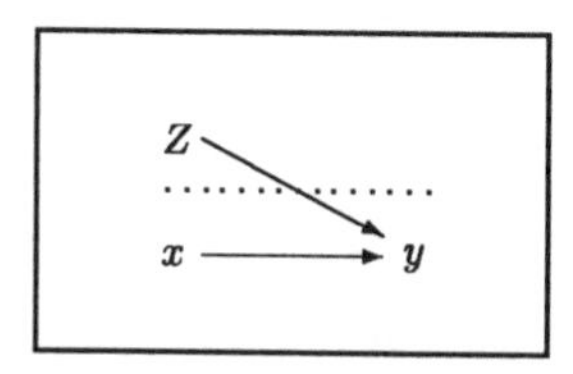

Figure 2.2 The structure of a multilevel proposition.

In this example we are interested in the effect of the macro-level variable Z (e.g., teacher efficacy) on the micro-level variable y (e.g., pupil motivation) controlling for the micro-level variable x (e.g., pupil aptitude).

Micro-level propositions
Micro-level propositions are of the form indicated in Figure 2.3.

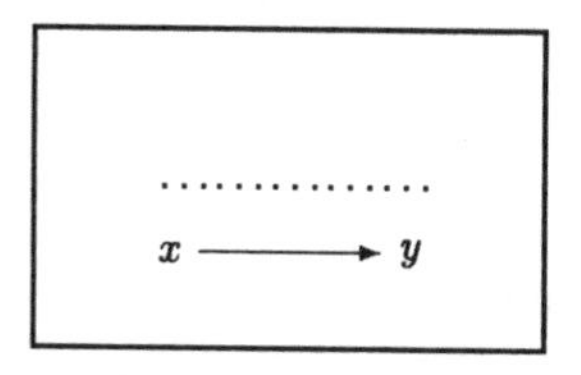

Figure 2.3 The structure of a micro-level proposition.

In this case the line indicates that there is a macro-level that is not referred to in the hypothesis that is put to the test, but that is used in the sampling design in the first stage. In assessing the strength of the relation between occupational status and income, for instance, respondents may have been selected for face-to-face interviews per zip-code area. This then may cause dependency (as a nuisance) in the data.

Macro-level propositions
Macro-level propositions are of the form of Figure 2.4.

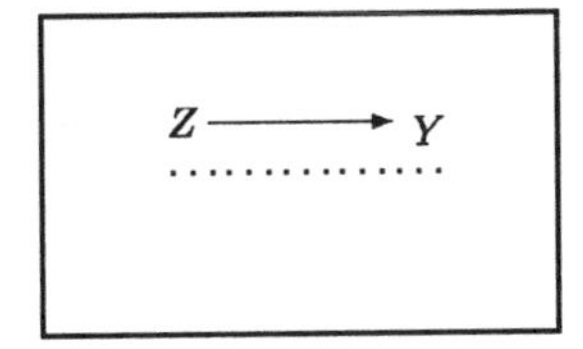

Figure 2.4 The structure of a macro-level proposition.

The line separating the macro-level from the micro-level seems to be superfluous here. When investigating the relation between long-range strategic planning policy of firms and their profits, there is no multilevel situation, and a simple random sample may have been taken. When either or both variables are not directly observable, however, and have to be measured at the micro-level (e.g., organizational climate measured as the average satisfaction of employees), then a two-stage sample is needed nevertheless. This is the case *a fortiori* for variables defined as aggregates of micro-level variables (e.g., the crime rate in a neighborhood).

Macro–micro relations
The most common situation in social research is that macro-level variables are supposed to have a relation with micro-level variables. There are three obvious instances of macro-to-micro relations, all of which are typical examples of the multilevel situation.

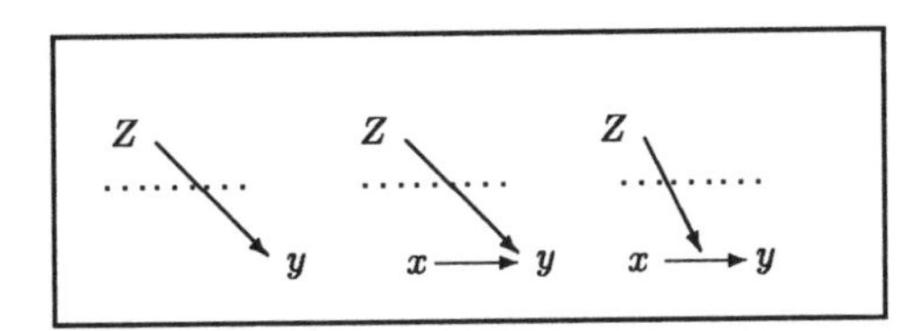

Figure 2.5 The structure of macro–micro propositions.

The first case is the macro-to-micro proposition. The more explicit the religious norms in social networks, for example, the more conservative the views that individuals have on contraception. The second proposition is a special case of this. It refers to the case where there is a relation between Z and y, given that the effect of x on y is taken into account. The example given may be modified to: 'for individuals of a given educational level'. The last case in the figure is the *macro–micro-interaction*, also known as the cross-level interaction: the relation between x and y is dependent on Z. Or stated otherwise: the relation between Z and y is dependent on x. The effect of aptitude on achievement, for instance, may be small in case of ability grouping of pupils within classrooms but large in ungrouped classrooms.

Next to these three situations there is the so-called emergent, or micro–macro, proposition (Figure 2.6).

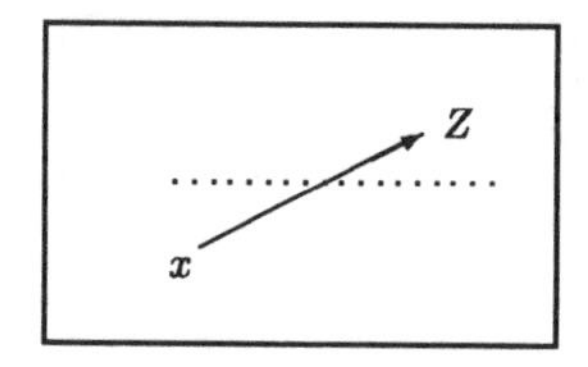

Figure 2.6 The structure of a micro–macro proposition.

In this case, a micro-level variable x affects a macro-level variable Z (student achievement may affect teacher's stress experience).

There are of course combinations of the various examples given. Figure 2.7 contains a causal chain that explains through which micro-variables there is an association between the macro-level variables W and Z (cf. Coleman, 1990).

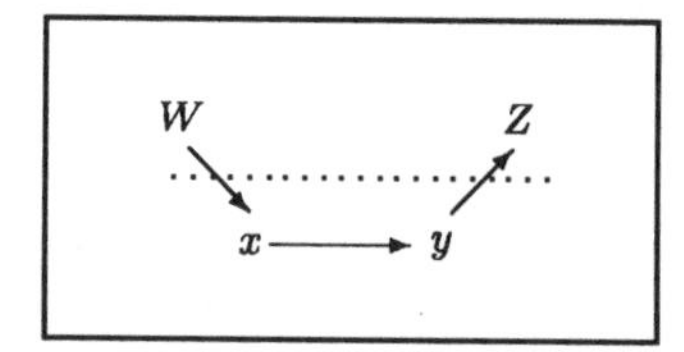

Figure 2.7 A causal macro–micro–micro–macro chain.

An example of this chain: why do the qualities of a football coach affect his social prestige? The reason is that good coaches are capable of motivating their players, thus leading the players to good performance, thus to winning games, and this of course leads to more social prestige for the coach. Another instance of a complex multilevel proposition is the contextual effects proposition. An example: 'low socio-economic status pupils achieve less in classrooms with a low average aptitude'. This is also a cross-level interaction effect, but the macro-level variable, average aptitude in the classroom, now is an aggregate of a micro-level variable. In the next chapters the statistical tools to handle multilevel structures will be introduced for outcome variables defined at the micro-level.

3 Statistical Treatment of Clustered Data

Before proceeding in the next chapters to explain ways for the statistical modeling of data with a multilevel structure, we focus attention in this chapter on the question: what will happen if we ignore the multilevel structure of the data? Are there any instances where one may proceed with single-level statistical models although the data stem from a multistage sampling design? What kind of errors may occur when this is done?

Next to this, we present some statistical methods for multilevel data that do not use the hierarchical linear model (which receives ample treatment in following chapters). First, we describe the intraclass correlation coefficient, a basic measure for the degree of dependency in clustered observations. Second, some simple statistics (mean, standard error of the mean, variance, correlation, reliability of aggregates) are treated for two-stage sampling designs. The relations are spelled out between within-group, between-group, and total regressions, and similarly for correlations. Finally, we mention some simple methods for combining evidence within groups into an overall test.

3.1 Aggregation

A common procedure in social research with two-level data is to aggregate the micro-level data to the macro-level. The simplest way to do this is to work with the averages for each macro-unit.

There is nothing wrong with aggregation in cases where the researcher is only interested in macro-level propositions, although it should be borne in mind that the reliability of an aggregated variable depends, among others, on the number of micro-level units in a macro-level unit (see later in this chapter), and thus will be larger for the larger macro-units than for the smaller ones. In cases where the researcher is interested in macro–micro or micro-level propositions, however, aggregation may result in gross errors.

The first potential error is the 'shift of meaning' (cf. Hüttner, 1981). A variable that is aggregated to the macro level refers to the macro-units, not directly to the micro-units. The firm average of a rating of employees on their working conditions, e.g., may be used as an index for 'organizational climate'. This variable refers to the firm, not directly to the employees.

The second potential error with aggregation is the ecological fallacy (Robinson, 1950). A correlation between macro-level variables cannot be used to make assertions about micro-level relations. The percentage of black inhabitants in a neighborhood could be related to average political views in the neighborhood, e.g., the higher the percentage of blacks in a neighborhood, the higher might be the proportion of people with extreme right-wing political views. This, however, does not give us any clue about the micro-level relation between race and political conviction. (The shift of meaning plays a role here, too. The percentage of black inhabitants is a variable that means something for the neighborhood, and this meaning is distinct from the meaning of ethnicity as an individual-level variable.) The ecological and other related fallacies are extensively discussed by Alker (1969).

The third potential error is the neglect of the original datastructure, especially when some kind of analysis of covariance is to be used. Suppose one is interested in assessing between-school differences in pupil achievement after correcting for intake differences, and that Figure 3.1 depicts the true situation. The figure depicts the situation for five groups, for each of which we have five observations. The groups are indicated by the shapes $\square, \times, +, \diamond$, and $*$. The five group means are indicated by $\bullet$.

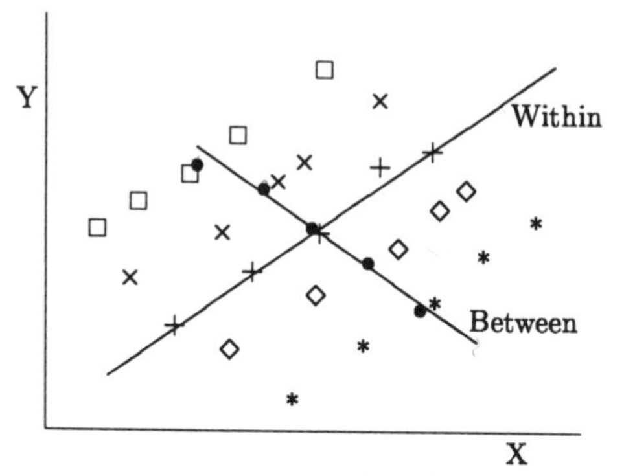

Figure 3.1 micro-level versus macro-level adjustments. (X, Y) values for five groups indicated by $*, \diamond, +, \times, \square$; group averages by $\bullet$.

Now suppose the question is: do the differences between the groups on the variable Y, after adjusting for differences on the variable X, have a substantial size? The micro-level approach, which adjusts for the within-group regression of Y on X, will lead to the regression line that has a positive slope. In this picture, the micro-units from the group that have the $\square$-symbol are all above the line, whereas the micro-units from the $*$-group are all under the regression line. The micro-level regression-approach thus will lead us to conclude that the five groups do differ given that an adjustment for X has been made. Now suppose that we would aggregate the data, and regress the average $\bar{Y}$ on the average $\bar{X}$. The averages are depicted by •. This situation is represented in the graph by the regression line with a negative slope. The averages of all groups are almost perfectly on the regression-line (the observed average $\bar{Y}$ can almost perfectly be predicted from the observed average $\bar{X}$), thus leading us to the conclusion that there are almost no differences between the five groups after adjusting for the average $\bar{X}$. Although the situation depicted in the graph is an idealized example, it clearly shows that working with aggregate data 'is dangerous at best, and disastrous at worst' (Aitkin and Longford, 1986, p. 42). When analysing multilevel data, without aggregation, the problem described in this paragraph can be dealt with by distinguishing between the within-groups and the between-groups regressions. This is worked out in Sections 3.6, 4.5, and 9.2.1.

The last objection against aggregation is, that it prevents from examining potential cross-level interaction effects of a specified micro-level variable with an as yet unspecified macro-level variable. Having aggregated the data to the macro-level one cannot examine relations like: is the sentence differential between black and white suspects different between judges, when allowance is made for differences in seriousness of crimes? Or, to give another example: is the effect of aptitude on achievement, present in the case of whole class instruction, smaller or even absent in case of ability grouping of pupils within classrooms?

3.2 Disaggregation

Now suppose that we treat our data at the micro level. There are two situations:

1. we also have a measure of a variable at the macro level, next to the measures at the micro level;

2. we only have measures of micro-level variables.

In situation (1), disaggregation leads to 'the miraculous multiplication of the number of units'. To illustrate what is meant: suppose a researcher is interested in the question whether older judges give more lenient sentences than younger judges. A two-stage sample is taken: in the first stage ten judges are sampled, and in the second stage per judge ten trials are sampled (in total there are thus $10 \times 10 = 100$ trials). One might disaggregate the data to the level of the trials and estimate the relation between the experience of the judge and the length of the sentence, without taking into account that some trials involve the same judge. This is like pretending that there are 100 independent observations, whereas in actual fact there are only 10 independent observations (the 10 judges). This shows that disaggregation and treating the data as if they are independent implies that the sample size is dramatically exaggerated. For the study of between-group differences, disaggregation often leads to serious risks of committing type I errors (asserting on the basis of the observations that there is a difference between older and younger judges whereas in the population of judges there is no such relation). On the other hand, for studying within-group differences, disaggregation often leads to unnecessarily conservative tests (i.e., too low type I error probabilities); this is elaborated in Moerbeek et al. (1997).

If only measures are taken at the micro level, analysing the data at the micro level is a correct way to proceed, as long as one takes into account that observations within a macro-unit may be correlated. In sampling theory, this phenomenon is known as the design effect for two-stage samples. If one wants to estimate the average management capability of young managers, while in the first stage a limited number of organizations (say 10) are selected and within each organization five managers are sampled, one runs the risk of making an error if (as is usually the case) there are systematic differences between organizations. In general, two-stage sampling leads to the situation that the 'effective' sample size that should be used to calculate standard errors is less than the total number of cases, the latter being given here by the 50 managers. The formula will be presented in one of the next sections.

Starting with Robinson's (1950) paper about the ecological fallacy, many papers have been written about the possibilities and dangers of cross-level inference, i.e., methods to conclude something about relations between micro-units on the basis of relations between data at the aggregate level, or conclude something about relations between macro-units on the basis of relations between disaggregated data. A concise discussion and many references are given by Pedhazur (1982, Chapter 13) and by Aitkin and Longford (1986). Our conclusion is that if the macro-units have any meaningful relation with the phenomenon under study, analysing only aggregated or only disaggregated data is apt to lead to misleading and erroneous conclusions. A multilevel approach, in which within-group and between-group relations are combined, is more difficult but much more productive. This approach requires, however, to specify assumptions about the way in which macro- and micro-effects are put together. The present chapter presents some multilevel procedures that are based on only a minimum of such assumptions (e.g., the additive model of equation (3.1)). The further chapters of this book are based on a more elaborate model, the so-called hierarchical linear model, since about 1990 the most widely accepted basis for multilevel analysis.

3.3 The intraclass correlation

The degree of resemblance between micro-units belonging to the same macro-unit can be expressed by the *intraclass correlation coefficient*. The term 'class' is conventionally used here and refers to the macro-units in the classification system under consideration. There are, however, several definitions of this coefficient, depending on the assumptions about the sampling design. In this section we assume a two-stage sampling design, and infinite populations at either level. The macro-units will also be referred to as *groups*.

A relevant model here is the *random effects ANOVA* model.[1] Indicating by Y_{ij} the outcome value observed for micro-unit i within macro-unit j, this model can be expressed as

$$Y_{ij} = \mu + U_j + R_{ij}\,, \tag{3.1}$$

where μ is the population grand mean, U_j is the specific effect of macro-unit j, and R_{ij} is the residual effect for micro-unit i within this macro-unit. In other words, macro-unit j has the 'true mean' $\mu + U_j$, and each measurement of a micro-unit within this macro-unit deviates from this true mean by some value, called R_{ij}. Units differ randomly from one another, which is reflected by the fact that U_j is a random variable and the name 'random effects model'. Some units have a high true mean, corresponding to a high value of U_j, others have a close to average, still others a low true mean. It is assumed that all variables are independent, the group effects U_j having population mean 0 and population variance τ^2 (the ***population between-group variance***), and the residuals having mean 0 and variance σ^2 (the ***population within-group variance***). For example, if micro-units are pupils and macro-units are schools, then the within-group variance is the variance within the schools about their true means, while the between-group variance is the variance between the schools' true means. The total variance of Y_{ij} is then equal to the sum of these two variances,

$$\text{var}(Y_{ij}) = \tau^2 + \sigma^2\,.$$

The number of micro-units within the j'th macro-unit is denoted by n_j. The number of macro-units is N, and the total sample size is $M = \sum_j n_j$.

In this situation, the intraclass correlation coefficient ρ_I can be defined as

$$\rho_I = \frac{\text{population variance } \textit{between} \text{ macro-units}}{\text{total variance}} = \frac{\tau^2}{\tau^2 + \sigma^2}\,. \tag{3.2}$$

It is the proportion of variance that is accounted for by the group level. This parameter is called a correlation coefficient, because it is equal to the correlation between values of two randomly drawn micro-units in the same, randomly drawn, macro-unit.

It is important to note that the population variance between macro-units is not directly reflected by the observed variance between the means of the macro-units (the observed between macro-units variance). The reason is that in a two-stage sample, variation between micro-units will also show up as extra observed variance between macro-units. It is indicated below how the observed variance between cluster means must be adjusted to yield a good estimator for the population variance between macro-units.

[1] This model is also known in the statistical literature as the one-way random effects ANOVA model and as Eisenhart's Type II ANOVA model. In multilevel modeling it is known as the empty model, which is treated further in Section 4.3.

Example 3.1 *Random data.*
Suppose we have a series of 100 observations as in the random digits Table 3.1.

Table 3.1 Data grouped into macro-units
(random digits from Glass and Stanley, 1970, p. 511).

j	Scores Y_{ij} for micro-units (random digits)										Average $\bar{Y}_{.j}$
01	60	36	59	46	53	35	07	53	39	49	43.7
02	83	79	94	24	02	56	62	33	44	42	51.9
03	32	96	00	74	05	36	40	98	32	32	44.5
04	19	32	25	38	45	57	62	05	26	06	31.5
05	11	22	09	47	47	07	39	93	74	08	35.7
06	31	75	15	72	60	68	98	00	53	39	51.1
07	88	49	29	93	82	14	45	40	45	04	48.9
08	30	93	44	77	44	07	48	18	38	28	42.7
09	22	88	84	88	93	27	49	99	87	48	68.5
10	78	21	21	69	93	35	90	29	12	86	53.4

The core part of the table contains the random digits. Now suppose that each row in the table is a macro-unit, so that for each macro-unit we have observations on 10 micro-units. The averages of the scores for each macro-unit are in the last column. There seem to be large differences between the randomly constructed macro-units, if we look at the variance in the macro-unit averages (which is 105.7). The total observed variance between the 100 micro-units is 814.0. Suppose the macro-units were schools, the micro-units pupils, and the random digits test scores. According to these two observed variances we might conclude that the schools differ considerably with respect to their average testscores. We know in this case, however, that in 'reality' the macro-units differ only by chance.

The following subsections show how the intraclass correlation can be estimated and tested. For a review of various inference procedures for the intraclass correlation we refer to Donner (1986). An extensive overview of many methods for estimating and testing the within-group and between-group variances is given by Searle, Casella, and McCulloch (1992).

3.3.1 *Within-group and between-group variance*

We continue referring to the macro-units as groups. To disentangle the information contained in the data about the population between-group variance and the population within-group variance, we consider the ***observed variance between groups*** and the ***observed variance within groups***. These are defined in the following way. The mean of macro-unit j is denoted

$$\bar{Y}_{.j} = \frac{1}{n_j} \sum_{i=1}^{n_j} Y_{ij} \; ,$$

and the overall mean is

$$\bar{Y}_{..} = \frac{1}{M} \sum_{j=1}^{N} \sum_{i=1}^{n_j} Y_{ij} = \frac{1}{M} \sum_{j=1}^{N} n_j \bar{Y}_{.j} \; .$$

The observed variance within group j is given by

$$S_j^2 = \frac{1}{n_j - 1} \sum_{i=1}^{n_j} (Y_{ij} - \bar{Y}_{.j})^2 .$$

This number will vary from group to group. To have one parameter that expresses the within-group variability for all groups jointly, one uses the observed within-group variance, or pooled within-group variance. This is a weighted average of the variances within the various macro-units, defined as

$$\begin{aligned} S^2_{\text{within}} &= \frac{1}{M - N} \sum_{j=1}^{N} \sum_{i=1}^{n_j} (Y_{ij} - \bar{Y}_{.j})^2 \qquad (3.3) \\ &= \frac{1}{M - N} \sum_{j=1}^{N} (n_j - 1) S_j^2 . \end{aligned}$$

If model (3.1) holds, the expected value of the observed within-group variance is exactly equal to the population within-group variance:

$$\text{Expected variance } within = \mathcal{E} S^2_{\text{within}} = \sigma^2 . \qquad (3.4)$$

The situation for the between-group variance is a bit more complicated. For equal group sizes n_j, the observed between-group variance is defined as the variance between the group means,

$$S^2_{\text{between}} = \frac{1}{(N-1)} \sum_{j=1}^{N} (\bar{Y}_{.j} - \bar{Y}_{..})^2 . \qquad (3.5)$$

For unequal group sizes, the contributions of the various groups need to be weighted. The following formula uses weights that are useful for estimating the population between-group variance:

$$S^2_{\text{between}} = \frac{1}{\tilde{n}(N-1)} \sum_{j=1}^{N} n_j (\bar{Y}_{.j} - \bar{Y}_{..})^2 . \qquad (3.6)$$

In this formula, $\tilde{n}$ is defined by

$$\tilde{n} = \frac{1}{N-1} \left\{ M - \frac{\sum_j n_j^2}{M} \right\} = \bar{n} - \frac{s^2(n_j)}{N \bar{n}} , \qquad (3.7)$$

where $\bar{n} = M/N$ is the mean sample size and

$$s^2(n_j) = \frac{1}{N-1} \sum_{j=1}^{N} (n_j - \bar{n})^2$$

is the variance of the sample sizes. If all n_j have the same value, then $\tilde{n}$ also has this value. In this case, S^2_{between} is just the variance of the group means, given by (3.5).

It can be shown that the total observed variance is a combination of the within-group and the between-group variances, expressed as follows:

$$\text{observed } total \text{ variance} = \frac{1}{M-1} \sum_{j=1}^{N} \sum_{i=1}^{n_j} (Y_{ij} - \bar{Y}_{..})^2$$

$$= \frac{M-N}{M-1} S^2_{\text{within}} + \frac{\tilde{n}(N-1)}{M-1} S^2_{\text{between}} \,. \tag{3.8}$$

The complications with respect to the between-group variance arise from the fact that the micro-level residuals R_{ij} also contribute, although to a minor extent, to the observed between-group variance. Statistical theory tells us that the expected between-group variance is given by

Expected observed variance *between* =

True variance *between* + Expected sampling error variance.

More specifically, the formula is (cf. Hays (1988, Section 13.3) for the case with constant n_j and Searle, Casella, and McCulloch (1992, Section 3.6) for the general case)

$$\mathcal{E} S^2_{\text{between}} = \tau^2 + \frac{\sigma^2}{\tilde{n}} \,, \tag{3.9}$$

which holds provided that model (3.1) is valid. The second term in this formula becomes small when $\tilde{n}$ becomes large. Thus for large group sizes, the expected observed between variance is practically equal to the true between variance. For small group sizes, however, it tends to be larger than the true between variance due to the random differences that also exist between the group means.

In practice, we do not know the population values of the between and within macro-unit variances; these have to be estimated from the data. One way of estimating these parameters is based on formulae (3.4) and (3.9). From the first it follows that the population within-group variance, σ^2, can be estimated unbiasedly by the observed within-group variance:

$$\hat{\sigma}^2 = S^2_{\text{within}} \,. \tag{3.10}$$

From the combination of the last two formulae it follows that the population between-group variance, τ^2, can be estimated unbiasedly by taking the observed between-groups variance and subtracting the contribution that true within-group variance makes, on average, according to (3.9), to the observed between-group variance:

$$\hat{\tau}^2 = S^2_{\text{between}} - \frac{S^2_{\text{within}}}{\tilde{n}} \,. \tag{3.11}$$

(Another expression is given in (3.14).) This expression can take negative values. This happens when the difference between group means is less than would be expected on the basis of the within-group variability, even if the true between-group variance τ^2 would be 0. In such a case, it is natural to estimate τ^2 as being 0.

It can be concluded that the split between observed within-group variance and observed between-group variance does not correspond precisely to the split between the within-group and between-group variances in the population: the observed between-group variance reflects the population between-group variance plus a bit of the population within-group variance.

The intraclass correlation is estimated according to formula (3.2) by

$$\hat{\rho}_{\text{I}} = \frac{\hat{\tau}^2}{\hat{\tau}^2 + \hat{\sigma}^2} \,. \tag{3.12}$$

(Formula (3.15) gives another, equivalent, expression.) The standard error of this estimator in the case where all group sizes are constant, $n_j = n$, is given by

$$\text{S.E.}(\hat{\rho}_I) = (1 - \rho_I)(1 + (n-1)\rho_I)\sqrt{\frac{2}{n(n-1)(N-1)}} \; . \tag{3.13}$$

This formula was given by Donner (1986, equation (6.1)), who also gives the (quite complicated) formula for the standard error for the case of variable group sizes.

The estimators given above are so-called analysis of variance (ANOVA) estimators. They have the advantage that they can be represented by explicit formulae. Other much used estimators are those produced by the maximum likelihood (ML) and residual maximum likelihood (REML) methods (cf. Section 4.6). For equal group sizes, the ANOVA estimators are the same as the REML estimators (Searle, Casella and McCulloch, 1992). For unequal group sizes, the ML and REML estimators are slightly more efficient than the ANOVA estimators. Multilevel software can be used to calculate the ML and REML estimates.

Example 3.2 *Within- and between-group variability for random data.*
For our random digits table of the earlier example the observed between variance is $S^2_{\text{between}} = 105.7$. The observed variance within the macro-units can be computed from formula (3.8). The observed total variance is known to be 814.0 and the observed between variance is given by 105.7. Solving (3.8) for the observed within variance yields $S^2_{\text{within}} = (99/90) \times (814.0 - (10/11) \times 105.7) = 789.7$. The estimated true variance within the macro-units then also is $\hat{\sigma}^2 = 789.7$. The estimate for the true between macro-units variance is computed from (3.11) as $\hat{\tau}^2 = 105.7 - (789.7/10) = 26.7$.

Finally, the estimate of the intraclass correlation is $\hat{\rho}_I = 26.7/(789.7 + 26.7) = 0.03$. Its standard error, computed from (3.13), is 0.06.

3.3.2 Testing for group differences

The intraclass correlation as defined by (3.2) can be zero or positive. A statistical test can be performed to investigate if a positive value for this coefficient could be attributed to chance. If it may be assumed that the within-group deviations R_{ij} are normally distributed, one can use an exact test for the hypothesis that the intraclass correlation is 0, which is the same as the null hypothesis that there are no group differences, or the true between-group variance is 0. This is just the F-test for a group effect in the one-way analysis of variance (ANOVA), which can be found in any textbook on ANOVA. The test statistic can be written as

$$F = \frac{\tilde{n} S^2_{\text{between}}}{S^2_{\text{within}}} \; ,$$

and it has an F distribution with $N - 1$ and $M - N$ degrees of freedom if the null hypothesis holds.

Example 3.3 *The F-test for the random data set.*
For the data of Table 3.1, $F = (10 \times 106)/789.7 = 1.34$ with 9 and 90 degrees of freedom. This value is far from significant ($p > 0.10$). Thus, there is no evidence of true between-group differences.

Statistical computer packages usually give the F statistic and the within-group variance, S^2_{within}. From this output, the estimated population between-group variance can be calculated by

$$\hat{\tau}^2 = \frac{S^2_{\text{within}}}{\tilde{n}}(F - 1) \tag{3.14}$$

and the estimated intraclass correlation coefficient by

$$\hat{\rho}_I = \frac{F - 1}{F + \tilde{n} - 1}, \tag{3.15}$$

where $\tilde{n}$ is given by (3.7). If $F < 1$, it is natural to replace both of these expressions by 0. These formulae show that a high value for the F statistic will lead to large estimates for the between-group variance as well as the intraclass correlation, but that the group sizes, as expressed by $\tilde{n}$, moderate the relation between the test statistic and the parameter estimates.

If there are covariates, it often is relevant to test whether there are group differences in addition to those accounted for by the effect of the covariates. This is achieved by the usual F-test for the group effect in an analysis of covariance (ANCOVA). Such a test is relevant because it is possible that the ANOVA F-test does not demonstrate any group effects, but that such effects do emerge when controlling for the covariates (or vice versa). Another check on whether the groups make a difference can be carried out by testing the group-by-covariate interaction effect. These tests can be found in textbooks on ANOVA and ANCOVA, and they are contained in the well-known general purpose statistical computer programs.

So, to test whether a given nesting structure in a data set calls for multilevel analysis, one can use standard techniques from the analysis of variance. In addition to testing for the main group effect, it is also advisable to test for group-by-covariate interactions. If there is neither evidence for a main effect nor for interaction effects involving the group structure, then the researcher may leave aside the nesting structure and analyse the data by unilevel methods such as ordinary least squares (OLS) regression analysis. This approach to test for group differences can be taken whenever the number of groups is not too large for the computer program being used. If there are too many groups, however, the program will refuse to do the job. In such a case it will still be possible to carry out the tests for group differences that are treated in the following chapters, following the logic of the hierarchical linear model. This will require the use of statistical multilevel software.

3.4 Design effects in two-stage samples

In the design of empirical investigations, the determination of sample sizes is an important decision. For two-stage samples, this is more complicated than for simple ('one-stage') random samples. An elaborate treatment of this question is given in Cochran (1977). This section gives a simple approach to the precision of estimating a population mean, indicating the basic role played by the intraclass correlation. We return to this question in Chapter 10.

Large samples are preferable to increase the precision of parameter estimates, i.e., to obtain tight confidence intervals around the parameter estimates. In a simple random sample the standard error of the mean is related to the sample size by the formula

$$\text{standard error} = \frac{\text{standard deviation}}{\sqrt{\text{sample size}}} . \tag{3.16}$$

This formula can be used to indicate the required sample size (in a simple random sample) if a given standard error is desired.

When using two-stage samples, however, the clustering of the data should be taken into account when determining the sample size. Let us suppose that all group sizes are equal, $n_j = n$ for all j. The (total) sample size then is Nn. The ***design effect*** is a number that indicates how much the sample size in the denominator of (3.16) is to be adjusted because of the sampling design used. It is the ratio of the variance obtained with the given sampling design to the variance obtained for a simple random sample from the same population, supposing that the total sample size is the same. A large design effect implies a relatively large variance, which is a disadvantage that may be offset by the cost reductions implied by the design. The design effect of a two-stage sample with equal group sizes is given by

$$\text{design effect} = 1 + (n-1)\,\rho_{\text{I}} . \tag{3.17}$$

This formula expresses that, from a purely statistical point of view, a two-stage sample becomes less attractive as ρ_{I} increases (clusters become more homogeneous) and as the group size n increases (the two-stage nature of the sampling design becomes stronger).

Suppose, e.g., we were studying the satisfaction of patients with their doctors' treatments. Furthermore, let us assume that some doctors have more satisfied patients than others, leading to a ρ_{I} of 0.30. The researchers used a two-stage sample, since that is far cheaper than selecting patients simply at random. They first randomly selected 100 doctors, from each chosen doctor selected five patients at random, and then interviewed each of these. In this case the design effect is $1 + (5 - 1) \times 0.30 = 2.20$. When estimating the standard error of the mean, we no longer can treat the observations as independent from each other. The effective sample size, i.e., the equivalent total sample size that we should use in estimating the standard error, is equal to

$$N_{\text{effective}} = \frac{Nn}{\text{design effect}} , \tag{3.18}$$

in which N is the number of selected macro-units. For our example we find $N_{\text{effective}} = (100 \times 5)/2.20 = 227$. So the two-stage sample with a total of 500 patients here is equivalent to a simple random sample of 227 patients.

One can also derive the total sample size using a two-stage sampling design on the basis of a desired level of precision, assuming that ρ_I is known, and fixing n because of budgetary or time-related considerations. The general rule is: this required sample size increases as ρ_I increases and it increases with the number of micro-units one wishes to select per macro-unit. Using (3.17) and (3.18) this can be derived numerically from the formula

$$N_{ts} = N_{srs} + N_{srs}\,(n-1)\,\rho_I \ .$$

The quantity N_{ts} in this formula refers to the total desired sample size when using a two-stage sample, whereas N_{srs} refers to the desired sample size if one would have used a simple random sample.

In practice, ρ_I is unknown. However, it often is possible to make an educated guess about it on the basis of earlier research.

In Figure 3.2, N_{ts} is graphed as a function of n and ρ_I (0.1, 0.2, 0.4, and 0.6, respectively), and taking $N_{srs} = 100$ as the desired sample size for an equally informative simple random sample.

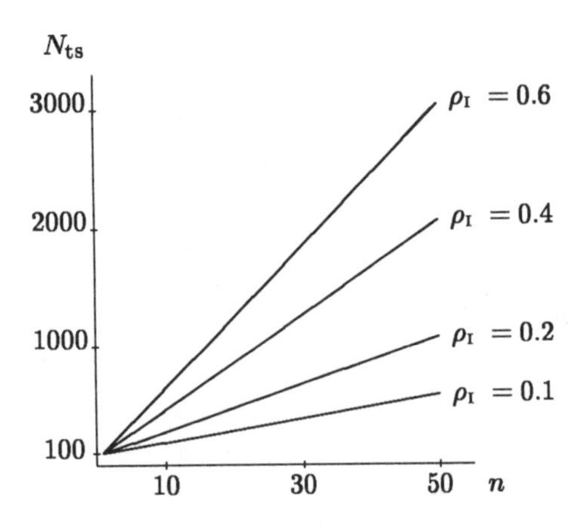

Figure 3.2 The total desired sample size in two-stage sampling.

3.5 Reliability of aggregated variables

Reliability, as conceived in psychological test theory (e.g., Lord and Novick, 1968) and in generalizability theory (e.g., Shavelson and Webb, 1991), is closely related to clustered data – although this may not be obvious at first sight. Classical psychological test theory considers a subject (an individual or other observational unit) with a given *true score*, of which imprecise, or unreliable, observations may be made. The observations can be considered to be nested within the subjects. If there is more than one observation per subject, the data are clustered. Whether there is only one observation or several, equation (3.1) is the model for this situation: the true score of subject j is $\mu + U_j$ and the i'th observation on this subject is Y_{ij}, with associated measurement error R_{ij}. If several observations are taken, these can be aggregated to the mean value $\bar{Y}_{.j}$ which then is the measurement for the true score of subject j.

The same idea can be used when it is not an individual subject who is to be measured, but some collective entity: a school, a firm, or in general any macro-level unit such as those mentioned in Table 2.2. For example, when the school climate is measured on the basis of questions posed to pupils of the school, then Y_{ij} could refer to the answer by pupil i in school j to a given question, and the opinion of the pupils about this school would be measured by the mean value $\bar{Y}_{.j}$. In terms of psychological test theory, the micro-level units i are regarded as parallel items for measuring the macro-level unit j.

The reliability of a measurement is defined generally as

$$\text{reliability} = \frac{\text{variance of true scores}}{\text{variance of observed scores}} .$$

It can be proved that this is equal to the correlation between independent replications of measuring the same subject. (This means in the mathematical model that the same value U_j is measured, but with independent realizations of the random error R_{ij}.) The reliability is indicated by the symbol λ_j.[2]

For measurement on the basis of a single observation according to model (3.1), reliability is just the intraclass correlation coefficient:

$$\lambda_j = \rho_{\text{I}} = \frac{\tau^2}{\tau^2 + \sigma^2} \qquad (\text{ if } n_j = 1) . \tag{3.19}$$

When several measurements are made for each macro-level unit, these constitute a cluster or group of measurements which are aggregated to the group mean $\bar{Y}_{.j}$. To apply to $\bar{Y}_{.j}$ the general definition of reliability, note that the observed variance is the variance between the observed means $\bar{Y}_{.j}$, while the true variance is the variance between the true scores $\mu + U_j$. Therefore the reliability of the aggregate is

$$\text{reliability of } \bar{Y}_{.j} = \frac{\text{variance between } \mu + U_j}{\text{variance between } \bar{Y}_{.j}} . \tag{3.20}$$

Example 3.4 *Reliability for random data.*
If in our previous random digits example the digits represented, e.g., the perceptions by teachers in schools of their working conditions, then the aggregated variable, an indicator for organizational climate, has an estimated reliability of $26.7/105.7 = 0.25$. (The population value of this reliability is 0, however, as the data are random, so true variance is nil ...)

It can readily be demonstrated that the reliability of aggregated variables increases as the number of micro-units per macro-unit increases, since the true variance of the group mean (with group size n_j) is τ^2 while the expected

[2] In the literature the reliability of a measurement X is frequently denoted by the symbol ρ_{XX}, so that the reliability coefficient λ_j could also be denoted by ρ_{YY}.

observed variance of the group mean is $\tau^2 + \sigma^2/n_j$. Hence the reliability can be expressed by

$$\lambda_j = \frac{\tau^2}{\tau^2 + \sigma^2/n_j} = \frac{n_j \rho_{\mathrm{I}}}{1 + (n_j - 1)\rho_{\mathrm{I}}} \,. \tag{3.21}$$

It is quite clear that if n_j is very large, then λ_j is almost 1. If $n_j = 1$ we are not able to distinguish between within- and between-group variance. Figure 3.3 presents a graph where the reliability of an aggregate is depicted as a function of n_j (denoted by n) and ρ_{I} (0.1 and 0.4, respectively).

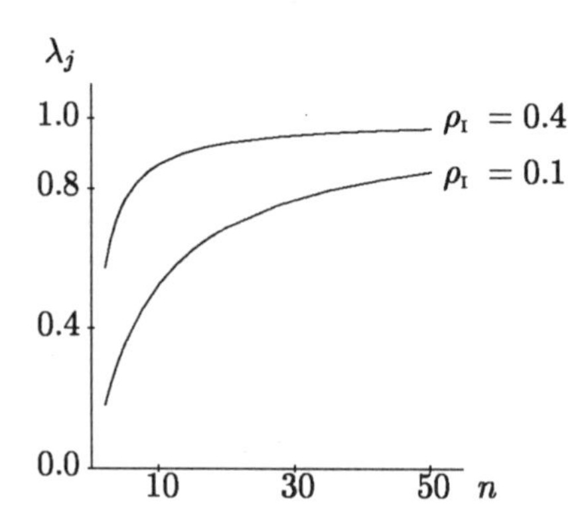

Figure 3.3 Reliability of aggregated variables.

3.6 Within- and between-group relations

We saw in Section 3.1 that regressions at the macro-level between aggregated variables $\bar{X}$ and $\bar{Y}$ can be completely different from the regressions between the micro-level variables X and Y. This section considers in more detail the interplay between macro-level and micro-level relations between two variables. First the focus is on regression of Y on X, then on the correlation between X and Y.

The main point of this section is that within-group relations can be, in principle, completely different from between-group relations. This is natural, because the processes at work within groups may be different from the processes at work between groups (see Section 3.1). Total relations, i.e., relations at the micro-level when the clustering into macro-units is disregarded, are mostly a kind of average of the within-group and between-group relations. Therefore it is necessary to consider within- and between-group relations jointly, whenever the clustering of micro-units in macro-units is meaningful for the phenomenon being studied.

3.6.1 Regressions

The linear regression of a 'dependent' variable Y on an 'explanatory' or 'independent' variable X is the linear function of X that yields the best[3] prediction of Y. When the bivariate distribution of (X, Y) is known and the data structure has only a single level, the expression for this regression function is

$$Y = \beta_0 + \beta_1 X + R \,,$$

where the regression coefficients are given by

$$\beta_0 = \mathcal{E}(Y) - \beta_1 \mathcal{E}(X) \,,$$

$$\beta_1 = \frac{\text{cov}(X, Y)}{\text{var}(X)} \,.$$

The constant term β_0 is called the ***intercept***, while β_1 is called the regression coefficient. The term R is the ***residual*** or ***error*** component, and expresses the part of the dependent variable Y that cannot be approximated by a linear function of Y. Recall from Section 1.2.2 that $\mathcal{E}(X)$ and $\mathcal{E}(Y)$ denote the population means (expected values) of X and Y, respectively.

In a multilevel data structure, this principle can be applied in various ways, depending on which population of X and Y values is being considered.

Let us consider the artificial dataset of Table 3.2. The first two columns in the table contain the identification numbers of the macro-unit (j) and the micro-unit (i). The other four columns contain the data. By X_{ij} is denoted the variable observed for micro-unit i in macro-unit j, and by $\bar{X}_{.j}$ the average of the X_{ij} values for group j. The analogous notation holds for the dependent variable Y.

Table 3.2 Artificial data on 5 macro-units each with 2 micro-units.

j	i	X_{ij}	$\bar{X}_{.j}$	Y_{ij}	$\bar{Y}_{.j}$
1	1	1	2	5	6
1	2	3	2	7	6
2	1	2	3	4	5
2	2	4	3	6	5
3	1	3	4	3	4
3	2	5	4	5	4
4	1	4	5	2	3
4	2	6	5	4	3
5	1	5	6	1	2
5	2	7	6	3	2

One might be interested in the relation between Y_{ij} and X_{ij}. The linear regression line of Y_{ij} on X_{ij} at the micro-level for the total group of 10 observations is

$$Y_{ij} = 5.33 - 0.33\, X_{ij} + R \,. \qquad \textit{(Total regression)}$$

[3]'Best prediction' means here the prediction that has the smallest mean squared error: the so-called least squares criterion.

This is the disaggregated relation, since the nesting of micro-units in macro-units is not taken into account. The regression coefficient is -0.33.

The aggregated relation is the linear regression relationship at the macro-level of the group means $\bar{Y}_{.j}$ on the group means $\bar{X}_{.j}$. This regression line is

$$\bar{Y}_{.j} = 8.00 - 1.00\,\bar{X}_{.j} + R\,. \qquad \textit{(Regression between group means)}$$

The regression coefficient now is -1.00.

A third option is to describe the relation between Y_{ij} and X_{ij} within each single group. Assuming that the regression coefficient has the same value in each group, this is the same as the regression of the within-group Y-deviations $(Y_{ij} - \bar{Y}_{.j})$ on the X-deviations $(X_{ij} - \bar{X}_{.j})$. This within-group regression line is given by

$$Y_{ij} = \bar{Y}_{.j} + 1.00\,(X_{ij} - \bar{X}_{.j}) + R\,, \qquad \textit{(Regression within groups)}$$

with a regression coefficient of $+1.00$.

Finally, and that is how the artificial dataset was constructed, Y_{ij} can be written as a function of the within-group and between-group relations between Y and X. This amounts to putting together the between-group and the within-group regression equations. The result is

$$\begin{aligned} Y_{ij} &= 8.00 - 1.00\,\bar{X}_{.j} + 1.00\,(X_{ij} - \bar{X}_{.j}) + R & (3.22)\\ &= 8.00 + 1.00\,X_{ij} - 2.00\,\bar{X}_{.j} + R\,. & \textit{(Multilevel regression)} \end{aligned}$$

Figure 3.4 graphically depicts the total, within-group, and between-group relations between the variables. The five parallel ascending lines represent

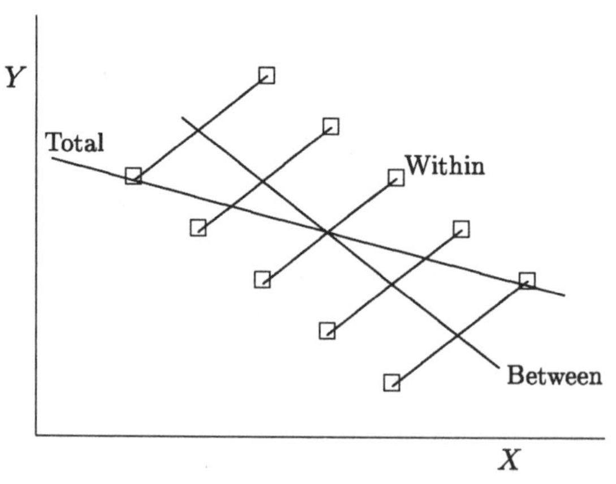

Figure 3.4 Within, between, and total relations.

the within-groups relation between Y and X. The steep descending line represents the relation at the aggregate level (i.e., between the group means), whereas the almost horizontal descending line represents the total relationship, i.e., the micro-level relation between X and Y ignoring the hierarchical structure.

The within-group regression coefficient is +1 whereas the between-group coefficient is −1. The total regression coefficient, −0.33, is in between these two. This illustrates that within-group and between-group relations can be completely different, even have opposite signs. The true relation between Y and X is revealed only when the within- and between-group relations are considered jointly, i.e., by the multilevel regression. In the multilevel regression, both the between-groups and the within-group regression coefficients play a role. Thus there are many different ways to describe the data, of which one is the best, because it describes how the data were generated. In this artificial data set, the residual R is 0; real data have, of course, non-zero residuals.

A population model

The interplay of within-group and between-group relations can be better understood on the basis of a population model such as (3.1). Since this section is about two variables, X and Y, a bivariate version of the model is needed. In this model, group (macro-unit) j has specific main effects U_{xj} and U_{yj} for variables X and Y, and associated with individual (micro-unit) i are the within-group deviations R_{xij} and R_{yij}. The population means are denoted μ_x and μ_y and it is assumed that the U's and the R's have population means 0. The U's on one hand and the R's on the other are independent. The formula for X and Y then reads

$$\begin{aligned} X_{ij} &= \mu_x + U_{xj} + R_{xij} , \\ Y_{ij} &= \mu_y + U_{yj} + R_{yij} . \end{aligned} \tag{3.23}$$

For the formulae that refer to relations between group means $\bar{X}_{.j}$ and $\bar{Y}_{.j}$, it is assumed that each group has the same size, denoted by n.

The correlation between the group effects is defined as

$$\rho_{\text{between}} = \rho(U_{xj}, U_{yj})$$

while the correlation between the individual deviations is defined by

$$\rho_{\text{within}} = \rho(R_{xij}, R_{yij}) .$$

One of the two variables X and Y might have a stronger group nature than the other, so that the intraclass correlation coefficients for X and Y may be different. These are denoted by $\rho_{\text{I}x}$ and $\rho_{\text{I}y}$, respectively.

The *within-group regression coefficient* is the regression coefficient within each group of Y on X, assumed to be the same for each group. This coefficient is denoted by β_{within} and defined by the within-group regression equation,

$$Y_{ij} = \mu_y + U_{yj} + \beta_{\text{within}} (X_{ij} - \mu_x - U_{xj}) + R . \tag{3.24}$$

This equation may be regarded as an analysis of covariance (ANCOVA) model for Y. Hence the within-group regression coefficient also is the effect of X in the ANCOVA approach to this multilevel data.

The within-group regression coefficient is also obtained when the Y-deviation values $(Y_{ij} - \bar{Y}_{.j})$ are regressed on the X-deviation values

$(X_{ij} - \bar{X}_{.j})$. In other words, it is also the regression coefficient obtained in the disaggregated analysis of the within-group deviation scores.

The population *between-group regression coefficient* is defined as the regression coefficient for the group effects U_x on U_y. This coefficient is denoted by $\beta_{\text{between } U}$ and is defined by the regression equation

$$U_{xj} = \beta_{\text{between } U}\, U_{yj} + R \,,$$

where R now is the group-level residual.

The *total regression coefficient* of X on Y is the regression coefficient in the disaggregated analysis, i.e., when the data are treated as single level data:

$$Y_{ij} = \mu_y + \beta_{\text{total}}\,(X_{ij} - \mu_x) + R \,.$$

The total regression coefficient can be expressed as a weighted mean of the within- and the between-groups coefficients, where the weight for the between-groups coefficient is just the intraclass correlation for X. The formula is

$$\beta_{\text{total}} = \rho_{\text{I}x}\, \beta_{\text{between } U} + (1 - \rho_{\text{I}x})\, \beta_{\text{within}} \,. \tag{3.25}$$

This expression implies that if X is a pure macro-level variable (so that $\rho_{\text{I}x} = 1$), the total regression coefficient is equal to the between-group coefficient. Conversely, if X is a pure micro-level variable we have $\rho_{\text{I}x} = 0$, and the total regression coefficient is just the within-group coefficient. Usually X will have both a within-group and a between-group component and the total regression coefficient will be somewhere between the two level-specific regression coefficients.

Regressions between observed group means[4]

At the macro-level, the regression of the *observed* group means $\bar{Y}_{.j}$ on $\bar{X}_{.j}$ is not the same as the regression of the 'true' group effects U_y on U_x. This is because the observed group averages, $\bar{X}_{.j}$ and $\bar{Y}_{.j}$, can be regarded as the 'true' group means to which some error, $\bar{R}_{x.j}$ and $\bar{R}_{y.j}$, has been added.[5] Therefore the regression coefficient for the observed group means is not exactly equal to the (population) between-group regression coefficient, but it is given by

$$\beta_{\text{between group means}} = \lambda_{xj}\, \beta_{\text{between } U} + (1 - \lambda_{xj})\, \beta_{\text{within}} \,, \tag{3.26}$$

where λ_{xj} is the reliability of the group means $\bar{X}_{.j}$ for measuring $\mu_x + U_{xj}$, given by equation (3.20) applied to the X variable. If n is large the reliability will be close to unity, and the regression coefficient for the group means will be close to the between-group regression coefficient at the population level.

Combining equations (3.25) and (3.26) leads to another expression for the total regression coefficient. This expression uses the correlation ratio η_x^2 which is defined as the ratio of the intraclass correlation coefficient to the reliability of the group mean,

$$\eta_x^2 = \frac{\rho_{\text{I}x}}{\lambda_{xj}} = \frac{\tau^2 + \sigma^2/n}{\tau^2 + \sigma^2} \,. \tag{3.27}$$

[4]The remainder of this subsection may be skipped by the cursory reader.

[5]The same phenomenon is at the basis of formulae (3.9) and (3.11).

For large group sizes the reliability approaches unity, so the correlation ratio approaches the intraclass correlation.

In the data, the correlation ratio η_x^2 is the same as the proportion of variance in X_{ij} explained by the group means, and it can be computed as the ratio of the between-group sum of squares relative to the total sum of squares in an analysis of variance, i.e.,

$$\hat{\eta}_x^2 = \frac{\sum_j n_j \, (\bar{X}_{.j} - \bar{X}_{..})^2}{\sum_{i,j} (X_{ij} - \bar{X}_{..})^2} \, .$$

The combined expression indicates how the total regression coefficient depends on the within-group regression coefficient and the regression coefficient between the group means:

$$\beta_{\text{total}} = \eta_x^2 \, \beta_{\text{between group means}} + (1 - \eta_x^2) \, \beta_{\text{within}} \, . \tag{3.28}$$

Expression (3.28) was first given by Duncan et al. (1961) and can be found also, e.g., in Pedhazur (1982, p. 538). A multivariate version was given by Maddala (1971). To apply this equation to an unbalanced data set, the regression coefficient between group means must be calculated in a weighted regression, group j having weight n_j.

Example 3.5 *Within- and between-group regressions for artificial data.*
In the artificial example given, the total sum of squares of X_{ij} as well as Y_{ij} is 30 and the between-group sums of squares for X and Y are 20. Hence the correlation ratios are $\hat{\eta}_x^2 = \hat{\eta}_y^2 = 20/30 = 0.667$. If we use this value and plug it into formula (3.28), we find

$$\hat{\beta}_{\text{total}} = 0.667 \times (-1.00) + (1 - 0.667) \times 1.00 = -0.33 \, ,$$

which is indeed what we found earlier.

3.6.2 Correlations

The quite extreme nature of the artificial data set of Table 3.2 becomes apparent when we consider the correlations.

The group means $(\bar{X}_{.j}, \bar{Y}_{.j})$ lie on a decreasing straight line, so the *observed between-group correlation*, which is defined as the correlation between the group means, is $R_{\text{between}} = -1$. The *within-group correlation* is defined as the correlation within the groups, assuming that this correlation is the same within each group. This can be calculated as the correlation coefficient between the within-group deviation scores $\tilde{X}_{ij} = (X_{ij} - \bar{X}_{.j})$ and $\tilde{Y}_{ij} = (Y_{ij} - \bar{Y}_{.j})$. In this data set the deviation scores $(\tilde{X}_{ij}, \tilde{Y}_{ij})$ are $(-1, -1)$ for $i = 1$ and $(+1, +1)$ for $i = 2$, so the within-group correlation here is $R_{\text{within}} = +1$. Thus, we see that the within-group as well as the between-group correlations are perfect, but of opposite signs. The disaggregated correlation, i.e., the correlation computed without taking the nesting structure into account, is $R_{\text{total}} = -0.33$. (This is the same as the value for the regression coefficient in the total (disaggregated) regression equation, because X and Y have the same variance.)

The population model again

Recall that in the population model mentioned above, the correlation coefficient between the group effects U_x and U_y was defined as ρ_{between} and the correlation between the individual deviations R_x and R_y was defined as ρ_{within}. The intraclass correlation coefficients for X and Y were denoted by $\rho_{\text{I}x}$ and $\rho_{\text{I}y}$.

How do these correlations between unobservable variables relate to correlations between observables? The population within-group correlation is also the correlation between the within-group deviation scores $(\tilde{X}_{ij}, \tilde{Y}_{ij})$:

$$\rho(\tilde{X}_{ij}, \tilde{Y}_{ij}) = \rho_{\text{within}} \; . \tag{3.29}$$

For the between-group coefficient the relation is, as always, a bit more complicated. The correlation coefficient between the group means is equal to

$$\rho(\bar{X}_{.j}, \bar{Y}_{.j}) = \sqrt{\lambda_{xj}\, \lambda_{yj}}\; \rho_{\text{between}} + \sqrt{(1-\lambda_{xj})\,(1-\lambda_{yj})}\; \rho_{\text{within}} , \tag{3.30}$$

where λ_{xj} and λ_{yj} are the reliability coefficients of the group means (see equation (3.20)). For large group sizes the reliabilities will be close to 1 (provided the intraclass correlations are larger than 0), so that the correlation between the group means will then be close to ρ_{between}. This formula shows that the correlation between group means is higher than total correlation, i.e., aggregation will increase correlation, *only if* the between-groups correlation coefficient is larger than the within-groups correlation coefficient. Therefore, the reason that correlations between group means are often higher than correlations between individuals is not the mathematical consequence of aggregation, but the consequence of the processes at the group level (determining the value of ρ_{between}) being different from the processes at the individual level (which determine the value of ρ_{within}).

The total correlation (i.e., the correlation in the disaggregated analysis) is a combination of the within-group and the between-group correlation coefficients. The combination depends on the intraclass correlations, as shown by the formula

$$\rho(X_{ij}, Y_{ij}) = \sqrt{\rho_{\text{I}x}\, \rho_{\text{I}y}}\; \rho_{\text{between}} + \sqrt{(1-\rho_{\text{I}x})\,(1-\rho_{\text{I}y})}\; \rho_{\text{within}} \; . \tag{3.31}$$

If the intraclass correlations are low, then X and Y have primarily the nature of level-one variables, and the total correlation will be close to the within-group correlation; on the other hand, if the intraclass correlations are close to 1, then X and Y have almost the nature of level-two variables and the total correlation is close to the between-group correlation.

If the intraclass correlations of X and Y are equal and denoted by ρ_{I}, then (3.31) can be formulated more simply as

$$\rho(X_{ij}, Y_{ij}) = \rho_{\text{I}}\, \rho_{\text{between}} + (1-\rho_{\text{I}})\, \rho_{\text{within}} \; .$$

In this case the weights ρ_{I} and $(1-\rho_{\text{I}})$ add up to 1 and the total regression coefficient is necessarily between the within-group and the between-group regression coefficient. In general, however, this is not always true, because the sum of the weights in (3.31) is smaller than 1 if the intraclass correlations for X and Y are different. For example, if one of the intraclass correlations is close to 0 and the other is close to 1, then one variable is mainly a level-one variable and the other mainly a level-two variable. Formula (3.31) then implies that the total correlation coefficient is close to 0, no matter how large the within-group and the between-group correlations. This is rather obvious, since a level-one variable with hardly any between-group variability cannot be substantially correlated with a variable with hardly any within-group variability.

Correlations between observed group means[6]
Analogous to the regression coefficients, also for the correlation coefficients we can combine the equations to see how the total correlation depends on the within-group correlation and the correlation between the group means. This yields

$$\rho(X_{ij}, Y_{ij}) = \eta_x\, \eta_y\, \rho(\bar{X}_{.j}, \bar{Y}_{.j}) + \sqrt{(1-\eta_x^2)\,(1-\eta_y^2)}\, \rho_{\text{within}}\,. \tag{3.32}$$

This expression was given by Knapp (1977) and can also be found, e.g., in Pedhazur (1982, p. 536). When it is applied to an unbalanced data set, the correlation between the group means should be calculated with weights n_j.

It may be noted that many texts do not make the explicit distinction between population and data. If the population and the data are equated, then the reliabilities are unity, the correlation ratios are the same as the intraclass correlations, and the population between-group correlation is equal to the correlation between the group means. The equation for the total correlation then becomes

$$R_{\text{total}} = \hat{\eta}_x\, \hat{\eta}_y\, R_{\text{between}} + \sqrt{(1-\hat{\eta}_x^2)\,(1-\hat{\eta}_y^2)}\, R_{\text{within}}\,. \tag{3.33}$$

When parameter estimation is being considered, however, confusion may be caused by neglecting this distinction.

Example 3.6 *Within- and between-group correlations for artificial data.*
The correlation ratios in the artificial data example are $\eta_x^2 = \eta_y^2 = 0.667$ and we also saw above that $R_{\text{within}} = +1$ and $R_{\text{between}} = -1$. Filling in these numbers in formula (3.33) yields

$$R_{\text{total}} = \sqrt{0.667^2} \times (-1.00) + \sqrt{(1-0.667)^2} \times 1.00 = -0.33\,,$$

which indeed is the value found earlier for the total correlation.

3.6.3 Estimation of within- and between-group correlations

There are several ways for obtaining estimates for the correlation parameters treated in this section.

A quick method is based on the intraclass correlations, estimated as in Section 3.3.1 or from the output of a multilevel computer program, and the observed within-group and total correlations. The observed within-group correlation is just the ordinary correlation coefficient between the within-group deviations $(X_{ij} - \bar{X}_{.j})$ and $(Y_{ij} - \bar{Y}_{.j})$, and the total correlation is the ordinary correlation coefficient between X and Y in the whole data set.

[6] The remainder of Section 3.6 may also be skipped by the cursory reader.

The quick method then is based on (3.29) and (3.31). This leads to the estimates

$$\hat{\rho}_{\text{within}} = R_{\text{within}} \,, \tag{3.34}$$

and

$$\hat{\rho}_{\text{between}} = \frac{R_{\text{total}} - \sqrt{(1-\hat{\rho}_{\text{I}x})\,(1-\hat{\rho}_{\text{I}y})}\, R_{\text{within}}}{\sqrt{\hat{\rho}_{\text{I}x}\,\hat{\rho}_{\text{I}y}}} \,. \tag{3.35}$$

This is not the statistically most efficient method, but it is straightforward and leads to good results if sample sizes are not too small.

The ANOVA method (Searle, 1956) goes via the variances and covariances, based on the definition

$$\rho(X,Y) = \frac{\text{cov}(X,Y)}{\sqrt{\text{var}(X)\,\text{var}(Y)}} \,.$$

Estimating the within- and between-group variances was discussed in Section 3.3.1. The within- and between-group covariances between X and Y can be estimated by formulae analogous to (3.3), (3.6), (3.10) and (3.11), replacing the squares $(Y_{ij} - \bar{Y}_{.j})^2$ and $(\bar{Y}_{.j} - \bar{Y}_{..})^2$ by the cross-products $(X_{ij} - \bar{X}_{.j})(Y_{ij} - \bar{Y}_{.j})$ and $(\bar{X}_{.j} - \bar{X}_{..})(\bar{Y}_{.j} - \bar{Y}_{..})$. It is shown in Searle, Casella, and McCulloch (1992, Section 11.1.a) how these calculations can be replaced by a calculation involving only sums of squares.

Finally, the maximum likelihood (ML) and residual maximum likelihood (REML) methods can be used. These are the most often used estimation methods (cf. Section 4.6) and are implemented in multilevel software. Chapter 13 describes multivariate multilevel models; the correlation coefficient between two variables refers to the simplest multivariate situation, viz., bivariate data. Formula (13.5) represents the model which allows the estimation of within-group and between-group correlations.

Example 3.7 *Within- and between-group correlations for school tests.*
In many examples in Chapters 4 and 5, a data set is used about school performance of $M = 2287$ pupils in $N = 131$ schools. The present example considers the correlation between the scores on an arithmetic test (X) and a language test (Y). Within-school correlations reflect the correspondence between the pupils' language and arithmetic capabilities (attenuated by the unreliability of the tests). Between-school correlations reflect the processes that determine the schools' populations (intake policy, social segregation of neighborhoods) and the influence of teachers and school policy. Thus the within-school correlations and the between-school correlations are caused by quite different processes.

The indices 'within', 'between', and 'total' will be abbreviated to 'w', 'b', and 't'. The observed correlations are $R_{\text{w}} = 0.626, R_{\text{b}} = 0.876$, and $R_{\text{t}} = 0.694$.

The ANOVA estimators are calculated along the principles of Section 3.3.1. The observed within-group variances and covariance, which are also the estimated within-group population variances and covariance (cf. (3.10)), are $\hat{\sigma}_x^2 = S^2_{\text{w x}} = 32.233, \hat{\sigma}_y^2 = S^2_{\text{w y}} = 64.319, \hat{\sigma}_{xy} = S_{\text{w xy}} = 28.516$.

The observed between-group variances and covariance are $S^2_{\text{b x}} = 13.433$, $S^2_{\text{b y}} = 20.580, S_{\text{b xy}} = 14.558$. For this data set, $\tilde{n} = 17.435$. According to

(3.11), the estimated population between-group variances and covariance are $\hat{\tau}_x^2 = 11.584, \hat{\tau}_y^2 = 16.890, \hat{\tau}_{xy} = 12.923$.

From these estimated variances, the intraclass correlations are computed as $\rho_{Ix} = 11.584/(11.584 + 32.233) = 0.2644$ and $\rho_{Iy} = 16.890/(16.890 + 64.319) = 0.2080$.

The 'quick' method uses the observed within and total correlations and the intraclass correlations. The resulting estimates are $\hat{\rho}_w = 0.626$ from (3.34) and $\hat{\rho}_b = 0.922$ from (3.35).

The ANOVA estimates for the within- and between-group correlations uses the estimated within- and between-group population variances and covariance. The results are $\hat{\rho}_w = 28.516/\sqrt{32.233 \times 64.319} = 0.626$ and $\hat{\rho}_b = 12.923/\sqrt{11.584 \times 16.890} = 0.924$.

The ML estimates for the within-group variances and covariance are obtained in Chapter 13 as $\hat{\sigma}_x^2 = 32.25, \hat{\sigma}_y^2 = 64.64, \hat{\sigma}_{xy} = 28.62$. For the between-group variances and covariance they are $\hat{\tau}_x^2 = 12.76, \hat{\tau}_y^2 = 19.00, \hat{\tau}_{xy} = 14.54$. This leads to estimated correlations $\hat{\rho}_w = 28.62/\sqrt{32.25 \times 64.64} = 0.627$ and $\hat{\rho}_b = 14.54/\sqrt{12.76 \times 19.00} = 0.934$.

It can be concluded that, for this large data set, the three methods all yield practically the same results. The within-school (pupil level) correlation, 0.62, is substantial. Thus pupils' language and arithmetic capacities are closely correlated. The between-school correlation, 0.92 or 0.93, is very high. This demonstrates that the school policies, the teaching quality, and the processes that determine the composition of the school population have practically the same effect on the pupils' language as on their arithmetic performance. Note that the observed between-school correlation, 0.88, is not quite as high because of the attenuation caused by unreliability that follows from (3.30).

3.7 Combination of within-group evidence

When research focuses on within-group relations and several groups (macro-units) were investigated, it is often desired to combine the evidence gathered in the various groups. For example, consider a study where a relation between work satisfaction (X) and sickness leave (Y) is studied in several organizations. If the organizations are sufficiently similar, or if they can be considered as a sample from some population of organizations, then the data can be analysed according to the hierarchical linear model of the next chapters. If, however, the organizations are too diverse and not representative of any population, then it still can be relevant to conduct one test for the relation between X and Y in which the evidence from all organizations is combined. Another example is meta-analysis, the statistically based combination of several studies. There exist many texts about meta-analysis; e.g., Hedges and Olkin (1985), Rosenthal (1991), Hedges (1992). A number of publications may contain information about the same phenomenon, and it can be important to combine this information in a single test. If the studies leading to the publications may be regarded as a sample from some population of studies, then again methods based on the hierarchical linear model can be used. The hierarchical linear model is treated in the following chapters; applications to meta-analysis are given, e.g., by Bryk and Raudenbush (1992, Chapter 7). If studies collected cannot be regarded as a sample from a population, then still the methods mentioned below may be used.

There exist various methods for combining evidence from several studies, based only on the assumption that this evidence is statistically independent. They can be applied already if the number of independent studies is at least two. The least demanding method is Fisher's combination of p-values (Fisher, 1932; Hedges and Olkin, 1985). This method assumes that in each of N studies a null hypothesis is tested, which results in independent p-values $p_1, ..., p_N$. The combined null hypothesis is that in *all* of the studies, the null hypothesis holds; the combined alternative hypothesis is that in *at least* one of the studies, the alternative hypothesis holds. It is not required that the N independent studies used the same operationalizations or methods of analysis, only that it is meaningful to test this combined null hypothesis. This hypothesis can be tested by minus twice the sum of the natural logarithms of the p-values,

$$\chi^2 = -2 \sum_{j=1}^{N} \ln(p_j) , \tag{3.36}$$

which under the combined null hypothesis has a chi-squared distribution with $2N$ degrees of freedom. Because of the shape of the logarithmic function, this combined statistic will already have a large value if at least one of the p-values is very small.

A stronger combination procedure can be achieved if the several studies all lead to estimates of theoretically the same parameter, denoted here by θ. Suppose that the j'th study yields a parameter estimate $\hat{\theta}_j$ with standard error s_j and that all the studies are statistically independent. Then the combined estimate with smallest standard error is the weighted average with weights inversely proportional to s_j^2

$$\hat{\theta} = \frac{\sum_j s_j^{-2}\, \hat{\theta}_j}{\sum_j s_j^{-2}} \tag{3.37}$$

with standard error

$$\text{S.E.}(\hat{\theta}) = \sqrt{\frac{1}{\sum_j s_j^{-2}}} \; . \tag{3.38}$$

For example, if standard errors are inversely proportional to the square root of sample size, $s_j = \sigma / \sqrt{n_j}$ for some value σ, then the weights are directly proportional to the sample sizes and the standard error of the combined estimate is $\sigma / \sqrt{\sum n_j}$. If the individual estimates are approximately normally distributed, and also (even when the estimates are not nearly normally distributed) when N is large, the t-ratio,

$$\frac{\hat{\theta}}{\text{S.E.}(\hat{\theta})} \; ,$$

can be tested in a standard normal distribution.

The choice between these two combination methods can be made as follows. The combination of estimates, expressed by (3.37), is more suitable if the true parameter value (the estimated θ) is approximately the same in each of the combined studies, while Fisher's method (3.36) for combining p-values is more suitable if it is possible that the effect sizes are very different between the N studies. More combination methods can be found in the literature about meta-analysis, e.g., Hedges and Olkin (1985).

Example 3.8 *Gossip behavior in six organizations.*
Wittek and Wielers (1998) investigated effects of informal social network structures on gossip behavior in six work organizations. One of the hypotheses tested was that individuals tend to gossip more if they are involved in more coalition triads. An individual A is involved in a coalition triad with two others, B and C, if he has a positive relation with B while A and B both have a negative relation with C. Six organizations were studied which were so different that an approach following the lines of the hierarchical linear model was not considered appropriate. For each organization separately, a multiple regression was carried out to estimate the effect of the number of coalition triads in which a person was involved on a measure for gossip behavior, controlling for some relevant other variables.

The p-values obtained were 0.015, 0.42, 0.19, 0.13, 0.25, and 0.43. Only one of these is significant (i.e., less than 0.05), and the question is whether this combination of six p-values would be unlikely under the combined null hypothesis which states that in *all* six organizations the effect of coalition triads on gossip is absent. Equation (3.36) yields the test statistic $\chi^2 = 22.00$ with $d.f. = 2 \times 6 = 12, p < 0.05$. Thus the result is significant, which shows that indeed in the combined data there is evidence that there is an effect of coalition triads on gossip behavior, even though this effect is significant in only one of the organizations considered separately.

In this chapter we have presented some statistics to describe hierarchically structured data. We mostly gave examples with balanced data, i.e., an equal number of micro-units per macro-unit. In practice, most data sets are unbalanced (except mainly for some experimental designs with longitudinal data on a fixed set of occasions). Our main purpose was to demonstrate how clustering in the data, i.e., dependent observations, is not only a nuisance that should be taken care of statistically, but can also be a very interesting phenomenon, worth further study. Finally, before proceeding with the introduction of the hierarchical linear model of multilevel analysis, it should be borne in mind that usually there are explanatory variables and statistically it is the independence not of the observations but of the residuals (the unexplained part of the dependent variable) which is the basic assumption of single-level linear models.

4 The Random Intercept Model

In the preceding chapters it was argued that the best way to analyse multilevel data is an approach that represents within-group as well as between-group relations within a single analysis, where 'group' refers to the units at the higher levels of the nesting hierarchy. Very often it makes sense to use probability models to represent the variability within and between groups, in other words, to conceive of the unexplained variation within groups and the unexplained variation between groups as random variability. For a study of pupils within schools, e.g., this means that not only unexplained variation between pupils, but also unexplained variation between schools is regarded as random variability. This can be expressed by statistical models with so-called random coefficients.

The hierarchical linear model is such a random coefficient model for multilevel, or hierarchically structured, data and is by now the main tool for multilevel analysis. Chapters 4 and 5 treat the definition of this model and the interpretation of the model parameters. The present chapter discusses the simpler case of the random intercept model; Chapter 5 treats the general hierarchical linear model, which also has random slopes. Testing the various components of the model is treated in Chapter 6. The later chapters treat various elaborations and other aspects of the hierarchical linear model. The focus of this treatment is on the two-level case, but Chapters 4 and 5 also contain sections on models with more than two levels of variability.

For the sake of concreteness, we refer to the level-one units as 'individuals', and to the level-two units as 'groups'. The reader may fill in other names for the units, if she has a different application in mind; e.g., if the application is to repeated measurements, 'measurement occasions' for level-one units and 'subjects' for level-two units. The nesting situation of 'measurement occasions within individuals' is given special attention in Chapter 12. The number of groups in the data is denoted N; the number of individuals in the groups may vary from group to group, and is denoted n_j for group j ($j = 1, 2, \ldots, N$). The total number of individuals is denoted $M = \sum_j n_j$.

The hierarchical linear model is a type of *regression model* that is particularly suitable for multilevel data. It differs from the usual multiple regression model in the fact that the equation defining the hierarchical linear model contains more than one error term: one (or more) for each level. As in all regression models, there is a distinction between *dependent* and *explanatory* variables: the aim is to construct a model that expresses how the dependent variable depends on, or is explained by, the explanatory variables. Instead of explanatory variable, the names predictor variable and independent variable are also in use. The dependent variable must be a variable at level one: the hierarchical linear model is a model for explaining something that happens at the lowest, most detailed level.

In this section, we assume that one explanatory variable is available at either level. In the notation, we distinguish the following types of indices and variables:

j is the index for the groups ($j = 1, \ldots, N$);
i is the index for the individuals within the groups ($i = 1, \ldots, n_j$).

The indices can be regarded as case numbers; note that the numbering for the individuals starts again in every group. For example, individual 1 in group 1 is different from individual 1 in group 2.

For individual i in group j, we have the following variables:

Y_{ij} is the dependent variable;
x_{ij} is the explanatory variable at the individual level;

for group j, we have that

z_j is the explanatory variable at the group level.

To understand the notation, it is essential to realize that the indices, i and j, indicate precisely on what the variables depend. The notation Y_{ij}, e.g., indicates that the value of variable Y depends on the group j and also on the individual i. (Since the individuals are nested within groups, the index i makes sense only if it is accompanied by the index j: to identify individual $i = 1$, we must know to which group we refer!) The notation z_j, on the other hand, indicates that the value of Z depends only on the group j, and not on the individual, i.

The basic idea of multilevel modeling is that the outcome variable Y has an individual as well as a group aspect. This carries through also for other level-one variables. The X variable, although it is a variable at the individual level, may also contain a group aspect. The mean of X in one group may be different from the mean in another group. In other words, X may (and often will) have a positive between-group variance. Stated more generally, the compositions of the various groups with respect to X may differ from one another. It should be kept in mind that explanatory variables that are defined at the individual level often also contain some information about the groups.

4.1 A regression model: fixed effects only

The simplest model is one without the random effects that are characteristic for multilevel models; it is the classical model of multiple regression. This model states that the dependent variable, Y_{ij}, can be written as the sum of a systematic part (a linear combination of the explanatory variables) and a random residual,

$$Y_{ij} = \beta_0 + \beta_1 x_{ij} + \beta_2 z_j + R_{ij} \,. \tag{4.1}$$

In this model equation, the β's are the regression parameters: β_0 is the intercept (i.e., the value obtained if x_{ij} as well as z_j are 0), β_1 is the coefficient for the individual variable X, while β_2 is the coefficient for the group variable Z. The variable R_{ij} is the residual (sometimes called *error*); an essential requirement in regression model (4.1) is that all residuals are mutually independent and have a zero mean; a convenient assumption is that in all groups they have the same variances (the homoscedasticity assumption) and are normally distributed. This model has a multilevel nature only to the extent that one of the explanatory variables refers to the lower and the other to the higher level.

Model (4.1) can be extended to a regression model where not only main effects of X and Z, but also the cross-level interaction effect is present. This type of interaction is discussed more elaborately in the following chapter. It means that the product variable $ZX = Z \times X$ is added to the list of explanatory variables. The resulting regression equation is

$$Y_{ij} = \beta_0 + \beta_1 x_{ij} + \beta_2 z_j + \beta_3 z_j x_{ij} + R_{ij} \,. \tag{4.2}$$

These models pretend, as it were, that all the multilevel structure in the data is fully explained by the group variable Z and the individual variable X. If two individuals are being considered and their X- and Z-values are given, then for their Y-value it is immaterial whether they belong to the same, or to different groups.

Models of the type (4.1) and (4.2), and their extensions with more explanatory variables at either or both levels, have in the past been widely used in research on data with a multilevel structure. They are convenient to handle for anybody who knows multiple regression analysis. Is anything wrong with them? YES! For data with a meaningful multilevel structure, it is practically always unfounded to make the *a priori* assumption that all of the group structure is represented by the explanatory variables. Given that there are only N groups, it is unfounded to do as if one has $n_1+n_2+\ldots+n_N$ independent replications. There is one exception: when all group sample sizes n_j are equal to 1, the researcher does not need to have any qualms about using these models because the nesting structure, although it may be present in the population, is not present in the data. Designs with $n_j = 1$ can be used when the explanatory variables have been chosen on the basis of substantive theory, and the focus of the research is on the regression coefficients rather than on how the variability of Y is partitioned into within-group and between-group variability.

In designs with group sizes larger than 1, however, the nesting structure often cannot be represented completely in the regression model by the explanatory variables. Additional effects of the nesting structure can be represented by letting the regression coefficients vary from group to group. Thus, the coefficients β_0 and β_1 in equation (4.1) must depend on the group, denoted by j. This is expressed in the formula by an extra index j for these coefficients. This yields the model

$$Y_{ij} = \beta_{0j} + \beta_{1j}\,x_{ij} + \beta_2\,z_j + R_{ij} \; . \tag{4.3}$$

Groups j can have a higher (or lower) value of β_{0j}, indicating that, for any given value of X, they tend to have higher (or lower) values of the dependent variable Y. Groups can also have a higher or lower value of β_{1j}, which indicates that the effect of X on Y is higher or lower. Since Z is a group-level variable, it would not make much sense conceptually to let the coefficient of Z depend on the group. Therefore β_2 is left unaltered in this formula.

The multilevel models treated in the following sections and in Chapter 5 contain diverse specifications of the varying coefficients β_{0j} and β_{1j}. The simplest version of model (4.3) is the version where β_{0j} and β_{1j} are constant (do not depend on j), i.e., the nesting structure has no effect, and we are back at model (4.1). In this case, the OLS[1] regression models of type (4.1) and (4.2) offer a good approach to analysing the data. If, on the other hand, the coefficients β_{0j} and β_{1j} do depend on j, then these regression models may give misleading results. Then it is preferable to take into account how the nesting structure influences the effects of X and Z on Y. This can be done using the random coefficient model of this and the following chapters. In this chapter the case is treated where the intercept β_{0j} depends on the group; the next chapter treats the case where also the regression coefficient β_{1j} is group-dependent.

4.2 Variable intercepts: fixed or random parameters?

Let us first consider only the regression on the level-one variable X. A first step towards modeling between-group variability is to let the intercept vary between groups. This reflects that some groups tend to have, on average, higher responses Y and others tend to have lower responses. This model is halfway between (4.1) and (4.3) (but omitting the effect of Z), in the sense that intercept β_{0j} does depend on the group but the regression coefficient of X, β_1, is constant:

$$Y_{ij} = \beta_{0j} + \beta_1\,x_{ij} + R_{ij} \; . \tag{4.4}$$

This is pictured in Figure 4.1. The group-dependent intercept can be split into an average intercept and the group-dependent deviation:

$$\beta_{0j} = \gamma_{00} + U_{0j} \; .$$

For reasons that will become clear in Chapter 5, the notation for the regression coefficients is changed here, and the average intercept is called γ_{00} while the regression coefficient for X is called γ_{10}. Substitution now leads to the model

$$Y_{ij} = \gamma_{00} + \gamma_{10}\,x_{ij} + U_{0j} + R_{ij} \; . \tag{4.5}$$

[1] Ordinary Least Squares.

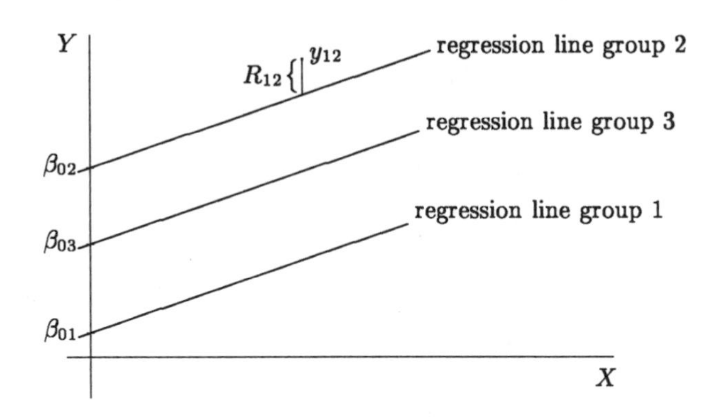

Figure 4.1 Different parallel regression lines.
The point y_{12} is indicated with its residual R_{12}.

The values U_{0j} are the main effects of the groups: conditional on an individual having a given X-value and being in group j, the Y-value is expected to be U_{0j} higher than in the average group. Model (4.5) can be understood in two ways:

(1) As a model where the U_{0j} are *fixed* parameters, N in number, of the statistical model. This is relevant if the groups j refer to categories each with its own distinct interpretation, e.g., a classification according to gender or religious denomination. In order to obtain identified parameters, the restriction that $\sum_j U_{0j} = 0$ can be imposed, so that effectively the groups lead to $N - 1$ regression parameters. This is the usual analysis of covariance model, in which the grouping variable is a factor. (Since we prefer to use Greek letters for statistical parameters and capital letters for random variables, we would prefer to use a Greek letter rather than U when we take this view of model (4.5).) In this specification it is impossible to use a group-level variable Z_j as an explanatory variable because it would be redundant given the fixed group effects.

(2) As a model where the U_{0j} are independent identically distributed *random* variables. Note that the U_{0j} are the unexplained group effects, which also may be called group residuals, controlling for the effects of variable X. These residuals now are assumed to be randomly drawn from a population with zero mean and an *a priori* unknown variance. This is relevant if the effects of the groups j (which can be neighborhoods, schools, companies, etc.), controlling for the explanatory variables, can be considered to be exchangeable. There is one parameter associated to the U_{0j} in this statistical model: their variance. This is the simplest random coefficient regression model. It is called the *random intercept model* because the group-dependent intercept, $\gamma_{00} + U_{0j}$, is a quantity which varies randomly from group to group. The groups are regarded as a sample from a population of groups. It is possible that there are group-level variables Z_j that express relevant

attributes of the groups (such variables will be incorporated in the model in Section 4.4 and, more extensively, in Section 5.2).

The next section discusses how to determine which of these specifications of model (4.5) is appropriate in a given situation.

Note that models (4.1) and (4.2) are OLS models, or fixed effects models, which do not take the nesting structure into account (except maybe by the use of a group-level variable Z_j), whereas models of type (1) above are OLS models that do take the nesting structure into account. The latter kind of OLS model has a much larger number of regression parameters, since in such models N groups lead to $N-1$ regression coefficients. It is important to distinguish between these two kinds of OLS models in discussions about how to handle data with a nested structure.

4.2.1 When to use random coefficient models?

These two different interpretations of equation (4.5) imply that multilevel data can be approached in two different ways, using models with fixed or with random coefficients. Which of these two interpretations is the most appropriate in a given situation depends on the focus of the statistical inference, the nature of the set of N groups, the magnitudes of the group sample sizes n_j, and the population distributions involved.

1. If the groups are regarded as unique entities and the researcher wishes primarily to draw conclusions pertaining to each of these N specific groups, then it is appropriate to use the analysis of covariance model. Examples are groups defined by gender or ethnic background.

2. If the groups are regarded as a sample from a (real or hypothetical) population and the researcher wishes to draw conclusions pertaining to this population, then the random coefficient model is appropriate. Examples are the groupings mentioned in Table 2.2.

3. If the researcher wishes to test effects of group-level variables, the random coefficient model should be used. The reason is that the fixed effects model already 'explains' all differences between group by the fixed effects, and there is no unexplained between-group variability left that could be explained by group-level variables. 'Random effects' and 'unexplained variability' are two ways of saying the same thing.

4. Especially for relatively small group sizes (in the range from 2 to 50 or 100), the random coefficient model has important advantages over the analysis of covariance model, provided that the assumptions about the random coefficients are reasonable. This can be understood as follows.

The random coefficient model includes the extra assumption of independent and identically distributed group effects U_{0j}. Stated less formally: the unexplained group effects are governed by 'mechanisms' that are roughly similar from one group to the next, and operate independently between the groups. The groups are said to be *exchangeable*. This assumption helps to counteract the paucity of the data that is implied by relatively small group sizes n_j. Since all group effects are assumed to come from the same population, the data from each group also have a bearing on inference with respect to the other groups, namely, through the information it provides about the population of groups.

In the analysis of covariance model, each of the U_{0j} is estimated as a separate parameter. If group sizes are small, then the data do not contain very much information about the values of the U_{0j} and there will be a considerable extent of overfitting in the analysis of covariance model: many parameters have large standard errors. This overfitting is avoided by using the random coefficient model, because the U_{0j} don't figure as parameters. If, on the other hand, the group sizes are large (say, 100 or more), then in the analysis of covariance the group-dependent parameters U_{0j} are estimated very precisely (with small standard errors), and the additional information that they come from the same population does not add much to this precision. In such a situation the difference between the results of the two approaches will be negligible.

5. The random coefficient model is mostly used with the additional assumption that the random coefficients, U_{0j} and R_{ij} in (4.5), are normally distributed. If this assumption is a very poor approximation, results obtained with this model may be unreliable. This can happen, e.g., when there are more outlying groups than can be accounted for by a normally distributed group effect U_{0j} with a common variance.

Other discussions about the choice between fixed and random coefficients can be found, e.g., in Searle et al. (1992; Section 1.4) and in Hsiao (1995, Section 8). An often mentioned condition for the use of random coefficient models is the restriction that the random coefficients should be independent of the explanatory variables. However, if there is a possible correlation between group-dependent coefficients and explanatory variables, this residual correlation can be removed, while continuing to use a random coefficient model, by also including effects of the group means of the explanatory variables. This is treated in Sections 4.5 and 9.2.1.

In order to choose between regarding the group-dependent intercepts U_{0j} as fixed statistical parameters and regarding them as random variables, a rule of thumb that often works in educational and social research is the following. This rule mainly depends on N, the number of groups in the data. If N is small, say, $N < 10$, then use the analysis of covariance approach: the problem with viewing the groups as a sample from a population is in this case, that the data will contain only scanty information about this population. If N is not small, say $N \geq 10$, while n_j is small or intermediate, say $n_j < 100$, then use the random coefficient approach: 10 or more groups is usually too large a number to be regarded as unique entities. If the group sizes n_j are large, say $n_j \geq 100$, then it does not matter much which view we take. However, this rule of thumb should be take with a large grain of salt and serves only to give a first hunch, not to determine the choice between fixed and random effects.

Populations and populations
When the researcher has indeed chosen to work with a random coefficient model, she must be aware that more than one population is involved in the multilevel analysis. Each level corresponds to a population! For a study of pupils in schools, there is a population of schools and a population of pupils. For voters in municipalities, there is a population of municipalities and a population of voters; etc. Recall that in this book we take a model-based view. This implies that the population are infinite hypothetical entities, that express 'what could be the case'. The random residuals and coefficients can be regarded as representing the effects of unmeasured variables and the approximate nature of the linear model. Randomness, in this sense, may be interpreted as unexplained variability.

Sometimes a random coefficient model can be used also when the population idea at the lower level is less natural. For example, in a study of longitudinal data where respondents are measured repeatedly, a multilevel model can be used with respondents at the second and measurements at the first level: measurements are nested within respondents. Then the population of respondents is an obvious concept. Measurements may be related to a population of time points. This will sometimes be natural, but not always. Another way of expressing the idea of random coefficient models in such a situation is to say that residual (non-explained) variability is present at level one as well as at level two, and this non-explained variability is represented by a probability model.

4.3 Definition of the random intercept model

In this text we treat the random coefficient view on model (4.5). This model, the *random intercept model*, is a simple case of the so-called ***hierarchical linear model.*** We shall not specifically treat the analysis of covariance model, and refer for this purpose to texts on analysis of variance and covariance. (For example, Cook and Campbell, 1979 or Stevens, 1996.) However, we shall encounter a number of considerations from the analysis of covariance that also play a role in multilevel modeling.

The empty model
Although this chapter follows an approach along the lines of regression analysis, the simplest case of the hierarchical linear model is the ***random effects analysis of variance model***, in which the explanatory variables, X and Z, do not figure. This model only contains random groups and random variation within groups. It can be expressed as a model[2] where the dependent variable is the sum of a general mean, γ_{00}, a random effect at the group level, U_{0j}, and a random effect at the individual level, R_{ij}:

$$Y_{ij} = \gamma_{00} + U_{0j} + R_{ij} \ . \tag{4.6}$$

[2]This is the same model encountered before in formula (3.1).

Groups with a high value of U_{0j} tend to have, on the average, high responses whereas groups with a low value of U_{0j} tend to have, on the average, low responses. The random variables U_{0j} and R_{ij} are assumed to have a mean of 0 (the mean of Y_{ij} is already represented by γ_{00}), to be mutually independent, and to have variances $\text{var}(R_{ij}) = \sigma^2$ and $\text{var}(U_{0j}) = \tau_0^2$. In the context of multilevel modeling (4.6) is called the *empty model*, because it contains not a single explanatory variable. It is important because it provides the basic partition of the variability in the data between the two levels. Given model (4.6), the total variance of Y can be decomposed as the sum of the level-two and the level-one variances,

$$\text{var}(Y_{ij}) = \text{var}(U_{0j}) + \text{var}(R_{ij}) = \tau_0^2 + \sigma^2 .$$

The covariance between two individuals (i and i', with $i \neq i'$) in the same group j is equal to the variance of the contribution U_{0j} that is shared by these individuals,

$$\text{cov}(Y_{ij}, Y_{i'j}) = \text{var}(U_{0j}) = \tau_0^2 ,$$

and their correlation is

$$\rho(Y_{ij}, Y_{i'j}) = \frac{\tau_0^2}{(\tau_0^2 + \sigma^2)} .$$

This parameter is just the *intraclass correlation coefficient* $\rho_I(Y)$ which we encountered already in Chapter 3. It can be interpreted in two ways: it is the correlation between two randomly drawn individuals in one randomly drawn group, and it is also the fraction of total variability that is due to the group level.

Example 4.1 *Empty model for language scores in elementary schools.*
In this example a data set is used that will be used in examples in many chapters of this book. The data set is concerned with grade 8 pupils (age about 11 years) in elementary schools in The Netherlands. After deleting pupils with missing values, the number of pupils is $M = 2287$, and the number of schools is $N = 131$. Class sizes in the original data set range from 10 to 42. In the data set reduced by deleting cases with missing data, the class sizes range from 4 to 35. The nesting structure is pupils within classes.

The dependent variable is the score on a language test. Most of the analyses of this data set in this book are concerned with investigating how the language test score depends on the pupil's intelligence and his or her family's social-economic status, and on a number of school or class variables.

Fitting the empty model yields the parameter estimates presented in Table 4.1. The 'deviance' in this table is given for the sake of completeness and later reference. This concept is explained in Chapter 6.

The estimates $\hat{\sigma}^2 = 64.57$ and $\hat{\tau}_0^2 = 19.42$ yield an intraclass correlation coefficient of $\hat{\rho} = 19.42/83.99 = 0.23$. This is rather high, compared to other results of educational research (values between 0.05 and 0.20 are common). This indicates that the grouping according to classes leads to an important similarity between the results of different pupils in the same class, although (as practically always) within-class differences are far larger than between-class differences.

For the overall distribution of the language scores, these estimates provide a mean of 40.36 and a standard deviation of $\sqrt{19.42 + 64.57} = 9.16$. The mean

Table 4.1 Estimates for empty model.

Fixed Effect	Coefficient	S.E.
γ_{00} = Intercept	40.36	0.43
Random Effect	**Variance Component**	**S.E.**
Level-two variance:		
$\tau_0^2 = \text{var}(U_{0j})$	19.42	2.92
Level-one variance:		
$\sigma^2 = \text{var}(R_{ij})$	64.57	1.97
Deviance	16253.2	

of 40.36 should be interpreted as the expected value of the language score for a random pupil in a randomly drawn class. This is close, but not identical, to the raw mean 40.94 and standard deviation 9.00 of the sample of 2287 pupils. The reason for this difference is that the estimation of model (4.6) implies a weighting of the various classes that is not taken into account in the calculation of the raw mean and standard deviation.

The estimates obtained from multilevel software for the two variance components, τ_0^2 and σ^2, will usually be slightly different from the estimates obtained from the formulae in Section 3.3.1. The reason is that different estimation methods are used: multilevel software uses the more efficient ML or REML method (cf. Section 4.6 below), which cannot be expressed in an explicit formula; the formulae of Section 3.3.1 are explicit but less efficient.

One explanatory variable

The following step is the inclusion of explanatory variables. These are used to try to explain part of the variability of Y; this refers to variability at level two as well as level one. With just one explanatory variable X, model (4.5) is produced:

$$Y_{ij} = \gamma_{00} + \gamma_{10}\, x_{ij} + U_{0j} + R_{ij} \; . \tag{4.5}$$

Essential assumptions are that all residuals, U_{0j} and R_{ij}, are mutually independent and have zero means given the values x_{ij} of the explanatory variable. For the U_{0j}, just as for the R_{ij}, it is assumed that they are drawn from normally distributed populations. The population variance of the lower level residuals R_{ij} is assumed to be constant across the groups, and is again denoted σ^2; the population variance of the higher level residuals U_{0j} is denoted τ_0^2. Thus, model (4.5) has four parameters: the regression coefficients γ_{00} and γ_{10} and the variance components σ^2 and τ_0^2.

The random variables U_{0j} can be regarded as residuals at the group level, or group effects that are left unexplained by X. Since residuals, or random errors, contain those parts of the variability of the dependent variable that are not modeled explicitly as a function of explanatory variables, this model contains unexplained variability at two nested levels. This partition of unexplained variability over the various levels is the essence of hierarchical random effects models.

The fixed intercept γ_{00} is the intercept for the average group. The regression coefficient γ_{10} can be interpreted as an unstandardized regression coefficient in the usual way: one unit increase in the value of X is associated with an average increase in Y of γ_{10} units. The residual variance, i.e., the variance conditional on the value of X, is

$$\mathrm{var}(Y_{ij} \mid x_{ij}) = \mathrm{var}(U_{0j}) + \mathrm{var}(R_{ij}) = \tau_0^2 + \sigma^2 ,$$

while the covariance between two different individuals (i and i', with $i \neq i'$) in the same group is

$$\mathrm{cov}(Y_{ij}, Y_{i'j} \mid x_{ij}, x_{i'j}) = \mathrm{var}(U_{0j}) = \tau_0^2 .$$

The fraction of residual variability that can be ascribed to level one is given by $\sigma^2/(\sigma^2 + \tau_0^2)$, and for level two this fraction is $\tau_0^2/(\sigma^2 + \tau_0^2)$.

Of the covariance or correlation between two individuals in the same group, a part may be explained by their X-values, and another part is unexplained. This unexplained, or residual, correlation between them is the *residual intraclass correlation coefficient*,

$$\rho_I(Y|X) = \frac{\tau_0^2}{\sigma^2 + \tau_0^2} .$$

This parameter is the correlation between the Y-values of two randomly drawn individuals in one randomly drawn group, controlling for variable X. It is analogous to the usual intraclass correlation coefficient, but now controlling for X. The formula for the (non-residual, or raw) intraclass correlation coefficient was just the same, but there the variance parameters τ_0^2 and σ^2 referred to the variances in the empty model whereas now they refer to the variances in model (4.5), which includes the effect of variable X.

If model (4.5) is valid while the intraclass correlation coefficient is 0, i.e., $U_{0j} = 0$ for all groups j, then the grouping is irrelevant for the Y-variable conditional on X, and one could have used ordinary linear regression, i.e., a model such as (4.1). If the residual intraclass correlation coefficient, or equivalently, τ_0^2, is positive, then the hierarchical linear model is a better analysis method than ordinary least squares ('OLS') regression analysis, among others, because the standard errors of the estimated coefficients produced by ordinary regression analysis are not to be trusted. This was discussed already in Section 3.2.[3]

The residual intraclass correlation coefficient is positive if and only if the intercept variance τ_0^2 is positive. Therefore testing the residual intraclass correlation coefficient amounts to the same as testing the intercept variance. Tests for this parameter are presented in Sections 6.2 and 6.3.

---

[3]**Mind the word** *valid*! You may have fitted model (4.5) and obtained an estimated value for $\rho_I(Y)$ that is quite low or even 0. This does not exclude the possibility that there are group differences of the kind treated in Chapter 5: random slopes. If there is an important random slope variance, even without an intercept variance, ordinary regression analysis may yield incorrect results and less insight than the hierarchical linear model.

Example 4.2 *Random intercept and one explanatory variable: IQ.*
As a variable at the pupil level that is essential for explaining the language score, we use the measure for verbal IQ taken from the ISI test. The IQ score has been centered, so that its mean is 0. This facilitates interpretation of various parameters. Its standard deviation in this data set is 2.07 (this is calculated as a descriptive statistic, without taking the grouping into account). The results are presented in Table 4.2.

Table 4.2 Estimates for random intercept model with effect for IQ.

Fixed Effect	Coefficient	S.E.
γ_{00} = Intercept	40.61	0.31
γ_{10} = Coefficient of IQ	2.488	0.070
Random Effect	**Variance Component**	**S.E.**
Level-two variance:		
$\tau_0^2 = \text{var}(U_{0j})$	9.50	1.52
Level-one variance:		
$\sigma^2 = \text{var}(R_{ij})$	42.23	1.29
Deviance	15251.8	

In the model presented in Table 4.2 each class, indexed by the letter j, has its own regression line, given by

$$Y = 40.61 + U_{0j} + 2.488\,\text{IQ} .$$

The U_{0j} are class-dependent deviations of the intercept and have a mean of 0 and a variance of 9.50, hence, a standard deviation of $\sqrt{9.50} = 3.08$. Figure 4.2 is a picture of fifteen such random regression lines. This figure can be regarded as a random sample from the population of schools defined by Table 4.2.

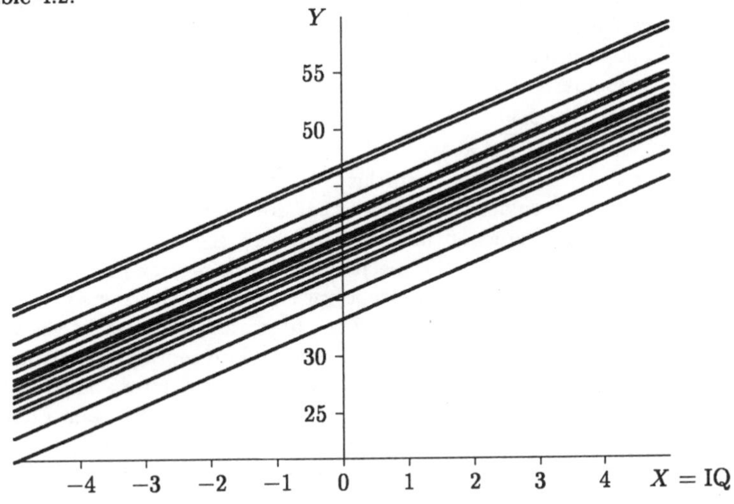

Figure 4.2 Fifteen randomly chosen regression lines according to the random intercept model of Table 4.2.

The scatter around the regression lines (i.e., the vertical distances R_{ij} between the observations and the regression line for the class under consideration) has a variance of 42.23 and therefore a standard deviation of $\sqrt{42.23} = 6.50$. These distances between observations and regression lines therefore tend to be much bigger than the vertical distances between the regression lines. However, the distances between the regression lines are not negligible. A school with a typical low average achievement (bottom 2.5 percent) will have a value of U_{0j} of about two standard deviations below the mean, so that it will have a regression line

$$Y = 40.61 - 2 \times 3.08 + 2.488 \, \text{IQ} = 34.45 + 2.488 \, \text{IQ} \; ,$$

whereas a school with a typical high achievement (top 2.5 percent) will have a regression line

$$Y = 40.61 + 2 \times 3.08 + 2.488 \, \text{IQ} = 46.77 + 2.488 \, \text{IQ} \; .$$

There appears to be a strong effect of IQ. Each additional measurement unit of IQ leads, on average, to 2.488 additional measurement units of the language score. To obtain a scale for effect that is independent of the measurement units, one can calculate ***standardized coefficients***, i.e., coefficients expressed in standard deviations as scale units. These are the coefficients that would be obtained if all variables were rescaled to unit variances. They are given by

$$\frac{\text{S.D.}(X)}{\text{S.D.}(Y)} \, \gamma \; ,$$

in this case estimated by $(2.07/9.16) \times 2.488 = 0.56$. In other words, each additional standard deviation on IQ leads, on average, to an increase in language score of 0.56 standard deviations.

The residual variance σ^2 as well as the random intercept variance τ_0^2 are much lower in this model than in the empty model (cf. Table 4.1). The residual variance is lower because between-pupil differences are partially explained. The intercept variance is lower because classes differ in average IQ score, so that this pupil-level variable also explains part of the differences between classes. The residual intraclass correlation is estimated by

$$\hat{\rho}_I(Y|X) = \frac{9.50}{42.23 + 9.50} = 0.18 \; ,$$

somewhat smaller than the raw intraclass correlation of 0.23 (see Table 4.1). These results may be compared to those that are obtained from an ordinary least squares (OLS) regression analysis, in which the nesting of pupils in classes is not taken into account. This analysis can be regarded as an analysis using model (4.5) in which the intercept variance τ_0^2 is constrained to be 0. The results are displayed in Table 4.3. The parameter estimates for the OLS method seem rather close to those for the random intercept model. However, the regression coefficient for IQ differs about 2 standard errors between the two models. This implies that, although the numerical values seem similar, they are nevertheless rather different from a statistical point of view. Further, the standard error of the intercept is twice as large in the results from the random intercept model as in those from the OLS analysis. This indicates that the OLS analysis produces an over-optimistic impression of the precision of this estimate, and illustrates the lack of trustworthiness of OLS estimates for multilevel data.

Table 4.3 Estimates for ordinary least squares regression.

Fixed Effect	Coefficient	S.E.
γ_{00} = Intercept	40.93	0.15
γ_{10} = Coefficient of IQ	2.654	0.072
Random Effect	Variance Component	S.E.
Level-one variance:		
$\sigma^2 = \text{var}(R_{ij})$	50.90	1.51
Deviance	15477.7	

4.4 More explanatory variables

Just like in multiple regression analysis, more than one explanatory variable can be used in the random intercept model. When the explanatory variables at the individual level are denoted $X_1, \ldots, X_p$, and those at the group level $Z_1, \ldots, Z_q$, adding their effects to the random intercept model leads to the following formula[4]

$$Y_{ij} = \gamma_{00} + \gamma_{10}\, x_{1ij} + \ldots + \gamma_{p0}\, x_{pij} + \gamma_{01}\, z_{1j} + \ldots + \gamma_{0q}\, z_{qj} + U_{0j} + R_{ij} \, . \tag{4.7}$$

The regression parameters, γ_{h0} $(h = 1, ..., p)$ and γ_{0h} $(h = 1, ..., q)$ for level-one and level-two explanatory variables, respectively, again have the same interpretation as unstandardized regression coefficients in multiple regression models: one unit increase in the value of X_h (or Z_h) is associated with an average increase in Y of γ_{h0} (or γ_{0h}) units. Just like in multiple regression analysis, some of the variables X_h and Z_h may be interaction variables, or non-linear (e.g., quadratic) transforms of basic variables.

The first part of the right-hand side of (4.7), incorporating the regression coefficients,

$$\gamma_{00} + \gamma_{10}\, x_{1ij} + \ldots + \gamma_{p0}\, x_{pij} + \gamma_{01}\, z_{1j} + \ldots + \gamma_{0q}\, z_{qj} \; ,$$

is called the *fixed part* of the model, because the coefficients are fixed (i.e., not stochastic). The remaining part,

$$U_{0j} + R_{ij} \; ,$$

is called the *random part* of the model.

It is again assumed that all residuals, U_{0j} and R_{ij}, are mutually independent and have zero means given the values of the explanatory variables. A somewhat less crucial assumption is that these residuals are drawn from normally distributed populations. The population variance of the level-one residuals R_{ij} is denoted σ^2 while the population variance of the level-two residuals U_{0j} is denoted τ_0^2.

Due to the two-level nature of the model under consideration, two special kinds of variables deserve special attention. These are group-level variables

[4]The subscripts of the regression coefficients γ seem a bit baroque. The reason is to obtain consistency with the notation in the next chapter, cf. (5.12).

that arise as aggregates of individual-level variables; and cross-level interactions. The latter will be treated in Chapter 5; the former are treated here and in the next section.

Some level-two variables are directly defined for the units of this level; others are defined through the subunits (units at level one) that are comprised in this unit. For example, when the nesting structure refers to children within families, the type of dwelling is directly a family-dependent variable, but the average age of the children is based on aggregation of a variable (age) that is itself a level-one variable. As another example, referring to a longitudinal study of primary school children, where the nesting structure refers to repeated measurements within individual children, the gender of the child is a direct level-two variable, whereas the average reading score over the age period of 6–12 years, or the slope of the increase in reading score over this period, is a level-two variable that is an aggregate of a level-one variable.

Varying group sizes

In most research, the group sizes n_j are variable between groups. Even in situations such as repeated measures designs, where level two corresponds to individual subjects and level one to time points of measurement of these subjects, and where it is often the researcher's intention to measure all subjects at the same moments, there often occur missing data which again leads to a data structure with variable n_j. This does not constitute a problem for the application of the hierarchical linear model in any way. The hierarchical linear model can even be applied if some groups have size $n_j = 1$, as long as some other groups have greater sizes. Of course, the smaller groups will have a smaller influence on the results than the larger groups.

However, it can be worthwhile to give some thought to the substantive meaning of the group size. The number of pupils in a school class may have effects on teaching and learning processes, the size of a department in a firm may have influence on the performance of the department and the interpersonal relations. In such cases it can be advisable to include group size in the model as an explanatory variable with a fixed effect. When the fraction of missing data differs between the groups it can be useful to check whether there is a systematic bias in data being missing by testing whether there is an effect of the fraction of missing data per group, including for this purpose this variable in the fixed part of the model.

4.5 Within- and between-group regressions

Especially group means are an important type of explanatory variable. A group mean for a given level-one explanatory variable is defined as the mean over all individuals, or level-one units, within the given group, or level-two unit.[5] This can be an important contextual variable. The group mean of a level-one explanatory variable allows to express the difference between *within-group* and *between-group* regressions, as was proposed already by Davis et al. (1961). We already saw in Section 3.6 that the coefficients for these two types of regression can be completely different. The within-group regression coefficient expresses the effect of the explanatory variable within a given group; the between-group regression coefficient expresses the effect of the group mean of the explanatory variable on the group mean of the dependent variable. In other words, the between-group regression coefficient is just the coefficient in a regression analysis for data that are aggregated (by averaging) to the group level.

Pursuing the example of children within families, suppose that we are interested in the amount of pocket money the children receive. This will depend on the child's age, but it could also depend on the average age of the children in the family. The within-group regression coefficient measures the effect of age differences within a given family; the between-group regression coefficient measures the effect of average age on the average pocket money received by the children in the family. In a simple model, not taking other variables into account, denote age (in years) of child i in family j by x_{ij}, and the average age of all children in family j by $z_j = \bar{x}_{.j}$. In the model

$$Y_{ij} = \gamma_{00} + \gamma_{10}\, x_{ij} + U_{0j} + R_{ij} \ ,$$

the within-group and between-group regression coefficients are forced to be equal. If we add the family mean, $z_j = \bar{x}_{.j}$, as an explanatory variable, we obtain

$$Y_{ij} = \gamma_{00} + \gamma_{10}\, x_{ij} + \gamma_{01}\bar{x}_{.j} + U_{0j} + R_{ij} \ . \tag{4.8}$$

This model is more flexible in that the within-group regression coefficient is allowed to differ from the between-group regression coefficient. This can be seen as follows.

If model (4.8) is considered within a given group, the terms can be reordered as

$$Y_{ij} = (\gamma_{00} + \gamma_{01}\bar{x}_{.j} + U_{0j}) + \gamma_{10}\, x_{ij} + R_{ij} \ .$$

The part between parentheses is the (random!) intercept for this group, and the regression coefficient of X within this group is γ_{10}. The systematic (i.e., non-random) part is the within-group regression line

$$Y = (\gamma_{00} + \gamma_{01}\bar{x}_{.j}) + \gamma_{10}\, x \ .$$

On the other hand, taking the group average at both sides of the equality sign in (4.8) yields the between-group regression model,

$$\begin{aligned} \bar{Y}_{.j} &= \gamma_{00} + \gamma_{10}\bar{x}_{.j} + \gamma_{01}\bar{x}_{.j} + U_{0j} + \bar{R}_{.j} \\ &= \gamma_{00} + (\gamma_{10} + \gamma_{01})\bar{x}_{.j} + U_{0j} + \bar{R}_{.j} \ . \end{aligned}$$

[5] If cases with missing data are deleted for doing the multilevel analysis, then one should calculate the group mean as the average over all level-one units for which this particular variable is available, and before deleting cases because they have missing values on other variables!

The systematic part of this model is represented by the regression line

$$Y = \gamma_{00} + (\gamma_{10} + \gamma_{01})\, x \ .$$

This shows that the between-group regression coefficient is $\gamma_{10} + \gamma_{01}$.

The difference between the within-group and the between-group regression coefficients can be tested in this model by testing the null hypothesis that $\gamma_{01} = 0$ by the method of Section 6.1. (This test is a version of what is known in econometrics as the Hausman specification test; see Hausman and Taylor, 1981, or Baltagi, 1995, Section 4.3.) Figure 4.3 gives a sketch of within-group regression lines that differ from the between-group regression line in the sense that the between-group regression is stronger ($\gamma_{01} > 0$). The reverse, where the between-group regression line is less steep than the within-group regression lines, can also be the case.

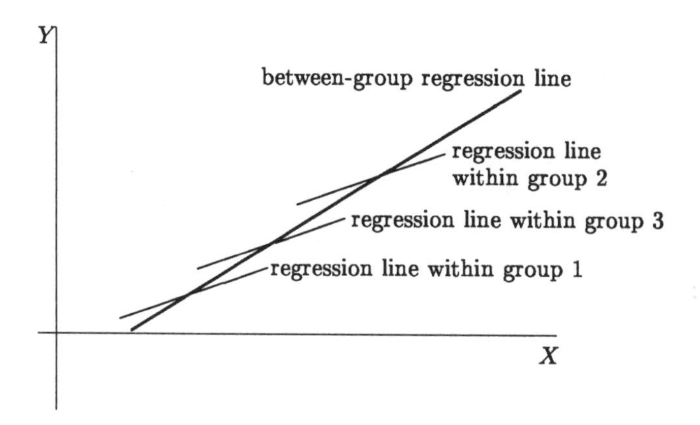

Figure 4.3 Different between-group and within-group regression lines.

If the within- and between-group regression coefficients are different, then it often is convenient to replace x_{ij} in (4.8) by the ***within-group deviation score***, defined as $x_{ij} - \bar{x}_{.j}$. To distinguish the corresponding parameters from those in (4.8), they are denoted by $\tilde{\gamma}$. The resulting model is

$$Y_{ij} = \tilde{\gamma}_{00} + \tilde{\gamma}_{10}\,(x_{ij} - \bar{x}_{.j}) + \tilde{\gamma}_{01}\,\bar{x}_{.j} + U_{0j} + R_{ij} \ . \qquad (4.9)$$

This model is statistically equivalent to model (4.8) but it has a more convenient parametrization because the between-group regression coefficient now is

$$\tilde{\gamma}_{01} = \gamma_{10} + \gamma_{01} \ , \qquad (4.10)$$

while the within-group regression coefficient is

$$\tilde{\gamma}_{10} = \gamma_{10} \ . \qquad (4.11)$$

The use of the within-group deviation score is called ***within-group centering***; some computer programs for multilevel analysis have special facilities for this.

Example 4.3 *Within- and between-group regressions for IQ.*
We continue example 2 by allowing differences between the within-group and between-group regressions of the language score on IQ. The results are displayed in Table 4.4. IQ here is the raw variable, i.e., without group-centering. In other words, the results refer to model (4.8).

Table 4.4 Estimates for random intercept model with different within- and between-group regressions.

Fixed Effect	Coefficient	S.E.
γ_{00} = Intercept	40.74	0.28
γ_{10} = Coefficient of IQ	2.415	0.072
γ_{01} = Coefficient of $\overline{\text{IQ}}$ (group mean)	1.589	0.313
Random Effect	**Variance Component**	**S.E.**
Level-two variance:		
$\tau_0^2 = \text{var}(U_{0j})$	7.73	1.30
Level-one variance:		
$\sigma^2 = \text{var}(R_{ij})$	42.15	1.28
Deviance	15227.5	

The within-group regression coefficient is 2.415 and the between-group regression coefficient is 2.415 + 1.589 = 4.004. A pupil with a given IQ obtains, on average, a higher language test score if he or she is in a class with a higher average IQ. In other words, the context effect of mean IQ gives an additional contribution over and above the effect of individual IQ. A figure for these results is qualitatively similar to Figure 4.3 in the sense that the within-group regression lines are less steep than the between-group regression line.
This table represents within each class, denoted j, a linear regression equation

$$Y = 40.74 + U_{0j} + 2.415\,\text{IQ} + 1.589\,\overline{\text{IQ}}$$

where U_{0j} is a class-dependent deviation with mean 0 and variance 7.73 (standard deviation 2.78). The within-class deviations about this regression equation, R_{ij}, have a variance of 42.15 (standard deviation 6.49). Within each class, the effect (regression coefficient) of IQ is 2.415, so the regression lines are parallel. Classes differ in two ways: they may have different mean IQ values, which affects the expected results Y through the term $1.589\,\overline{\text{IQ}}$; this is an explained difference between the classes; and they have randomly differing values for U_{0j}, which is an unexplained difference. These two ingredients contribute to the class-dependent intercept, given by $40.74 + U_{0j} + 1.589\,\overline{\text{IQ}}$.

The within-group and between-group regression coefficients would be equal if, in formula (4.8), the coefficient of average IQ would be 0, i.e., $\gamma_{01} = 0$. This null hypothesis can be tested (see Section 6.1) by the t-ratio defined as

$$t = \frac{\text{estimate}}{\text{standard error}},$$

given here by 1.589/0.313 = 5.08, a highly significant result. In other words, we may conclude that the within- and between-group regression coefficients are different indeed.

If the individual IQ variable had been replaced by within-group deviation scores $\text{IQ}_{ij} - \overline{\text{IQ}}_{.j}$, i.e., model (4.9) had been used, then the estimates obtained would have been $\hat{\gamma}_{10} = 2.415$ and $\hat{\gamma}_{01} = 4.004$, cf. formulae (4.10) and (4.11). Indeed, the regression equation given above can be described equivalently by

$$Y = 40.74 + U_{0j} + 2.415\,(\text{IQ} - \overline{\text{IQ}}) + 4.004\,\overline{\text{IQ}}\ ,$$

which indicates explicitly that the within-group regression coefficient is 2.415 while the between-group regression coefficient, i.e., the coefficient of the group means $\overline{Y}$ on the group means $\overline{\text{IQ}}$, is 4.004.

When interpreting the results of a multilevel analysis, it is important to keep in mind that the conceptual interpretation of within-group and between-group regression coefficients usually is completely different. These two coefficients may express quite contradictory mechanisms. This is related to the shift of meaning and the ecological fallacy discussed in Section 3.1. For theoretically important variables in multilevel studies, it is the rule rather than the exception that within-group regression coefficients differ from between-group regression coefficients (although the statistical significance of this difference may be another matter, depending as it is on sample sizes, etc.).

4.6 Parameter estimation

The random intercept model (4.7) is defined by its statistical parameters: the regression parameters, γ, and the variance components, σ^2 and τ_0^2. Note that the random effects, U_{0j}, are not parameters in a statistical sense, but latent (i.e., not directly observable) variables. The literature (e.g., Longford, 1993a) contains two major estimation methods for estimating the statistical parameters, under the assumption that the U_{0j} as well as the R_{ij} are normally distributed: maximum likelihood (ML) and residual (or restricted) maximum likelihood (REML).

The two methods differ little with respect to estimating the regression coefficients, but they do differ with respect to estimating the variance components. A very brief indication of the difference between the two estimation methods is, that the REML method estimates the variance components while taking into account the loss of degrees of freedom resulting from the estimation of the regression parameters, whereas the ML method does not take this into account. The result is that the ML estimators for the variance components have a downward bias, and the REML estimators don't. For example, the usual variance estimator for a single-level sample, in which the sum of squared deviations is divided by sample size minus 1, is a REML estimator; the corresponding ML estimator divides instead by the total sample size. The difference can be important especially when the number of groups is small. For a large number of groups (as a rule of thumb, 'large' here means larger than 30), the difference between the ML and the REML estimates is immaterial. The literature suggests that the REML method is preferable with respect to the estimation of the variance parameters (and the covariance parameters for the more general models treated in Chapter 5). When one wishes to carry out deviance tests (see Section 6.2), it sometimes is required to use ML rather than REML estimates.[6]

Various algorithms are available to determine these estimates. They have names such as EM (Expectation - Maximization), Fisher scoring, IGLS (Iterative Generalized Least Squares), and RIGLS (Residual or Restricted IGLS). They are iterative, which means that a number of steps are taken in which a provisional estimate comes closer and closer to the final estimate. When all goes well, the steps converge to the ML or REML estimate. Technical details can be found, e.g., in Bryk and Raudenbush (1992), Goldstein (1995), or Longford (1993a, 1995). In principle, the algorithms all yield the same estimates for a given estimation method (ML or REML). The differences are that for some complicated models, the algorithms may vary in the amount of computational problems (sometimes one algorithm may converge and the other not), and that computing time may be different. For the practical user, the differences between the algorithms are hardly worth thinking about.

An aspect of the estimation of hierarchical linear model parameters that surprises some users of this model is the fact that it is possible that the variance parameters, in model (4.7) notably parameter τ_0^2, can be estimated to be exactly 0. The value of 0 is then also reported for the standard error by many computer programs! Even when the estimate is $\hat{\tau}_0^2 = 0$, this does not mean that the data imply absolute certainty that the population value of τ_0^2 is equal to 0. Such an estimate can be understood as follows. For simplicity, consider the empty model, (4.6). The level-one residual variance σ^2 is estimated by the pooled within-group variance. The parameter τ_0^2 is estimated by comparing this within-group variability to the between-group variability. The latter is determined not only by τ_0^2 but also by σ^2, since

$$\text{var}(\overline{Y}_{.j}) = \tau_0^2 + \frac{\sigma^2}{n_j} \,. \tag{4.12}$$

Note that τ_0^2, being a variance, cannot be negative. This implies that, even if $\tau_0^2 = 0$, a positive between-group variability is expected. If observed between-group variability is equal to or smaller than what is expected from (4.12) in case $\tau_0^2 = 0$, then the estimate $\hat{\tau}_0^2 = 0$ is reported (cf. the discussion following (3.11)).

If the group sizes n_j are variable, the larger groups will, naturally, have a larger influence on the estimates than the smaller groups. The influence of group size on the estimates is, however, mediated by the intraclass correlation coefficient. Consider, e.g., the estimation of the mean intercept, γ_{00}. If the residual intraclass correlation is 0, the groups have an influence on the estimated value of γ_{00} that is proportional to their size. In the extreme case that the residual intraclass correlation is 1, each group has an equally large influence, independent of its size. In practice, where the residual intraclass

[6] When models are compared with different fixed parts, deviance tests should be based on ML estimation. Deviance tests with REML estimates may be used for comparing models with different random parts and the same fixed part. Different random parts will be treated in the next chapter.

correlation is between 0 and 1, larger groups will have a larger influence, but less than proportionately.

4.7 'Estimating' random group effects: posterior means

The random group effects U_{0j} are latent variables rather than statistical parameters, and therefore are not estimated as an integral part of the statistical parameter estimation. However, there can be many reasons why it can nevertheless be desirable to 'estimate' them.[7] This can be done by a method known as *empirical Bayes estimation* which produces so-called *posterior means*, see, e.g., Efron and Morris (1975). The basic idea of this method is that U_{0j} is 'estimated' by combining two kinds of information:

(1) the data from group j,
(2) the fact (or, rather, the model assumption) that the unobserved U_{0j} is a random variable just like all other random group effects, and therefore has a normal distribution with mean 0 and variance τ_0^2.

In other words, data information is combined with population information.

The formula is given here only for the empty model, i.e., the model without explanatory variables. The idea for more complicated models is analogous; formulae can be found in the literature, e.g. Longford (1993a, Section 2.10).

The empty model was formulated in (4.6) as

$$Y_{ij} = \beta_{0j} + R_{ij} = \gamma_{00} + U_{0j} + R_{ij} \ .$$

Since γ_{00} is already an estimated parameter, an estimate for β_{0j} will be the same as an estimate for U_{0j} plus γ_{00}. Therefore, estimating β_{0j} and estimating U_{0j} are equivalent problems given that an estimate for γ_{00} is available.

If we used only group j, β_{0j} would be estimated by the group mean, which is also the OLS estimate,

$$\hat{\beta}_{0j} = \overline{Y}_{.j} \ . \tag{4.13}$$

If we looked only at the population, we would estimate β_{0j} by its population mean, γ_{00}. This parameter is estimated by the overall mean,

$$\hat{\gamma}_{00} = \overline{Y}_{..} = \sum_{j=1}^{N} \frac{n_j}{M} \overline{Y}_{.j} \ ,$$

where $M = \sum_j n_j$ denotes the total sample size. Another possibility is to combine the information from group j with the population information. The optimal combined 'estimate' for β_{0j} is a weighted average of the two previous estimates:

$$\hat{\beta}_{0j}^{\text{EB}} = \lambda_j \, \hat{\beta}_{0j} + (1 - \lambda_j) \, \hat{\gamma}_{00} \ , \tag{4.14}$$

[7] The word estimate is put between quotation marks because the proper statistical term for finding likely values of the U_{0j}, being random variables, is ***prediction***. The term of estimation is reserved for finding likely values for statistical parameters. Since prediction is associated in everyday speech, however, with determining something about the future, we prefer to speak here about 'estimation' between parentheses.

where EB stands for 'empirical Bayes' and the weight λ_j is defined as the reliability of the mean of group j (see equation (3.21)),

$$\lambda_j = \frac{\tau_0^2}{\tau_0^2 + \sigma^2/n_j} .$$

The ratio of the two weights, $\lambda_j/(1-\lambda_j)$, is just the ratio of true variance τ_0^2 to error variance σ^2/n_j. In practice we do not know the true values of the parameters σ^2 and τ_0^2, and we substitute estimated values to calculate (4.14).

Formula (4.14) is called the posterior mean, or the empirical Bayes estimate, for β_{0j}. This term comes from Bayesian statistics. It refers to the distinction between the *prior* knowledge about the group effects, which is based only on the population from which they are drawn, and the *posterior* knowledge which is based also on the observations made about this group. There is an important parallel between random coefficient models and Bayesian statistical models, because the random coefficients used in the hierarchical linear model are analogous to the random parameters that are essential in the Bayesian statistical paradigm. This empirical Bayes estimate is treated from a Bayesian standpoint, e.g., in Press (1989, p. 43).

Formula (4.14) can be regarded as follows: the OLS estimate (4.13) for group j is pushed a bit toward the general mean $\hat{\gamma}_{00}$. This is an example of *shrinkage to the mean* just like is being used, e.g., in psychometrics, for the estimation of true scores. The corresponding estimator sometimes is called the Kelley estimator; see, e.g., Kelley (1927), Lord and Novick (1968), or other textbooks on classical psychological test theory. From the definition of the weight λ_j it is apparent that the influence of the data of group j itself becomes larger as group size n_j becomes larger. For large groups, the posterior mean is practically equal to $\hat{\beta}_{0j}$, the intercept that would be estimated from data on group j alone.

In principle, the OLS estimate (4.13) and the empirical Bayes estimate (4.14) both are sensible procedures for estimating the mean of group j. The former does not need the assumption that group j is a random element from the population of groups, and is an unbiased estimate.[8] The latter is biased toward the population mean, but for a randomly drawn group it has a smaller mean squared error. The squared error averaged over all groups will be smaller for the empirical Bayes estimate, but the price is a conservative (drawn to the average) appraisal of the groups with truly very high or very low values of β_{0j}. The estimation variance of the empirical Bayes estimate is

$$\text{var}\left(\hat{\beta}_{0j}^{\text{EB}} - \beta_{0j}\right) = (1-\lambda_j)\,\tau_0^2 , \tag{4.15}$$

if the uncertainty due to the estimation of γ_{00} (which is of secondary importance anyway) is neglected. This formula also is well-known from classical psychological test theory (e.g., Lord and Novick, 1968).

[8] Unbiasedness means that the average of many – hypothetical – independent replications of this estimate for this particular group j would be very close to the true value β_{0j}.

The same principle can be applied (but with more complicated formulae) to the 'estimation' of the group-dependent intercept $\beta_{0j} = \gamma_{00} + U_{0j}$ in random intercept models that do include explanatory variables, such as (4.7). This intercept can be 'estimated' again by γ_{00} plus the posterior mean of U_{0j}, and is then also referred to as the ***posterior intercept.***

Instead of being primarily interested in the intercept as defined by $\gamma_{00} + U_{0j}$, which is the value of the regression equation for *all* explanatory variables having the value 0, one may also be interested in the value of the regression line for group j for the case where only the level-one variables x_{1ij} to x_{pij} are 0 while the level-two variables have the values proper to this group. To 'estimate' this version of the intercept of group j, we use

$$\hat{\gamma}_{00} + \hat{\gamma}_{01}\, z_{1j} + \ldots + \hat{\gamma}_{0q}\, z_{qj} + \hat{U}^{\mathrm{EB}}_{0j} \,, \tag{4.16}$$

where the values $\hat{\gamma}$ indicate the (ML or REML) estimates of the regression coefficients. These values also are sometimes called posterior intercepts.

The posterior means (4.14) can be used, e.g., to see which groups have unexpectedly high or low values on the outcome variable, given their values on the explanatory variables. They can also be used in a residual analysis, for checking the assumption of normality for the random group effects, and for detecting outliers, cf. Chapter 9. The posterior intercepts (4.16) indicate the total main effect of group j, controlling for the level-one variables X_1 to X_p, but including the effects of the level-two variables Z_1 to Z_q. For example, in a study of pupils in schools where the dependent variable is a relevant indicator of scholastic performance, these posterior intercepts could be valuable information for the parents indicating the contribution of the various schools to the performance of their beloved children.

Example 4.4 *Posterior means for random data.*
We can illustrate the 'estimation' procedure by returning to the random digits table (Chapter 3, Table 3.1). Macro unit 04 in that table has an average of $\overline{Y}_{.j} = 31.5$ over its 10 random digits. The grand mean of the total 100 random digits is $\overline{Y}_{..} = 47.2$. The average of macro unit 04 thus seems to be far below the grand mean. But the reliability of this mean is only $\lambda_j = 26.7/\{26.7 + (789.7/10)\} = 0.25$. Applying (4.14), the posterior mean is calculated as

$$0.25 \times 31.5 + (1 - 0.25) \times 47.2 = 43.3 \,.$$

In words: the posterior mean for macro-unit 04 is determined for 75 percent (i.e., $1 - \lambda_j$) by the grand mean of 47.2 and by only 25 percent (i.e., λ_j) by its OLS mean of 31.5. The shrinkage to the grand mean is evident. Because of the low estimated intraclass correlation of $\hat{\rho}_I = 0.03$ and the low number of observations per macro unit, $n_j = 10$, the empirical Bayes estimate of the average of macro unit 04 is closer to the grand mean than to the group mean.

4.7.1 Posterior confidence intervals

Now suppose that parents have to choose a school for their children, and that they wish to do so on the basis of the value a school adds to abilities that pupils already have when entering the school (as indicated by an IQ test).

Let us focus on the language scores. 'Good' schools then are schools where pupils on average are 'over-achievers', that is to say, they achieve more than expected on the basis of their IQ. 'Poor' schools are schools where pupils on average have language scores that are lower than one would expect given their IQ scores.

In this case the level-two residuals U_{0j} from a two-level model with language as the dependent and IQ as the predictor variable convey the relevant information. But remember that each U_{0j} has to be estimated from the data, and that there is sampling error associated with each residual, since we work with a sample of students from each school. Of course we might argue that within each school the entire population of students is studied, but in general we should handle each parameter estimate with its associated uncertainty since we are now considering the performance of the school for a hypothetical new pupil at this school.

Therefore, instead of simply comparing schools on the basis of the level-two residuals it is better to compare these residuals taking account of the associated confidence intervals.

The standard error of the empirical Bayes estimate is smaller than the mean squared error of the OLS estimate based on the data only for the given macro-unit (the given school, in our example). This is just the point of using the empirical Bayes estimate. For the empty model the standard error is the square root of (4.15), which can also be expressed as

$$\text{S.E.}\left(\hat{\beta}_{0j}^{\text{EB}}\right) = \frac{1}{\sqrt{\tau_0^{-2} + n_j\,\sigma^{-2}}} \,. \tag{4.17}$$

This formula also was given by Longford (1993a, Section 1.7). Thus, the standard error depends on the within-group as well as the between-group variance and on the number of sampled pupils for the school. For models with explanatory variables, the standard error can be obtained from computer output of multilevel software (see Chapter 15). Denoting the standard error for school j shortly by SE_j, the corresponding ninety percent confidence intervals can be calculated as the intervals

$$\left(\hat{\beta}_{0j}^{\text{EB}} - 1.64 \times \text{SE}_j,\ \hat{\beta}_{0j}^{\text{EB}} + 1.64 \times \text{SE}_j\right) .$$

Two cautionary remarks are in order, however.

In the first place, the shrinkage construction of the empirical Bayes estimates implies a bias: 'good' schools (with a high U_{0j}) will tend to be represented too negatively, 'poor' schools (with a low U_{0j}) will tend to be represented too positively (especially if the sample sizes are small). The smaller standard error is bought at the expense of this bias! These confidence intervals have the property that, on average, the random group effects U_{0j} will be included in the confidence interval for ninety percent of the groups. But for close to average groups the coverage probability is higher, while for the groups with very low or very high group effects the coverage probability will be lower than ninety percent.

In the second place, users of such information generally wish to compare a series of groups.

This problem was addressed by Goldstein and Healy (1995). The parent in our example will make her or his own selection of schools, and (if the parent is a trained statistician) will compare the schools on the basis of whether the confidence intervals overlap. In that case, the parent is implicitly performing a series of statistical tests on the differences between the group effects U_{0j}. Goldstein and Healy (1995, p. 175) write: 'It is a common statistical misconception to suppose that two quantities whose 95% confidence intervals just fail to overlap are significantly different at the 5% significance level'. The reader is referred to their article on how to adjust the width of the confidence intervals in order to perform such significance testing. For example, testing the equality of a series of level-two residuals at the five percent significance level, requires confidence intervals that are constructed by multiplying the standard error given above by 1.39 rather than the well-known five percent value of 1.96. For a ten percent significance level, the factor is 1.24 rather than 1.64. So the 'comparative confidence intervals' are allowed to be narrower than the confidence intervals used for assessing single groups.

Example 4.5 *Comparing added value of schools.*
Table 4.2 presents the multilevel model where language scores are controlled for IQ. The posterior means $\hat{\beta}_{0j}^{\text{EB}}$, which can be interpreted as the estimated value added, are graphically presented in Figure 4.4. The figure also presents the confidence intervals for testing the equality of any pair of residuals at a significance level of five percent. For convenience the schools are ordered on the size of their posterior mean.

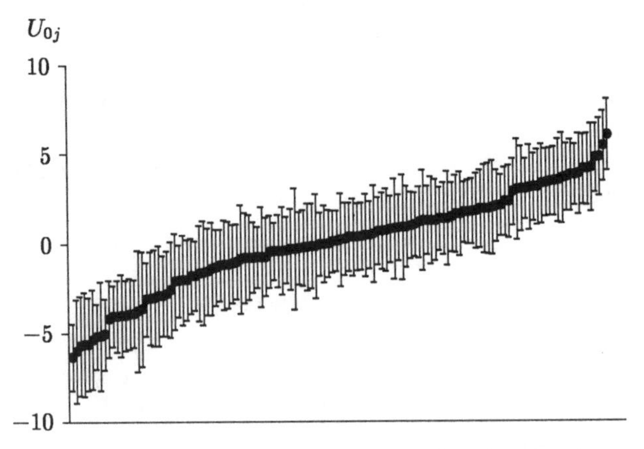

Figure 4.4 The added value scores for 131 schools with comparative posterior confidence intervals.

Note that approximately 30 schools have confidence intervals that overlap the confidence interval of the best school in this sample, implying that their value-added scores do not differ significantly. At the lower extreme, also about

30 schools do not differ significantly from the worst school in this sample. We can also deduce from the graph that about half of this sample of schools have approximately average scores that cannot be distinguished statistically.

Example 4.6 *Posterior confidence intervals for random data.*
Let us also look, once again, at the random digits example. In Figure 4.5 we graphed for each of the 10 macro-units (now in the original order) their OLS means and their posterior means with confidence intervals.

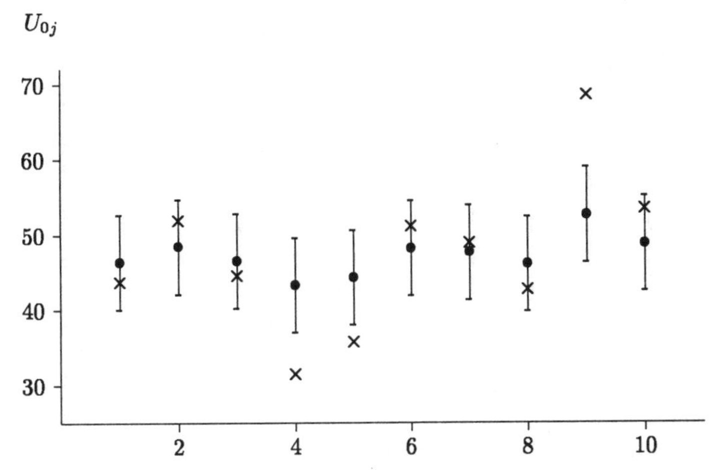

Figure 4.5 OLS means (×) and posterior means (•) with comparative posterior confidence intervals.

Once again we clearly observe the shrinkage since the OLS means (×) are further apart than the posterior means (•). Further, as we would expect from a random digits example, none of the pairwise comparisons results in any significant differences between the macro-units since all 10 confidence intervals overlap.

4.8 Three-level random intercept models

The three-level random intercept model is a straightforward extension of the two-level model. In previous examples, data were used where students were nested within schools. The actual hierarchical structure of educational data is, however: students nested within classes nested within schools. Other examples are: siblings within families within neighborhoods, or people within regions within states. Less obvious examples are: students within cohorts within schools, or longitudinal measurements within persons within groups. These latter cases will be illustrated in Chapter 12 on longitudinal data. For the time being we concentrate on 'simple' three-level hierarchical data structures. The dependent variable now is denoted by Y_{ijk}, referring to, e.g., pupil i in class j in school k. More generally, one can talk about level-one unit i in level-two unit j in level-three unit k. The three-level model for such data with one explanatory variable may be formulated as a regression model

$$Y_{ijk} = \beta_{0jk} + \beta_1 x_{ijk} + R_{ijk} \, , \tag{4.18}$$

where β_{0jk} is the intercept in level-two unit j within level-three unit k. For the intercept we have the level-two model,

$$\beta_{0jk} = \delta_{00k} + U_{0jk} \, , \tag{4.19}$$

where δ_{00k} is the average intercept in level-three unit k. For this average intercept we have the level-three model,

$$\delta_{00k} = \gamma_{000} + V_{00k} \, . \tag{4.20}$$

This shows that now there are three residuals, as there is variability on three levels. Their variances are denoted by

$$\text{var}(R_{ijk}) = \sigma^2, \quad \text{var}(U_{0jk}) = \tau^2, \quad \text{var}(V_{00k}) = \varphi^2 \, . \tag{4.21}$$

The total variance between all level-1 units now equals $\sigma^2 + \tau^2 + \varphi^2$, and the population variance between the level-two units is $\tau^2 + \varphi^2$. Substituting (4.20) and (4.19) into the level-one model (4.18) and using (in view of the next chapter) the triple indexing notation γ_{100} for the regression coefficient β_1 yields

$$Y_{ijk} = \gamma_{000} + \gamma_{100} \, x_{ijk} + V_{00k} + U_{0jk} + R_{ijk} \, . \tag{4.22}$$

Example 4.7 *A three-level model: students in classes in schools.*
For this example we use a dataset on 3792 students in 280 classes in 57 secondary schools with complete data (see Opdenakker and Van Damme, 1997). At school entrance students were administered tests on IQ, mathematics ability, achievement motivation, and furthermore data was collected on the educational level of the father and the students' gender.

The response variable is the score on a mathematics test administered at the end of the second grade of the secondary school (when the students were approximately 14 years old). Table 4.5 contains the results of the analysis of the empty three-level model (Model 1) and a model with a fixed effect of the students' intelligence.

Table 4.5 Estimates for three-level model.

	Model 1		Model 2	
Fixed Effects	Coefficient	S.E.	Coefficient	S.E.
γ_{000} = Intercept	7.96	0.23	− 4.55	0.50
γ_{100} = Coefficient of IQ			0.121	0.005
Random Effects	Var. Comp.	S.E.	Var. Comp.	S.E.
Level-three variance:				
$\varphi_0^2 = \text{var}(V_{00k})$	2.124	0.546	1.109	0.287
Level-two variance:				
$\tau_0^2 = \text{var}(U_{0jk})$	1.746	0.226	0.701	0.116
Level-one variance:				
$\sigma^2 = \text{var}(R_{ijk})$	7.816	0.186	6.910	0.165
Deviance	19009.7		18402.7	

The total variance is 11.686, the sum of the three variance components. Since this is a three-level model there are several kinds of intraclass correlation coefficient. Of the total variance, $2.124/11.686 = 18$ percent is situated at the school level while $(2.214 + 1.746)/11.686 = 33$ percent is situated at the class and school level. The level-three intraclass correlation expressing the likeness of students in the same schools thus is estimated to be 18 percent, while the intraclass correlation expressing the likeness of students in the same classes and the same schools thus is estimated to be 0.33. In addition, one can also estimate the intraclass correlation that expresses the likeness of classes in the same schools. This level-two intraclass correlation is estimated to be $2.124/(2.124 + 1.746) = 0.55$. This is more than 0.5: the school level contributes slightly more to variability than the class level. The interpretation is that if one randomly takes two classes within one school and calculates the average mathematics achievement level in one of the two, one can predict reasonably accurately the average achievement level in the other class. Of course we could have estimated a two-level model as well, ignoring the class level, but that would have led to a redistribution of the class-level variance to the two other levels, and it would affect the validity of hypothesis tests for added fixed effects.

Model 2 shows that the fixed effect of IQ is very strong, with a t-ratio (cf. Section 6.1) of $0.121/0.005 = 24.2$. (The intercept changes drastically because the IQ score does not have a zero mean; the conventional IQ scale, with a population mean of 100, was used.) Adding the effect of IQ leads to a stronger decrease in the class- and school-level variances than in the student-level variance. This suggests that schools and classes are rather homogeneous with respect to IQ and/or that intelligence may play its role partly at the school and class levels.

As in the two-level model, predictor variables at any of the three levels can be added. All features of the two-level model can be generalized to the three-level model quite straightforwardly: significance testing, model building, testing the model fit, centering of variables etc., although the researcher should be more careful now because of the more complicated formulation.

For example, for a level-one explanatory variable there can be three kinds of regressions. In the school example, these are the within-class regression, the within-school/between-class regression, and the between-school regression. Coefficients for these distinct regressions can be obtained by using the class means as well as the school means as explanatory variables with fixed effects.

Example 4.8 *Within-class, between-class, and between-school regressions.* Continuing the previous example, we now investigate whether indeed the effect of IQ is in part a class-level or school-level effect: in other words, whether the within-class, between-class/within-school, and between-school regressions are different. Table 4.6 presents the results.

In Model 3, the effects of the class mean, $\overline{\mathrm{IQ}}_{.jk}$, as well as the school mean $\overline{\overline{\mathrm{IQ}}}_{..k}$, have been added. The class mean has a clearly significant effect ($t = 0.106/0.013 = 8.15$), which indicates that between-class regressions are different from within-class regressions. The school mean does not have a significant effect ($t = 0.039/0.028 = 1.39$), so there is no evidence that the between-school regressions are different from the between-class regressions. It can be concluded that the composition with respect to intelligence plays a role at the class level, but not at the school level.

Table 4.6 Estimates for three-level model with distinct within-class, within-school, and between-school regressions.

	Model 3		Model 4	
Fixed Effects	Coefficient	S.E.	Coefficient	S.E.
Intercept	−18.16	2.66	−18.14	2.66
Coefficient of IQ_{ijk}	0.107	0.005		
Coefficient of $IQ_{ijk} - \overline{IQ}_{.jk}$			0.107	0.005
Coefficient of $\overline{IQ}_{.jk}$	0.106	0.013		
Coefficient of $\overline{IQ}_{.jk} - \overline{\overline{IQ}}_{..k}$			0.212	0.012
Coefficient of $\overline{\overline{IQ}}_{..k}$	0.039	0.028	0.252	0.025
Random Effects	Var. Comp.	S.E.	Var. Comp.	S.E.
Level-three variance:				
var(V_{00k})	0.798	0.211	0.798	0.211
Level-two variance:				
var(U_{0jk})	0.433	0.089	0.433	0.089
Level-one variance:				
var(R_{ijk})	6.893	0.164	6.893	0.164
Deviance	18324.3		18324.3	

Like in Section 4.5, replacing the variables by the deviation scores leads to an equivalent model formulation in which, however, the within-class, between-class, and between-school regression coefficients are given directly by the fixed parameters. In the three-level case, this means that we must use the following three variables:

$IQ_{ijk} - \overline{IQ}_{.jk}$, the within-class deviation score of the student from the class mean;

$\overline{IQ}_{.jk} - \overline{\overline{IQ}}_{..k}$, the within-school deviation score of the class mean from the school mean;

$\overline{\overline{IQ}}_{..k}$, the school mean itself.

The results are shown as Model 4.

We see here that the within-class regression coefficient is 0.107, equal to the coefficient of student-level IQ in Model 3; the between-class/within-school regression coefficient is 0.212, equal (up to rounding errors) to the sum of the student-level and the class-level coefficients in Model 3; while the between-school regression coefficient is 0.252, equal to the sum of all three coefficients in Model 3. From Model 3 we know that the difference between the last two coefficients is not significant.

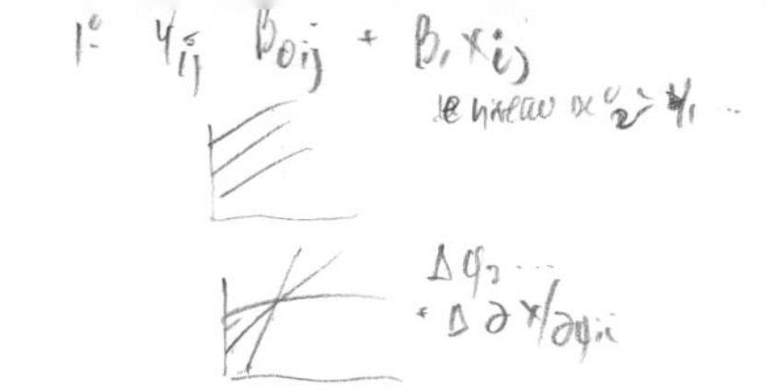

5 The Hierarchical Linear Model

In the previous chapter the simpler case of the hierarchical linear model was treated, where only intercepts are assumed to be random. In the more general case, slopes may also be random. For a study of pupils within schools, e.g., the effect of the pupil's intelligence or socio-economic status on scholastic performance could differ between schools. This chapter presents the general hierarchical linear model, which allows intercepts as well as slopes to vary randomly. The chapter follows the approach of the previous one: most attention is paid to the case of a two-level nesting structure, and the level-one units are called – for convenience only – 'individuals', while the level-two units are called 'groups'. The notation is also the same.

5.1 Random slopes

In the random intercept model of Chapter 4, the groups differ with respect to the average value of the dependent variable: the only random group effect is the random intercept. But the relation between explanatory and dependent variables can differ between groups in more ways. For example, in the educational field (nesting structure: pupils within classrooms), it is possible that the effect of socio-economic status of pupils on their scholastic achievement is stronger in some classrooms than in others. As an example in developmental psychology (repeated measurements within individual subjects), it is possible that some subjects progress faster than others. In the analysis of covariance, this phenomenon is known as heterogeneity of regressions across groups, or as group-by-covariate interaction. In the hierarchical linear model, it is modeled by *random slopes*.

Let us go back to a model with group-specific regressions of Y on one level-one variable X only, like model (4.3) but without the effect of Z,

$$Y_{ij} = \beta_{0j} + \beta_{1j}\, x_{ij} + R_{ij} \ . \tag{5.1}$$

The intercepts β_{0j} as well as the regression coefficients, or slopes, β_{1j} are group-dependent. These group-dependent coefficients can be split into an average coefficient and the group-dependent deviation:

$$\begin{aligned}\beta_{0j} &= \gamma_{00} + U_{0j} \\ \beta_{1j} &= \gamma_{10} + U_{1j} \ .\end{aligned} \tag{5.2}$$

Substitution leads to the model

$$Y_{ij} = \gamma_{00} + \gamma_{10}\, x_{ij} + U_{0j} + U_{1j}\, x_{ij} + R_{ij} \; . \tag{5.3}$$

It is assumed here that the level-two residuals U_{0j} and U_{1j} as well as the level-one residuals R_{ij} have means 0, given the values of the explanatory variable X. Thus, γ_{10} is the average regression coefficient just like γ_{00} is the average intercept. The first part of (5.3), $\gamma_{00} + \gamma_{10} x_{ij}$, is called the *fixed part* of the model. The second part, $U_{0j} + U_{1j} x_{ij} + R_{ij}$, is called the *random part.*

The term $U_{1j}\, x_{ij}$ can be regarded as a *random interaction between group and* X. This model implies that the groups are characterized by two random effects: their intercept and their slope. We say that X has a random slope, or a random effect, or a random coefficient. These two group effects will usually not be independent, but correlated. It is assumed that, for different groups, the pairs of random effects (U_{0j}, U_{1j}) are independent and identically distributed, that they are independent of the level-one residuals R_{ij}, and that all R_{ij} are independent and identically distributed. The variance of the level-one residuals R_{ij} is again denoted σ^2; the variances and covariance of the level-two residuals (U_{0j}, U_{1j}) are denoted as follows:

$$\begin{aligned} \text{var}(U_{0j}) &= \tau_{00} = \tau_0^2 \; ; \\ \text{var}(U_{1j}) &= \tau_{11} = \tau_1^2 \; ; \\ \text{cov}(U_{0j}, U_{1j}) &= \tau_{01} \; . \end{aligned} \tag{5.4}$$

Just like in the preceding chapter, one can say that the unexplained group effects are assumed to be exchangeable.

5.1.1 Heteroscedasticity

Model (5.3) implies not only that individuals within the same group have correlated Y-values (recall the residual intraclass correlation coefficient of Chapter 4), but also that this correlation as well as the variance of Y are dependent on the value of X. In an example, this can be understood as follows. Suppose that, in a study of the effect of socio-economic status (SES) on scholastic performance (Y), we have schools which do not differ in their effect on high-SES children, but which do differ in the effect of SES on Y (e.g., because of teacher expectancy effects). Then for children from a high SES background it does not matter which school they go to, but for children from a low SES background it does. The school then adds a component of variance for the low-SES children, but not for the high-SES children: as a consequence, the variance of Y (for a random child at a random school) will be larger for the former than for the latter children. Further, the within-school correlation between high-SES children will be nil, whereas between low-SES children it will be positive.

This example shows that model (5.3) implies that the variance of Y, given the value x on X, depends on x. This is called ***heteroscedasticity*** in the statistical literature. An expression for the variance of (5.3) is obtained as the sum of the variances of the random variables involved plus a term depending on the covariance between U_{0j} and U_{1j} (the other random variables are uncorrelated). Using (5.3) and (5.4), the result is

$$\text{var}(Y_{ij} \mid x_{ij}) = \tau_0^2 + 2\,\tau_{01}\,x_{ij} + \tau_1^2\,x_{ij}^2 + \sigma^2 , \qquad (5.5)$$

and, for two different individuals (i and i', with $i \neq i'$) in the same group,

$$\text{cov}(Y_{ij}, Y_{i'j} \mid x_{ij},\, x_{i'j}) = \tau_0^2 + \tau_{01}\,(x_{ij} + x_{i'j}) + \tau_1^2\,x_{ij}\,x_{i'j} \,. \qquad (5.6)$$

Formula (5.5) implies that the residual variance of Y is minimal for $x_{ij} = -\tau_{01}/\tau_{11}$. (This is deduced by differentiation with respect to x_{ij}.) When this value is within the range of possible X-values, the residual variance first decreases and then increases again; if this value is smaller than all X-values, then the residual variance is an increasing function of x; if it is larger than all X-values, then the residual variance is decreasing.

5.1.2 *Don't force* τ_{01} *to be 0!*

All of the preceding discussion implies that the group effects depend on x: according to (5.3), this effect is given by $U_{0j} + U_{1j}\,x$. This is illustrated by Figure 5.1. It gives a hypothetical graph of the regression of school achievement (Y) on intelligence (X).

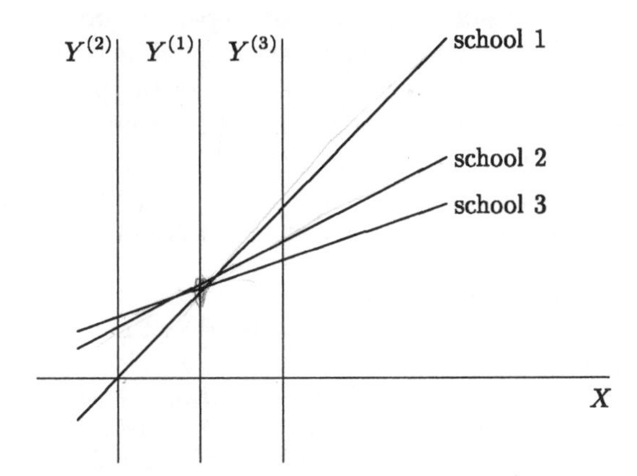

Figure 5.1 Different vertical axes.

It is clear that there are slope differences between the three schools. Looking at the $Y^{(1)}$-axis, there are almost no intercept differences between the schools. But if we add a value 10 to each intelligence score x, then the Y-axis is shifted to the left by 10 units: the $Y^{(2)}$ axis. Now school 3 is the best, school 1 the worst: there are strong intercept differences. If we would have subtracted 10 from the x-scores, we would have obtained the $Y^{(3)}$ axis, with again intercept differences but now in reverse order. This implies that the intercept variance τ_{00}, and also the intercept-by-slope covariance τ_{01}, depend on the origin (0-value) for the X-variable. From this we can learn two things:

(1) Since the origin of most variables in the social sciences is arbitrary, in random slope models the intercept-by-slope covariance should be a free

parameter estimated from the data, and not *a priori* constrained to the value 0 (i.e., left out of the model).
(2) In random slope models we should be careful with the interpretation of the intercept variance and the intercept-by-slope covariance, since the intercept refers to an individual with $x = 0$. For the interpretation of these parameters it is helpful when the scale for X is defined so that $x = 0$ has a well-interpretable meaning, preferably as a reference situation. For example, in repeated measurements when X refers to time, or measurement number, it can be convenient to let $x = 0$ correspond to the start, or the end, of the measurements. In nesting structures of individuals within groups, it is often convenient to let $x = 0$ correspond to the overall mean of the population or the sample – e.g., if X is IQ at the conventional scale with mean 100, it is advised to subtract 100 to obtain a population mean of 0.

5.1.3 Interpretation of random slope variances

For the interpretation of the variance of the random slopes, τ_1^2, it is illuminating to take also the average slope, γ_{10}, into consideration. Model (5.3) implies that the regression coefficient, or slope, for group j is $\gamma_{10} + U_{1j}$. This is a normally distributed random variable with mean γ_{10} and standard deviation $\tau_1 = \sqrt{\tau_1^2}$. Since about 95 percent of the probability of a normal distribution is within two standard deviations from the mean, it follows that approximately 95 percent of the groups have slopes between $\gamma_{10} - 2\tau_1$ and $\gamma_{10} + 2\tau_1$. Conversely, about one in forty groups has a slope less than $\gamma_{10} - 2\tau_1$ and one in forty has a slope steeper than $\gamma_{10} + 2\tau_1$.

Example 5.1 *A random slope for IQ.*
We continue the examples of Chapter 4, where the effect of IQ on a language test score was studied. Recall that IQ is here on a scale with mean 0 and its standard deviation in this data set is 2.07. A random slope of IQ is added to the model, i.e., it is allowed that the effect of IQ differs between classes. The model is an extension of model (5.3): a fixed effect for the class average on IQ is added. The model reads

$$Y_{ij} = \gamma_{00} + \gamma_{10}\,x_{ij} + \gamma_{01}\,\bar{x}_{.j} + U_{0j} + U_{1j}\,x_{ij} + R_{ij} \;.$$

The results can be read from Table 5.1. Note that the heading 'Level-two random effects' refers to the random intercept and random slope which are random effects associated to the level-two units (the class), but that the variable that has the random slope, IQ, is itself a level-one variable.

Figure 5.2 presents a sample of fifteen regression lines, randomly chosen according to the model of Table 5.1. (The values of the group mean $\overline{\text{IQ}}$ were chosen randomly from a normal distribution with mean −0.127 and standard deviation 1.005, which are the mean and standard deviation of the group mean of IQ in this data set.) This figure thus demonstrates the population of regression lines that characterizes, according to this model, the population of schools.

Table 5.1 Estimates for random slope model.

Fixed Effect	Coefficient	S.E.
γ_{00} = Intercept	40.75	0.29
γ_{10} = Coefficient of IQ	2.459	0.083
γ_{01} = Coefficient of $\overline{\text{IQ}}$ (group mean)	1.405	0.322
Random Effect	**Variance Component**	**S.E.**
Level-two random effects:		
$\tau_0^2 = \text{var}(U_{0j})$	7.92	1.32
$\tau_1^2 = \text{var}(U_{1j})$	0.200	0.098
$\tau_{01} = \text{cov}(U_{0j}, U_{1j})$	−0.820	0.267
Level-one variance:		
$\sigma^2 = \text{var}(R_{ij})$	41.35	1.29
Deviance	15213.5	

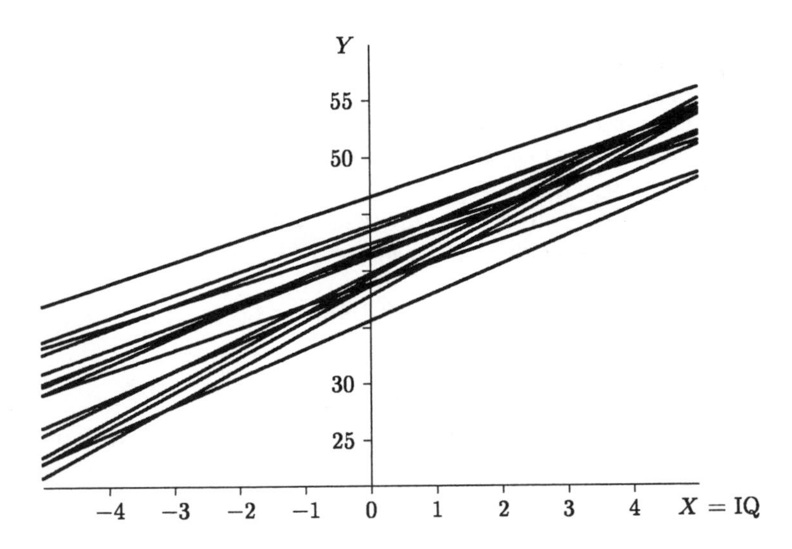

Figure 5.2 Fifteen random regression lines according to the model of Table 5.1 (with randomly chosen intercepts and slopes).

Should the value of 0.2 for the random slope variance be considered to be high? The slope standard deviation is $\sqrt{0.2} = 0.45$, and the average slope is $\gamma_{10} = 2.46$. The values of average slope ± two standard deviations range from 1.56 to 3.36. This implies that the effect of IQ is clearly positive in all classes, but high effects of IQ are more than twice as large as low effects. This may indeed be considered an important difference. (As indicated above, 'high' and 'low' are understood here as those values occurring in classes with, respectively, the top 2.5 percent, and the bottom 2.5 percent, of the class-dependent effects.)

Interpretation of intercept–slope covariance

The correlation between random slope and random intercept is $-0.82/\sqrt{7.92 \times 0.2} = -0.65$. Recall that all variables are centered (i.e., they have a zero mean), so that the intercept corresponds to the language test score for a pupil with an average intelligence in a class with an average mean intelligence. The negative correlation between slope and intercept means that classes with a higher performance for a pupil of average intelligence have a lower within-class effect of intelligence. Thus, the higher average performance tends to be achieved more by higher language scores of the less intelligent, than by higher scores of the more intelligent pupils.

In a random slope model, the within-group coherence cannot be simply expressed by the intraclass correlation coefficient or its residual version. The reason is that, in terms of the present example, the correlation between pupils in the same class depends on their intelligence. Thus, the extent to which a given classroom deserves to be called 'good' varies across pupils.

To investigate how the contribution of classrooms to pupils' performance depends on IQ, consider the equation implied by the parameter estimates:

$$Y_{ij} = 40.75 + 2.459\,\mathrm{IQ}_{ij} + 1.405\,\overline{\mathrm{IQ}}_{.j} + U_{0j} + U_{1j}\,\mathrm{IQ}_{ij} + R_{ij}\ .$$

Recall from example 2 that the standard deviation of the IQ score is about 2, and the mean is 0. Hence pupils with an intelligence among the bottom few percent or the top few percent have IQ scores of about ± 4. Substituting these values in the contribution of the random effects gives $U_{0j} \pm 4U_{1j}$. It follows from equations (5.5) and (5.6) that for pupils with IQ $= \pm 4$, we have

$$\begin{aligned}
&\mathrm{var}(Y_{ij} \mid \mathrm{IQ}_{ij} = -4) = \\
&\quad 7.92 + 2 \times (-0.820) \times (-4) + (-4)^2 \times 0.200 + 41.35 = 59.03,\\
&\mathrm{cov}(Y_{ij}, Y_{i'j} \mid \mathrm{IQ}_{ij} = -4, \mathrm{IQ}_{i'j} = 4) = 7.92 - 16 \times 0.20 = 4.72,\\
&\mathrm{var}(Y_{ij} \mid \mathrm{IQ}_{ij} = 4) = \\
&\quad 7.92 - 8 \times 0.820 + 16 \times 0.200 + 41.35 = 45.91,
\end{aligned}$$

and therefore

$$\rho(Y_{ij}, Y_{i'j} \mid \mathrm{IQ}_{ij} = -4, \mathrm{IQ}_{i'j} = 4) = \frac{4.72}{\sqrt{59.03 \times 45.91}} = 0.09.$$

Hence, the language test scores of the most intelligent and the least intelligent pupils in the same class are positively correlated over the population of classes: classes that have relatively good results for the less able tend also to have relatively good results for the more able students.

This positive correlation corresponds to the result that the value of IQ for which the variance given by (5.5) is minimal, is outside the range from -4 to $+4$. For the estimates in Table 5.1, this variance is

$$\mathrm{var}(Y_{ij} \mid \mathrm{IQ}_{ij} = x) = 7.92\ -\ 1.64\,x\ +\ 0.2\,x^2\ +\ \sigma^2\ .$$

Equating to 0 the derivative of this function of x yields that the variance is minimal for $x = 1.64/0.4 = 4.1$, just outside the IQ range from -4 to $+4$. This again implies that classes tend mostly to perform either higher, or lower, over the entire range of IQ. This is illustrated also by Figure 5.2 (which, however, also contains some regression lines that cross each other within the range of IQ).

5.2 Explanation of random intercepts and slopes

Regression analysis aims at explaining variability in the outcome (i.e., dependent) variable. Explanation is understood here in a quite limited way, viz., as being able to predict the value of the dependent variable from knowledge of the values of the explanatory variables. The unexplained variability in single-level multiple regression analysis is just the variance of the residual term. Variability in multilevel data, however, has a more complicated structure. This is related to the fact, mentioned in the preceding chapter, that several populations are involved in multilevel modeling: one population for each level. Explaining variability in a multilevel structure can be achieved by explaining variability between individuals but also by explaining variability between groups; if there are random slopes as well as random intercepts, at the group level one could try to explain the variability of slopes as well as intercepts.

In the model defined by (5.1) – (5.3), some variability in Y is explained by the regression on X, i.e., by the term $\gamma_{10}\, x_{ij}$; the random coefficients U_{0j}, U_{1j}, and R_{ij} each express different parts of the unexplained variability. In order to try to explain more of the unexplained variability, all three of these can be the point of attack. In the first place, one can try to find explanations in the population of individuals (at level one). The part of residual variance that is expressed by $\sigma^2 = \text{var}(R_{ij})$ can be diminished by including other level-one variables. Since group compositions with respect to level-one variables can differ from group to group, inclusion of such variables may also diminish residual variance at the group level. A second possibility is to try to find explanations in the population of groups (at level two). If we wish to reduce the unexplained variability associated with U_{0j} and U_{1j}, we can also say that we wish to expand equations (5.2) by predicting the group-dependent regression coefficients β_{0j} and β_{1j} from level-two variables Z. Supposing for the moment that we have one such variable, this leads to regression formulae for β_{0j} and β_{1j} on the variable Z,

$$\beta_{0j} = \gamma_{00} + \gamma_{01}\, z_j + U_{0j} \tag{5.7}$$

$$\beta_{1j} = \gamma_{10} + \gamma_{11}\, z_j + U_{1j}\ . \tag{5.8}$$

In words, the βs are treated as dependent variables in regression models for the population of groups; however, these are 'latent regressions', because the βs cannot be observed without error. Equation (5.7) is called an ***intercepts as outcomes*** model, and (5.8) a ***slopes as outcomes*** model.[1]

5.2.1 Cross-level interaction effects

This chapter started with the basic model (5.1), reading

$$Y_{ij} = \beta_{0j} + \beta_{1j}\, x_{ij} + R_{ij}\ .$$

Substituting (5.7) and (5.8) in this equation leads to the model

$$Y_{ij} = (\gamma_{00} + \gamma_{01}\, z_j + U_{0j}) + (\gamma_{10} + \gamma_{11}\, z_j + U_{1j})\, x_{ij} + R_{ij}$$

[1] In the older literature, these equations were applied to the estimated groupwise regression coefficients rather than the latent coefficients. The statistical estimation then was carried out in two stages: first ordinary least squares ('OLS') estimation within each group, then OLS estimation with the estimated coefficients as outcomes. This is statistically inefficient and does not differentiate the 'true score' variability of the latent coefficients from the sampling variability of the estimated groupwise regression coefficients. We do not treat this two-stage method.

$$= \gamma_{00} + \gamma_{01} z_j + \gamma_{10} x_{ij} + \gamma_{11} z_j x_{ij} + U_{0j} + U_{1j} x_{ij} + R_{ij} \; . \tag{5.9}$$

The last expression was rearranged so that first comes the fixed part and then the random part. Comparing this with model (5.3) shows that this explanation of the random intercept and slope leads to a different fixed part of the model, but does not change the formula for the random part, which remains $U_{0j} + U_{1j} x_{ij} + R_{ij}$. However, it is to be expected that the residual random intercept and slope variances, τ_0^2 and τ_1^2, will be less than their counterparts in model (5.3) because part of the variability of intercept and slopes now is explained by Z. In Chapter 7 we will see, however, that this is not necessarily so for the estimated values of these parameters.

In equation (5.9) we see that explaining the intercept β_{0j} by a level-two variable Z leads to a main effect of Z, while explaining the coefficient β_{1j} of X by the level-two variable Z leads to a product interaction effect of X and Z. Such an interaction between a level-one and a level-two variable is called a *cross-level interaction.*

For the definition of interaction variables such as the product $z_j x_{ij}$ in (5.9), it is advisable to use component variables Z and X for which the values $Z = 0$ and $X = 0$, respectively, have some interpretable meaning. For example, the variables Z and X could be centered around their means, so that $Z = 0$ means that Z has its average value, and analogous for X. Another possibility is that the zero values correspond to some kind of reference value. The reason is that, in the presence of the interaction term $\gamma_{11} z_j x_{ij}$, the main effect coefficient γ_{10} of X is to be interpreted as the effect of X *for cases with* $Z = 0$, while the main effect coefficient γ_{01} of Z is to be interpreted as the effect of Z *for cases with* $X = 0$.

Example 5.2 *Cross-level interaction between IQ and group size.*
The group size of the school classes yields a partial explanation of the class-dependent slopes. Group size ranges from 5 to 37, with a mean of 23.1. The school variable Z_2 is defined as group size minus 23.1. (The name Z_1 is implicitly used already for $\overline{\text{IQ}}$, the group mean of IQ.) When this variable is added to the model of Example 5.1, the parameter estimates presented in Table 5.2 are obtained.

The value of Z_2 ranges from about −18 to about +14. The class-dependent regression coefficient of IQ, cf. (5.8), is $\gamma_{10} + \gamma_{12} z_j + U_{1j}$, estimated as $2.443 - 0.022\, z_j + U_{1j}$. For z_j ranging between −18 and +14, the fixed part of this expression ranges (in reverse order) between about 2.1 and 2.8. This implies moderate differences in effect of intelligence on language score between classes with small and those with large groups, smaller group sizes being associated with higher effects of intelligence on language scores.

Cross-level interactions can be considered on the basis of two different kinds of argument. The above presentation is in line with an inductive argument: if a researcher finds a significant random slope variance, she may be led to think of level-two variables that could explain the random slope. An alternative approach is to base the cross-level interaction on substantive (theoretical) arguments formulated before looking at the data.

Table 5.2 Estimates for model with random slope and cross-level interaction.

Fixed Effect	Coefficient	S.E.
γ_{00} = Intercept	40.89	0.29
γ_{10} = Coefficient of IQ	2.443	0.082
γ_{01} = Coefficient of $\overline{\text{IQ}}$	1.246	0.326
γ_{02} = Coefficient of Z_2	0.057	0.037
γ_{12} = Coefficient of $Z_2 \times$ IQ	−0.022	0.011

Random Effect	Variance Component	S.E.
Level-two random effects:		
$\tau_0^2 = \text{var}(U_{0j})$	7.67	1.29
$\tau_1^2 = \text{var}(U_{1j})$	0.178	0.095
$\tau_{01} = \text{cov}(U_{0j}, U_{1j})$	−0.769	0.260
Level-one variance:		
$\sigma^2 = \text{var}(R_{ij})$	41.36	1.29
Deviance	15208.4	

The researcher then is led to estimate and test the cross-level interaction effect irrespective of whether a random slope variance was found. If a cross-level interaction effect exists, the power of the statistical test of this fixed effect is considerably higher than the power of the test for the corresponding random slope (assuming that the same model serves as the null hypothesis). Therefore it is not contradictory to look for a specific cross-level interaction even if no significant random slope was found. This is further elaborated in the last part of subsection 6.4.1.

More variables

The preceding models can be extended by including more variables that have random effects, and more variables explaining these random effects. Suppose that there are p level-one explanatory variables $X_1, \ldots, X_p$ and q level-two explanatory variables $Z_1, \ldots, Z_q$. Then, if the researcher is not afraid of a model with too many parameters, he can consider the model where all X-variables have varying slopes, and where the random intercept as well as all these slopes are explained by all Z-variables. At the within-group level, i.e., for the individuals, the model then is a regression model with p variables,

$$Y_{ij} = \beta_{0j} + \beta_{1j}\, x_{1ij} + \ldots + \beta_{pj}\, x_{pij} + R_{ij} \,. \tag{5.10}$$

The explanation of the regression coefficients β_{0j} to β_{pj} is based on the between-group model, which is a q-variable regression model for the group-dependent coefficient β_{hj},

$$\beta_{hj} = \gamma_{h0} + \gamma_{h1}\, z_{1j} + \ldots + \gamma_{hq}\, z_{qj} + U_{hj} \,. \tag{5.11}$$

Substitution of (5.11) in (5.10) and rearrangement of terms then yields the model

$$Y_{ij} = \gamma_{00} + \sum_{h=1}^{p} \gamma_{h0}\, x_{hij} + \sum_{k=1}^{q} \gamma_{0k}\, z_{kj} + \sum_{k=1}^{q}\sum_{h=1}^{p} \gamma_{hk}\, z_{kj}\, x_{hij}$$
$$+ U_{0j} + \sum_{h=1}^{p} U_{hj}\, x_{hij} + R_{ij} \; . \tag{5.12}$$

This shows that we obtain main effects of each X and Z variable as well as all cross-level product interactions. Further, we see the reason why, in formula (4.7), the fixed coefficients were called γ_{h0} for the level-one variable X_h and γ_{0k} for the level-two variable Z_k.

The groups are now characterized by $p+1$ random coefficients U_{0j} to U_{pj}. These random coefficients are independent between groups, but may be correlated within groups. It is assumed that the vector $(U_{0j}, \ldots, U_{pj})$ is independent of the level-one residuals R_{ij} and that all residuals have population means 0, given the values of all explanatory variables. It is also assumed that the level-one residual R_{ij} has a normal distribution with constant variance σ^2 and that $(U_{0j}, \ldots, U_{pj})$ has a multivariate normal distribution with a constant covariance matrix. Analogous to (5.4), the variances and covariances of the level-two random effects are denoted

$$\begin{aligned} \text{var}(U_{hj}) &= \tau_{hh} = \tau_h^2 \quad (h = 1, \ldots, p) \; ; \\ \text{cov}(U_{hj}, U_{kj}) &= \tau_{hk} \quad (h, k = 1, \ldots, p) \; . \end{aligned} \tag{5.13}$$

Example 5.3 *A model with many fixed effects.*
For this example, it must be noted that there is a distinction between class and group. The class is the set of pupils who are being taught by the same teacher in the same classroom. The group is the subset of those pupils in the class who are in grade 8. Some classes are combinations of grade 7 and grade 8 pupils. Only the grade 8 pupils are part of this data set. Accordingly, a variable COMB is defined that indicates whether a class is such a multi-grade class (COMB = 1; 53 classes), or entirely composed of grade 8 pupils (COMB = 0; 78 classes).

The model for the language test score in this example includes main effects and cross-level interactions between the following variables.

Pupil level

* IQ (as used in the preceding examples)
* SES = social-economic status of the pupil's family (a numerical variable with mean 0 and standard deviation 10.9).

Class level

* $\overline{\text{IQ}}$ = average IQ in the group
* GS = group size
* COMB = indicator of multi-grade classes.

(The class average of SES is not included in this example, because other analyses showed that this variable has no significant effect; in other words, there is no significant difference between the within-group and the between-group regressions on SES.)

The covariance of the random slopes of IQ and SES is not included in the table, because including this covariance in the model led to failure of convergence of the estimating algorithm. We may assume that this covariance is not different from 0 to an important extent.

Estimating model (5.12) (with $p = 2$, $q = 3$) leads to the results presented in Table 5.3. A discussion of this table is deferred to the next example.

Table 5.3 Estimates for model with random slopes and many effects.

Fixed Effect	Coefficient	S.E.
γ_{00} = Intercept	41.51	0.37
γ_{10} = Coefficient of IQ	2.125	0.102
γ_{20} = Coefficient of SES	0.154	0.020
γ_{01} = Coefficient of $\overline{\text{IQ}}$	0.833	0.325
γ_{02} = Coefficient of GS	−0.057	0.050
γ_{03} = Coefficient of COMB	−1.936	0.798
γ_{11} = Coefficient of IQ × $\overline{\text{IQ}}$	−0.049	0.082
γ_{12} = Coefficient of IQ × GS	−0.001	0.015
γ_{13} = Coefficient of IQ × COMB	0.374	0.243
γ_{21} = Coefficient of SES × $\overline{\text{IQ}}$	−0.020	0.019
γ_{22} = Coefficient of SES × GS	−0.001	0.003
γ_{23} = Coefficient of SES × COMB	0.017	0.046

Random Effect	Variance Component	S.E.
Level-two random effects:		
$\tau_0^2 = \text{var}(U_{0j})$	7.46	1.24
$\tau_1^2 = \text{var}(U_{1j})$	0.110	0.081
$\tau_{01} = \text{cov}(U_{0j}, U_{1j})$	−0.636	0.235
$\tau_2^2 = \text{var}(U_{2j})$	0.0	0.0
$\tau_{02} = \text{cov}(U_{0j}, U_{2j})$	0.0	0.0
Level-one variance:		
$\sigma^2 = \text{var}(R_{ij})$	39.37	1.23
Deviance	15089.1	

Unless p and q are quite small, model (5.12) entails a number of statistical parameters that usually is too large for comfort. Therefore, two simplifications are often used.

(1) Not all X-variables are considered to have random slopes. Note that the explanation of the variable slopes by the Z-variables may have led to some random residual coefficients that are not significantly different from 0 (testing the τ-parameters is discussed in Chapter 6) or that even are estimated to be 0.

(2) Given that the coefficients β_{hj} of a certain variable X_h are variable across groups, it is not necessary to use all variables Z_k for explaining their variability. The number of cross-level interactions can be restricted by explaining each β_{hj} by only a well-chosen subset of the Z_k.

Which variables to give random slopes, and which cross-level interaction variables to use, depends on subject-matter as well as empirical considerations. The statistical aspects of testing and model fitting are treated in later chapters.

Example 5.4 *A parsimonious model in the case of many variables.*
In Table 5.3, the random slope variance of SES is estimated by 0 (this happens occasionally, cf. section 5.4). Therefore, this random slope is excluded from the model.

We shall see in Chapter 6, that significance of fixed effects can be tested by applying a t-test to the ratio of estimate to standard error. This consideration led us to the model in which all cross-level interactions are excluded, but for the interaction between IQ and COMB. The resulting estimates are displayed in Table 5.4.

Table 5.4 Estimates for a more parsimonious model with a random slope and many effects.

Fixed Effect	Coefficient	S.E.
γ_{00} = Intercept	41.32	0.35
γ_{10} = Coefficient of IQ	2.113	0.092
γ_{20} = Coefficient of SES	0.156	0.015
γ_{01} = Coefficient of $\overline{\text{IQ}}$	0.876	0.324
γ_{03} = Coefficient of COMB	−1.396	0.574
γ_{13} = Coefficient of IQ × COMB	0.447	0.170

Random Effect	Variance Component	S.E.
Level-two random effects:		
$\tau_0^2 = \text{var}(U_{0j})$	7.56	1.25
$\tau_1^2 = \text{var}(U_{1j})$	0.128	0.084
$\tau_{01} = \text{cov}(U_{0j}, U_{1j})$	−0.588	0.239
Level-one variance:		
$\sigma^2 = \text{var}(R_{ij})$	39.34	1.23
Deviance	15093.4	

The estimates of the remaining effects do not change much compared to Table 5.3, but the standard errors of many fixed coefficients decrease considerably. This can be explained by the omission of the many non-significant effects.

Note that COMB is not a centered variable, but a 0–1 dummy with necessarily a positive mean. All other explanatory variables have 0 means. Therefore, the intercept γ_{00} is the mean of pupils with average characteristics in a single-grade class (COMB = 0), and the regression coefficient of IQ γ_{10} is the average effect of IQ in single-grade classes. The mean of pupils in multi-grade classes is $\gamma_{00} + \gamma_{03}$. The interaction effect γ_{13} is the additional effect of IQ in multi-grade classes. Hence the effect of IQ in multi-grade classes is $\gamma_{10} + \gamma_{13}$.

Compared to Example 5.2, it turns out that it is not group size but rather COMB, that seems to have an effect, both a main effect and an interaction effect with IQ. Partitioned classes (COMB = 1) lead to lower language test scores and to a higher effect of intelligence. The unexplained variance of class-dependent slopes of language score on IQ is considerably lower than in Table 5.2, when the interaction of IQ with GS (group size), instead of COMB, was included in the model.

The model found can be expressed as a model with variable intercepts and slopes by the formula

$$Y_{ij} = \beta_{0j} + \beta_{1j}\, x_{1ij} + \beta_{2j}\, x_{2ij} + R_{ij} \ ,$$

where X_1 is IQ and X_2 is SES. The intercept is

$$\beta_{0j} = \gamma_{00} + \gamma_{01}\, z_{1j} + \gamma_{03}\, z_{3j} + U_{0j} \ ,$$

where Z_1 is average IQ and Z_3 is COMB. The coefficient of X_1 is

$$\beta_{1j} = \gamma_{10} + \gamma_{13}\, z_{3j} + U_{1j} \ ,$$

while the coefficient of X_2 is not variable,

$$\beta_{2j} = \gamma_{20} \ .$$

5.2.2 A general formulation of fixed and random parts

Formally, and in many computer programs, these simplifications lead to a representation of the hierarchical linear model that is slightly different from (5.12). (For example, the HLM program uses the formulations (5.10) – (5.12) whereas MLn uses formulation (5.14).) Whether a level-one variable was obtained as a cross-level interaction or not is immaterial to the computer program. Even the difference between level-one variables and level-two variables, although possibly relevant for the way the data are stored, is not of any importance for the parameter estimation. Therefore, all variables – level-one and level-two variables, including product interactions – can be represented mathematically simply as x_{hij}. When there are r explanatory variables, ordered so that the first p have fixed *and* random coefficients, while the last $r - p$ have only fixed coefficients,[2] the hierarchical linear model can be represented as

$$Y_{ij} = \gamma_0 + \sum_{h=1}^{r} \gamma_h\, x_{hij} + U_{0j} + \sum_{h=1}^{p} U_{hj}\, x_{hij} + R_{ij} \ . \tag{5.14}$$

The two terms,

$$\gamma_0 + \sum_{h=1}^{r} \gamma_h\, x_{hij} \quad \text{and} \quad U_{0j} + \sum_{h=1}^{p} U_{hj}\, x_{hij} + R_{ij} \ ,$$

are the fixed and the random parts of the model, respectively.

In cases where the explanation of the random effects works extremely well, one may end up with models without any random effects at level 2. In other words, the random intercept U_{0j} and all random slopes U_{hj} in (5.14) have zero variance, and may just as well be omitted from the formula. In this case, the resulting model may be analysed just as well with OLS regression analysis, because the residuals are independent and have constant variance. Of course, this is known only after the multilevel analysis has been carried out. In such a case, the within-group dependence between measurements has been fully explained by the available explanatory variables (and their interactions). This underlines that whether the hierarchical linear model is a more adequate model for analysis than OLS regression depends not on the dependence of the measurements, but on the dependence of the residuals.

[2]It is mathematically possible that some variables have a random but not a fixed effect. This makes sense only in special cases.

5.3 Specification of random slope models

Given that random slope models are available, the researcher has many options to model his data. Each predictor may be assigned a random slope, and each random slope may covary with any other random slope. Parsimonious models, however, should be preferred, if only for the simple reason that a strong scientific theory is general rather than specific. A good guide for choosing between a fixed or a random slope for a given predictor variable should preferably be found in the theory that is being investigated. If the theory (whether this is a general scientific theory or a practical policy theory) does not give any clue with respect to a random slope for a certain predictor variable, then one may be tempted to refrain from using random slopes. However, this implies a risk of invalid statistical tests, because if some variable does have a random slope, then omitting this feature from the model could affect the estimated standard errors of the other variables. The specification of the hierarchical linear model, including the random part, is discussed more fully in Section 6.4 and Chapter 9.

In data exploration, one can try various specifications. Often it appears that the chance of detecting slope variation is high for variables with strong fixed effects. This, however, is an empirical rather than a theoretical assertion. Actually, it may well be that when a fixed effect is – almost – zero, there does exist slope variation. Consider, for instance, the case where male teachers treat boys advantageously over girls, whereas for female teachers the situation is reversed. If half of the sample consists of male and the other half of female teachers, then, all other things being equal, the main gender effect on achievement will be absent, since in half of the classes the gender effect will be positive and in the other half negative. The fixed effect of students' gender then is zero but varies across classes (depending on the teachers' gender). In this example, of course, the random effect would disappear if one should specify the cross-level interaction effect of teachers' gender with students' gender.

5.3.1 Centering variables with random slopes?

Recall from Figure 5.1 that the intercept variance and the meaning of the intercept in random slope models depend on the location of the X variable. Also the covariance between the intercepts and the slopes is dependent on this location. In the examples presented so far we have used an IQ score for which the grand mean was zero (the original score was transformed by subtracting the grand mean IQ). This facilitated interpretation since the intercept could be interpreted as the expected score for a student with average IQ. Making the IQ slope random did not have consequences for these meanings.

In Section 4.5 a model was introduced by which we could distinguish within- from between-group regression. Two models were discussed:

$$Y_{ij} = \gamma_{00} + \gamma_{10}\, x_{ij} + \gamma_{01}\, \overline{x}_{.j} + U_{0j} + R_{ij} \tag{4.8}$$

and

$$Y_{ij} = \tilde{\gamma}_{00} + \tilde{\gamma}_{10}\,(x_{ij} - \overline{x}_{.j}) + \tilde{\gamma}_{01}\,\overline{x}_{.j} + U_{0j} + R_{ij} \; . \tag{4.9}$$

It was shown that $\tilde{\gamma}_{01} = \gamma_{10} + \gamma_{01}$, $\tilde{\gamma}_{10} = \gamma_{10}$, so that the two models are equivalent.

Are the models also equivalent when the effect of X_{ij} or $(X_{ij} - \overline{X}_{.j})$ is random across groups? This was discussed by Kreft, de Leeuw, and Aiken (1995). Let us first consider the extension of (4.8). Define the level-one and level-two models

$$\begin{aligned} Y_{ij} &= \beta_{0j} + \beta_{1j}\,x_{ij} + \gamma_{01}\,\overline{x}_{.j} + R_{ij} \\ \beta_{0j} &= \gamma_{00} + U_{0j} \\ \beta_{1j} &= \gamma_{10} + U_{1j} \; ; \end{aligned}$$

substituting the level-two model into the level-one model leads to

$$Y_{ij} = \gamma_{00} + \gamma_{10}\,x_{ij} + \gamma_{01}\,\overline{x}_{.j} + U_{0j} + U_{1j}\,x_{ij} + R_{ij} \; .$$

Next we consider the extension of (4.9):

$$\begin{aligned} Y_{ij} &= \tilde{\beta}_{0j} + \tilde{\beta}_{1j}\,(x_{ij} - \overline{x}_{.j}) + \tilde{\gamma}_{01}\,\overline{x}_{.j} + R_{ij} \\ \tilde{\beta}_{0j} &= \tilde{\gamma}_{00} + U_{0j} \\ \tilde{\beta}_{1j} &= \tilde{\gamma}_{10} + U_{1j} \; ; \end{aligned}$$

substitution and rearrangement of terms now yields

$$\begin{aligned} Y_{ij} = \tilde{\gamma}_{00} &+ \tilde{\gamma}_{10}\,x_{ij} + (\tilde{\gamma}_{01} - \tilde{\gamma}_{10})\,\overline{x}_{.j} \\ &+ U_{0j} + U_{1j}\,x_{ij} - U_{1j}\,\overline{x}_{.j} + R_{ij} \; . \end{aligned}$$

This shows that the two models differ in the term $U_{1j}\,\overline{x}_{.j}$ which is included in the group-mean centered random slope model but not in the other model. Therefore in general there is no one-to-one relation between the γ and the $\tilde{\gamma}$ parameters, so the models are not statistically equivalent except for the extraordinary case where variable X has no between-group variability.

This implies that in constant slope models one can either use X_{ij} and $\overline{X}_{.j}$ or $(X_{ij} - \overline{X}_{.j})$ and $\overline{X}_{.j}$ as predictors, since this results in statistically equivalent models, but in random slope models one should carefully choose one or the other specification.

On which consideration should this choice be based? Generally one should be reluctant to use group-mean centered random slopes models unless there is a clear theory (or an empirical clue) that not in the first place the absolute score X_{ij} but rather the relative score $(X_{ij} - \overline{X}_{.j})$ is related to Y_{ij}. Now $(X_{ij} - \overline{X}_{.j})$ indicates the relative position of an individual in his or her group, and examples of instances where one may be particularly interested in this variable may be:

* research on normative or comparative social reference processes (e.g., Guldemond, 1994),
* research on relative deprivation,
* research on teachers' rating of student performance.

5.4 Estimation

What was mentioned in section (4.6) can be applied, with the necessary extensions, also to the estimation of parameters in the more complicated model (5.14). A number of iterative estimation algorithms have been proposed e.g., by Laird and Ware (1982), Goldstein (1986), and Longford (1987), and are now implemented in multilevel software.

The following may give some intuitive understanding of estimation methods. If the parameters of the random part, i.e., the parameters in (5.13) together with σ^2, were known, then the regression coefficients could be estimated straightforwardly with the so-called generalized least squares ('GLS') method. Conversely, if all regression coefficients γ_{hk} were known, the 'total residuals' (which seems an apt name for the second line of equation (5.12)) could be computed, and their covariance matrix could be used to estimate the parameters of the random part. These two partial estimation processes can be alternated: use provisional values for the random part parameters to estimate regression coefficients, use the latter estimates to estimate the random part parameters again (and now better), then go on to estimate the regression coefficients again, and so on ***ad libitum*** – or, rather, until convergence of this iterative process. This loose description is close to the iterated generalized least squares ('IGLS') method that is one of the algorithms to calculate the ML estimates.

There exist other methods (one called Fisher scoring, treated in Longford (1993a), the other nicknamed EM for Expectation – Maximization, see Bryk and Raudenbush (1992)) which calculate the same estimates, each with its own advantages.

Parameters can again be estimated with the ML or with the REML method; the REML method is preferable in the sense that it produces less biased estimates for the random part parameters in the case of small sample sizes, but the ML method is more convenient if one wishes to use deviance tests (see the next chapter). The IGLS algorithm produces ML estimates, whereas the so-called RIGLS ('restricted IGLS') algorithm yields the REML estimates. For the random slopes model also it is possible that estimates for the variance parameters τ_h^2 are exactly 0. The explanation is analogous to the explanation given for the intercept variances in Section 4.6.

The random group effects U_{hj} can again be 'estimated' by the empirical Bayes method, and the resulting 'estimates' are called posterior slopes (sometimes posterior means). This is analogous to what is treated in Section 4.7 about posterior means.

Usually the estimation algorithms do not allow to include an unlimited number of random slopes. Depending on the data set and the model specification, it is not uncommon that the algorithm refuses to converge for more than two or three variables with random slopes. Sometimes the convergence can be improved by linearly transforming the variables with random slopes so that they have (approximately) zero means, or by transforming them to have (approximately) zero correlations.

For some data sets, the estimation method can produce estimated variance and covariance parameters that correspond to impossible covariance matrices for the random effects at level two, e.g., τ_{01} sometimes is estimated larger than $\tau_0 \times \tau_1$. This would imply an intercept-slope correlation larger than 1. This is not an error of the estimation procedure, and it can be understood as follows. The estimation procedure is directed at the mean vector and covariance matrix of the vector of all observations. Some combinations of parameter values correspond to permissible structures of the latter covariance matrix that, nevertheless, cannot be formulated as a random effects model such as (5.3). Even if the estimated values for the τ_{hk} parameters do not combine into a positive definite matrix τ, the σ^2 parameter will still make the covariance matrix of the original observations (cf. equations (5.5) and (5.6)) positive definite. Therefore, such a strange result is in contradiction to the random effects formulation (5.3), but not to the more general formulation of a patterned covariance matrix for the observations.

Most computer programs give the standard errors of the variances of the random intercept and slopes, some give the standard errors of the estimated standard deviations $\hat{\tau}_h$ instead. The two standard errors can be transformed into each other by the approximation formula

$$\text{S.E.}(\hat{\tau}^2) \approx 2\,\hat{\tau}\,\text{S.E.}(\hat{\tau}) . \tag{5.15}$$

However, some caution is necessary in the use of these standard errors. The simplest way for making a confidence interval for τ^2 or τ, viz., taking the estimated value plus or minus twice the standard error, is a valid approximation only if the relative standard error of $\hat{\tau}$ (i.e., standard error divided by parameter estimate) is small, say, less than 1/4.

5.5 Three and more levels

When the data have a three-level hierarchy, slopes of level-one variables can be made random at level two and also at level three. In this case there will be at least two level-two and two level-three equations: one for the random intercept and one for the random slope. So, in the case of one explanatory variable, the model might be formulated as follows:

$$Y_{ijk} = \beta_{0jk} + \beta_{1jk}\, x_{ijk} + R_{ijk} \qquad \text{(Level-one model)}$$
$$\beta_{0jk} = \delta_{00k} + U_{0jk} \qquad \text{(Level-two model for intercept)}$$
$$\beta_{1jk} = \delta_{10k} + U_{1jk} \qquad \text{(Level-two model for slope)}$$
$$\delta_{0jk} = \gamma_{000} + V_{00k} \qquad \text{(Level-three model for intercept)}$$
$$\delta_{1jk} = \gamma_{100} + V_{10k}. \qquad \text{(Level-three model for slope)}$$

In the specification of such a model, for each level-one variable with random slope it has to be decided whether its slope must be random at level two, random at level three, or both. Generally one should have either strong *a priori* knowledge or a good theory to formulate models as complex as this one or even more complex models (i.e., with more random slopes). Further, for each level-two variable it must be decided whether its slope is random at level three.

Example 5.5 *A three-level model with a random slope.*
We continue with the example of Section 4.8, where we illustrated the three-level model using a data set about a math test administered to students in classes in schools. Now we include the available covariates (which are all centered around their grand means) and moreover the regression coefficient for the mathematics pretest is allowed to be random at level two and level three. The results are in Table 5.5.

Table 5.5 Estimates for three-level model with random slopes.

Fixed Effect	Coefficient	S.E.
γ_{000} = Intercept	8.41	0.16
γ_{100} = Coefficient of IQ	0.050	0.005
γ_{200} = Coefficient of pretest	0.146	0.011
γ_{300} = Coefficient of motivation	0.032	0.008
γ_{400} = Coefficient of father's education	0.039	0.015
γ_{500} = Coefficient of gender	0.221	0.106
Random Effect	**Variance Component**	**S.E.**
Level-three random effects:		
$\varphi_0^2 = \text{var}(V_{00k})$	0.971	0.254
$\varphi_2^2 = \text{var}(V_{20k})$	0.0024	0.0010
$\varphi_{02} = \text{cov}(V_{00k}, V_{20k})$	0.0381	0.0135
Level-two random effects:		
$\tau_0^2 = \text{var}(U_{0jk})$	0.439	0.089
$\tau_2^2 = \text{var}(U_{2jk})$	0.0019	0.0009
$\tau_{02} = \text{cov}(U_{0jk}, U_{2jk})$	0.0398	0.0068
Level-one variance:		
$\sigma^2 = \text{var}(R_{ijk})$	5.978	0.145
Deviance	17808.0	

The interpretation of the fixed part is straightforward as in conventional single-level regression models. The random part is more complicated. Since all predictor variables were grand mean centered, the intercept variances φ_0^2 (level three) and τ_0^2 (level two) have a clear meaning: they represent the amount of variation in mathematics achievement across schools and across classes within schools, respectively, for the average student whilst controlling for differences in IQ, mathematics ability, achievement motivation, educational level of the father, and gender. Comparing this table with Table 4.5 shows that much of the initial level-three and (especially) level-two variation has now been accounted for. Once there is control for initial differences, schools and classes within schools differ considerably less in the average mathematics achievement of their students at the end of grade two.

Now we turn to the slope variance. The fixed slope coefficient for the mathematics pretest is estimated to be 0.146. The variance at level three for this slope is 0.0024, and at level two 0.0019. So the variability between schools of the effect of the pretest is somewhat larger than the variability of this effect between classes. On one end of the distribution there are a few percent of the schools that have an effect of the pretest that is only $0.146 - 2 \times \sqrt{0.0024} =$

0.05, whereas in the most selective schools this effect is $0.146 + 2 \times \sqrt{0.0024} = 0.252$. Since there is also variation in this coefficient across classes within schools, the gap between initial low and initial high achievers (4 standard deviations apart; this standard deviation for the pretest is 8.21) within the same school can become as big as $(0.146 + 2 \times \sqrt{0.0024 + 0.0019}) \times (4 \times 8.21) = 9$ points, whereas on the other hand it can stay as low as $(0.146 - 2 \times \sqrt{0.0024 + 0.0019}) \times (4 \times 8.21) = 1$ point in the least selective schools. Given the standard deviation of 3.4 for the dependent variable, a difference of 1 is very low whereas 9 is quite a large difference.

6 Testing and Model Specification

(It is assumed for this chapter that the reader has a basic knowledge of statistical testing: null hypothesis, alternative hypothesis, errors of the first and the second kind, significance level, statistical power.)

6.1 Tests for fixed parameters

Suppose we are working with a model represented as (5.14). The null hypothesis that a certain regression parameter is 0, i.e.,

$$\mathrm{H}_0 : \gamma_h = 0 \ , \tag{6.1}$$

can be tested by a t-test. The statistical estimation leads to an estimate $\hat{\gamma}_h$ with associated standard error $\mathrm{S.E.}(\hat{\gamma}_h)$. Their ratio is a t-value:

$$T(\gamma_h) = \frac{\hat{\gamma}_h}{\mathrm{S.E.}(\hat{\gamma}_h)} \ . \tag{6.2}$$

One-sided as well as two-sided tests can be carried out on the basis of this test statistic.[1] Under the null hypothesis, $T(\gamma_h)$ has approximately a t-distribution, but the number of degrees of freedom ($d.f.$) is somewhat more complicated than in multiple linear regression, because of the presence of the two levels. The approximation by the t-distribution is not exact even if the normality assumption for the random coefficients holds. Suppose first that we are testing the coefficient of a level-one variable. (In accordance with the specification of model (5.14), a cross-level interaction variable is also considered a level-one variable.) If the total number of level-one units is M and the total number of explanatory variables is r, then we can take $d.f. = M - r - 1$. For testing the coefficient of a level-two variable when there are N level-two units and q explanatory variables at level 2, we take $d.f. = N - q - 1$. If the number of units minus the number of variables is large enough, say, larger than 40, the t-distribution can be replaced by a standard normal distribution.

Example 6.1 *Testing within- and between-group regressions.*
We wish to test if between-group and within-group regressions of language test score on IQ are different from one another, when controlling for social-economic status (SES). A model with a random slope for IQ is used. Two models are estimated and presented in Table 6.1. The first contains the raw

[1] This is one of the common principles for construction of a t-test. This type of test is called the ***Wald test***, after the statistician Abraham Wald (1902–1950).

IQ variable along with the group mean, the second contains the within-group deviation variable $\widetilde{\text{IQ}}$, defined as

$$\widetilde{\text{IQ}}_{ij} = \text{IQ}_{ij} - \overline{\text{IQ}}_{.j} \ ,$$

also together with the group mean. To test whether within- and between-group regression coefficients are different, the significance of the group mean $\overline{\text{IQ}}$ is tested, while controlling for the effect of the original variable, IQ. The difference of within- and between-group regressions is discussed further in Section 4.5.

Table 6.1 Estimates for two models with different between- and within-group regressions.

	Model 1		Model 2	
Fixed Effects	Coefficient	S.E.	Coefficient	S.E.
γ_{00} = Intercept	40.78	0.29	40.78	0.29
γ_{10} = Coefficient of IQ	2.249	0.082		
γ_{20} = Coefficient of $\widetilde{\text{IQ}}$			2.249	0.082
γ_{30} = Coefficient of SES	0.156	0.015	0.156	0.015
γ_{01} = Coefficient of $\overline{\text{IQ}}$	1.084	0.325	3.333	0.320
Random Effects	Var. Comp.	S.E.	Var. Comp.	S.E.
Level-two random effects:				
$\tau_0^2 = \text{var}(U_{0j})$	8.19	1.33	8.19	1.33
$\tau_1^2 = \text{var}(U_{1j})$	0.170	0.091	0.170	0.091
$\tau_{01} = \text{cov}(U_{0j}, U_{1j})$	−0.722	0.258	−0.722	0.258
Level-one variance:				
$\sigma^2 = \text{var}(R_{ij})$	39.29	1.22	39.29	1.22
Deviance	15103.7		15103.7	

The table shows that only the estimate for $\overline{\text{IQ}}$ differs between the two models. This is in accordance with Section 4.5: if IQ is variable 1 and $\widetilde{\text{IQ}}$ is variable 2, so that their regression coefficients are γ_{10} and γ_{20}, respectively, then the within-group regression coefficient is γ_{10} in Model 1 and γ_{20} in Model 2, while the between-group regression coefficient is $\gamma_{10}+\gamma_{01}$ in Model 1 and γ_{01} in Model 2. The models are equivalent representations of the data and differ only in the parametrization. The deviances indeed are exactly the same.

The within-group regression and between-group regressions are the same if $\gamma_{01} = 0$ in Model 1, i.e., there is no effect of the group mean given that the model controls for the raw variable (i.e., the variable without group-centering). The t-test statistic for testing $H_0 : \gamma_{01} = 0$ in Model 1 is equal to 1.084/0.325 = 3.33. This is strongly significant ($p < 0.001$). It may be concluded that within-group and between-group regressions are significantly different.

The results for Model 2 can be used to test if the within-group or between-group regressions are 0. The t-test statistic for testing the within-group regression is 2.249/0.082 = 27.4, the statistic for testing the between-group regression is 3.333/0.320 = 10.4. Both are extremely significant. Concluding, there are positive within-group as well as between-group regressions, and these are different from one another.

6.1.1 *Multi-parameter tests for fixed effects*

Sometimes we wish to test several regression parameters simultaneously. For example, consider testing the effect of a categorical variable with more than three categories. The effect of such a variable can be represented in the fixed part of the hierarchical linear model by $c-1$ dummy variables, where c is the number of categories, and this effect is nil if and only if all the corresponding $c-1$ regression coefficients are 0. Two types of test are much used for this purpose: the multivariate Wald test and the likelihood ratio test, also known as the ***deviance test***. The latter test is explained in the next section.

For the multivariate Wald test, we need not only the standard errors of the estimates but also the covariances among them. Suppose that we consider a certain vector γ of q regression parameters, for which we wish to test the null hypothesis

$$\mathrm{H}_0 : \gamma = 0 \, . \tag{6.3}$$

The statistical estimation leads to an estimate $\hat{\gamma}$ and an associated estimated covariance matrix $\hat{\Sigma}_\gamma$. From these, we can let a computer program calculate the test statistic represented in matrix form by

$$\hat{\gamma}' \, \hat{\Sigma}_\gamma^{-1} \, \hat{\gamma} \, . \tag{6.4}$$

Under the null hypothesis, the distribution of this statistic can be approximated by the chi-squared distribution with q degrees of freedom.[2]

The way of obtaining tests presented in this section is not applicable to tests of whether parameters (variances or covariances) in the random part of the model are 0. The reason is the fact that, if a population variance parameter is 0, its estimate divided by the estimated standard error does not approximately have a t-distribution. Tests for such hypotheses are discussed in the next section.

6.2 Deviance tests

The deviance test, or likelihood ratio test, is a quite general principle for statistical testing. In applications of the hierarchical linear model this test is mainly used for multiparameter tests and for tests about the random part of the model. The general principle is as follows.

When parameters of a statistical model are estimated by the maximum likelihood (ML) method, the estimation also provides the likelihood, which can be transformed into the ***deviance*** defined as minus twice the natural logarithm of the likelihood. This deviance can be regarded as a measure of lack of fit between model and data, but (in most statistical models) one cannot interpret the values of the deviance directly, but only differences in deviance values for several models fitted to the same data set.

---

[2] This approximation neglects the fact that $\hat{\Sigma}_\gamma$ is estimated, and not known exactly. It corresponds to using the standard normal distribution to test the value of (6.2). In principle, this could be taken into account by using the F rather than the chi-squared distribution. If the number of groups is large, the difference is not appreciable.

Suppose that two models are fitted to one data set, model M_0 with m_0 parameters and a larger model M_1 with m_1 parameters. So M_1 can be regarded as an extension of M_0, with $m_1 - m_0$ parameters added. Suppose that M_0 is tested as the null hypothesis and M_1 is the alternative hypothesis. Indicating the deviances by D_0 and D_1, respectively, their difference $D_0 - D_1$ can be used as a test statistic having a chi-squared distribution with $m_1 - m_0$ degrees of freedom. This type of test can be applied to parameters of the fixed as well as of the random part.

The deviance produced by the residual maximum likelihood (REML) method can be used in deviance tests only if the two models compared (M_0 and M_1) have the same fixed parts and differ only in their random parts.

Example 6.2 *Test of random intercept.*
In example 4.2, the random intercept model yields a deviance of $D_1 = 15251.8$, while the OLS regression model has a deviance of $D_0 = 15477.7$. There is $m_1 - m_0 = 1$ parameter added, the random intercept. The deviance difference is 225.9, immensely significant in a chi-squared distribution with $d.f. = 1$. This implies that, even when controlling for the effect of IQ, the differences between groups are strongly significant.

For example, suppose one is testing the fixed effect of a categorical explanatory variable with c categories. This categorical variable can be represented by $c-1$ dummy variables. Model M_0 is the hierarchical linear model with the effects of the other variables in the fixed part and with the given random part. Model M_1 also includes all these effects; in addition, the $c-1$ regression coefficients of the dummy variables have been added. Hence the difference in the number of parameters is $m_1 - m_0 = c - 1$. This implies that the deviance difference $D_0 - D_1$ can be tested in a chi-squared distribution with $d.f. = c - 1$. This test is an alternative for the multi-parameter Wald test treated in the preceding section. These tests will be very close to each other for intermediate and large sample sizes.

Example 6.3 *Effect of a categorical variable.*
In the data set used in Chapters 4 and 5, schools differ according to their denomination: public, catholic, protestant, or non-denominational private. To represent these four categories, three dummy variables are used, contrasting the last three against the first category. This means that all dummy variables are 0 for public schools, the first is 1 for catholic schools, the second is 1 for protestant schools, and the third is 1 for non-denominational private schools.

When the fixed effects of these $c - 1 = 3$ variables are added to the model presented in Table 5.2, which in this example has the role of M_0 with deviance $D_0 = 15208.4$, the deviance decreases to $D_1 = 15193.6$. The chi-squared value is $D_0 - D_1 = 14.8$ with $d.f. = 3$, $p < 0.005$. It can be concluded that, when controlling for IQ and group size (see the specification of model M_0 in Table 5.2), there are differences between the three types of school.

The estimated fixed effects (with standard errors) of the dummy variables are 1.70 (0.64) for the catholic, −0.80 (0.67) for the protestant, and 1.09 (1.30) for the non-denominational private schools. These effects are relative to the public schools. This implies that the catholic schools achieve higher, controlling for IQ and group size, than the public schools, while the other two categories do not differ significantly from the public schools.

6.2.1 *Halved p-values for variance parameters*

Variances are by definition non-negative. When testing the null hypothesis that a variance of the random intercept or of a random slope is zero, the alternative hypothesis is therefore one-sided. This observation can indeed be used to arrive at a sharpened version of the deviance test for variance parameters. This principle was derived by Miller (1977) and Self and Liang (1987).[3]

First consider the case that a random intercept is tested. The null model M_0 then is the model without a random part at level 2, i.e., all observations Y_{ij} are independent, conditional on the values of the explanatory variables. This is an ordinary linear regression model. The alternative model M_1 is the random intercept model with the same explanatory variables. There is $m_1 - m_0 = 1$ additional parameter, the random intercept variance τ_0^2. For the observed deviances D_0 of model M_0 (this model can be estimated by ordinary least squares) and D_1 of the random intercept model, the difference $D_0 - D_1$ is calculated. If $D_0 - D_1 = 0$, the random intercept variance is definitely not significant (it is estimated as being 0 ....). If $D_0 - D_1 > 0$, the tail probability of the difference $D_0 - D_1$ is looked up in a table of the chi-squared distribution with $d.f. = 1$. The p-value for testing the significance of the random intercept variance is half this tail value.

Second, consider the case of testing a random slope. Specifically, suppose that model (5.14) holds, and the null hypothesis is that the last random slope variance is zero: $\tau_p^2 = 0$. Under this null hypothesis, the p covariances, τ_{hp} for $h = 0, \ldots, p-1$, also are 0. The alternative hypothesis is the model defined by (5.14), and has $m_1 - m_0 = p + 1$ parameters more than the null model (one variance and p covariances). For example, if there are no other random slopes in the model ($p = 1$), $m_1 - m_0 = 2$. The same procedure is followed as for testing the random intercept: Both models are estimated, yielding the deviance difference $D_0 - D_1$. If $D_0 - D_1 = 0$, the random slope variance is not significant. If $D_0 - D_1 > 0$, the tail probability of the difference $D_0 - D_1$ is looked up in a table of the chi-squared distribution with $d.f. = p + 1$. The p-value for testing the significance of the random slope variance is half this tail value.

[3] The argumentation can be given loosely as follows. If the variance parameter is 0, then the probability is about 1/2 that the estimated value is 0, and also 1/2 that the estimated value is positive. For example, for estimating the level-two intercept variance τ^2 in an empty model, formula (3.11) can be used. If indeed $\tau^{@} = 0$, the probability is about 1/2 that this expression will be negative, and therefore truncated to 0. If the variance parameter is estimated as 0, the associated covariance parameters also are estimated at 0, with the result that all estimated parameters under model M_1 are the same as under M_0. This implies $D_1 = D_0$. The chi-squared distribution holds under the condition that the variance parameter is estimated at a positive value. It can be concluded that the null distribution of the deviance difference is a so-called mixed distribution, with probability 1/2 for the value 0, and probability 1/2 for a chi-squared distribution.

Example 6.4 *Test of random slope.*
When comparing Tables 4.4 and 5.1, it can be concluded that $m_1 - m_0 = 2$ parameters are added and the deviance diminishes by $D_0 - D_1 = 15227.5 - 15213.5 = 14.0$. Testing the value of 14.0 in a chi-squared distribution with $d.f. = 2$ yields $p < 0.001$. Halving the p-value leads to $p < 0.0005$. Thus, the significance probability of the random slope for IQ in the model of Table 5.1 is $p < 0.0005$.

As another example, suppose that one wishes to test the significance of the random slope for IQ in the model of Table 5.4. Then the model must be fitted in which this effect is omitted and all other effects remain. Thus, the omitted parameters are τ_1^2 and τ_{01} so that $d.f. = 2$. Fitting this reduced model leads to a deviance of $D_0 = 15101.5$, which is 8.5 more than the deviance in Table 5.4. The chi-squared distribution with $d.f. = 2$ gives a tail probability of $p < 0.02$, so that halving the p-value yields $p < 0.01$. Thus, testing the random slope for IQ in Table 5.4 yields a significant outcome with $p < 0.01$.

6.3 Other tests for parameters in the random part

The deviance tests are very convenient for testing parameters in the random part, but other tests for random intercepts and slopes also exist. In Section 3.3.2, the ANOVA F-test for the intraclass correlation was mentioned. This is effectively a test for randomness of the intercept. If it is desired to test the random intercept while controlling for explanatory variables, one may use the F-test from the ANCOVA model, using the explanatory variables as covariates.

Bryk and Raudenbush (1992) present chi-squared tests for random intercepts and slopes, i.e., for variance parameters in the random part. They are based on calculating OLS estimates for the values of the random effects within each group and testing these values for equality. For the random intercept, these are the large-sample chi-squared approximations to the F-tests in the ANOVA or ANCOVA model mentioned in Section 3.3.2.

Another test for variances parameters in the random part was proposed by Berkhof and Snijders (1998), and is explained in Section 9.2.2.

6.4 Model specification

Model specification is the choice of a satisfactory model. For the hierarchical linear model this amounts to the selection of relevant explanatory variables (and interactions) in the fixed part, and relevant random slopes (with their covariance pattern) in the random part. Model specification is one of the most difficult parts of statistical inference, because there are two steering wheels: substantive (subject-matter related) and statistical considerations. These steering wheels must be handled jointly. The purpose of model specification is to arrive at a model that describes the observed data to a satisfactory extent but without unnecessary complications. A parallel purpose is to obtain a model that is substantively interesting without wringing from the data drops that are really based on chance but interpreted as substance.

In linear regression analysis, model specification is already complicated (as is elaborated in many textbooks on regression, e.g., Ryan, 1997), and in multilevel analysis the number of complications is multiplied because of the complicated nature of the random part.

The complicated nature of the hierarchical linear model, combined with the two steering wheels for model specification, implies that there are no clear and fixed rules to follow. Model specification is a process guided by the following principles.

1. Considerations relating to the subject matter. These follow from field knowledge, existing theory, detailed problem formulation, and common sense.

2. The distinction between effects that are indicated *a priori* as effects to be tested, i.e., effects on which the research is focused, and effects that are necessary to obtain a good model fit. Often the tested effects are a subset of the fixed effects, and the random part is to be fitted adequately but of secondary interest. When there is no strong prior knowledge about which variables to include in the random part, one may follow a data-driven approach to select the variables for the random part.

3. A preference for 'hierarchical' models in the general sense (not the 'hierarchical linear model' sense) that if a model contains an interaction effect, then usually also the corresponding main effects should be included (even if these are not significant); and if a variable has a random slope, it normally should also have a fixed effect. The reason is that omitting such effects may lead to erroneous interpretations.

4. Doing justice to the multilevel nature of the problem. This is done as follows.

   (a) When a given level-one variable is present, one should be aware that the within-group regression coefficient may differ from the between-group regression coefficient, as described in Section 4.5. This can be investigated by calculating a new variable, defined as the group mean of the level-one variable, and testing the effect of the new variable.

   (b) When there is an important random intercept variance, there are important unexplained differences between group means. One may look for level-two variables (original level-two variables as well as aggregates of level-one variables) that explain part of these between-group differences.

   (c) When there is an important random slope variance of some level-one variable, say, X_1, there are important unexplained differences between the within-group effects of X_1 on Y. Here also one may look for level-two variables that explain part of these differences.

This leads to cross-level interactions as explained in Section 5.2. Recall from this section, however, that cross-level interactions can also be expected from theoretical considerations, even if no significant random slope variance is found.

5. Awareness of the necessity of including certain covariances of random effects. Including such covariances means that they are free parameters in the model, not constrained to 0 but estimated from the data.

   In Section 5.1.2, attention was given to the necessity to include in the model all covariances τ_{0h} between random slopes and random intercept. Another case of this point arises when a categorical variable with $c \geq 3$ categories has a random effect. This is implemented by giving random slopes to the $c - 1$ dummy variables that are used to represent the categorical variable. The covariances between these random slopes should then also be included in the model.

   Formulated generally, suppose that two variables X_h and $X_{h'}$ have random effects, and that the meaning of these variables is such that they could be replaced by two linear combinations, $aX_h + a'X_{h'}$ and $bX_h + b'X_{h'}$. (For the random intercept and random slope discussed in Section 5.1.2, the relevant type of linear combination would correspond to a change of origin of the variable with the random slope.) Then the covariance $\tau_{hh'}$ between the two random slopes should be included in the model.

6. Reluctance to include non-significant effects in the model; one could also say, a reluctance to overfitting. Each of points 1–4 above, however, could override this reluctance.

   An obvious example of this overriding is the case where one wishes to test for the effect of X_2, while controlling for the effect of X_1. The purpose of the analysis is a subject matter consideration, and even if the effect of X_1 is non-significant, one still should include this effect in the model.

7. The desire to obtain a good fit, and include all effects in the model that contribute to a good fit; in practice, this leads to the inclusion of all significant effects unless the data set is so large that certain effects, although significant, are deemed unimportant nevertheless.

8. Awareness of the following two basic statistical facts:

   (a) Every test of a given statistical parameter controls for all other effects in the model used as a null hypothesis (M_0 in Section 6.2). Since the latter set of effects has an influence on the interpretation as well as on the statistical power, test results may depend on the set of other effects included in the model.

   (b) We are constantly making errors of the first and the second kind. Especially the latter, since statistical power often is rather low.

This implies that an effect being non-significant does not mean that the effect is absent in the population. It also implies that a significant effect may be so by chance (but the probability of this is no larger than the level of significance – most often set at 0.05).

Multilevel research often is based on data with a limited number of groups. Since power for detecting effects of level-two variables depends strongly on the number of groups in the data, warnings about a low power are especially important for level-two variables.

These considerations are nice; but how to proceed in practice? To get an insight into the data, it is usually advisable to start with a descriptive analysis of the variables: an investigation of their means, standard deviations, correlations, and distributional forms. It is also helpful to make a preliminary ('quick and dirty') analysis with a simpler method such as OLS regression.

When starting with the multilevel analysis as such, in most situations (longitudinal data may provide an exception), it is advisable to start with fitting the empty model (4.6). This gives the raw within-group and between-group variances, from which the estimated intraclass correlation can be calculated. These parameters are useful as a general description and a starting point for further model fitting. The process of further model specification will include ***forward steps***: select additional effects (fixed or random), test their significance, and decide whether or not to include them in the model; and ***backward steps***: exclude effects from the model because they are not important from a statistical or substantive point of view. We mention two possible approaches in the following subsections.

6.4.1 Working upward from level one

In the spirit of Section 5.2, one may start with constructing a model for level one, i.e., first explain within-group variability, and explain between-group variability subsequently. This two-phase approach is advocated by Bryk and Raudenbush (1992) and is followed in the program HLM (Bryk, Raudenbush, and Congdon, 1996) (see Chapter 15).

Modeling within-group variability

Subject matter knowledge and availability of data leads to a number of level-one variables X_1 to X_p which are deemed important, or hypothetically important, to predict or explain the value of Y. These variables lead to equation (5.10) as a starting point:

$$Y_{ij} = \beta_{0j} + \beta_{1j}\,x_{1ij} + \ldots + \beta_{pj}\,x_{pij} + R_{ij} \;.$$

This equation represents the within-group effects of X_h on Y. The between-group variability is first modeled as random variability. This is represented by splitting the group-dependent regression coefficients β_{hj} in a mean coefficient γ_{h0} and a group-dependent deviation U_{0j}:

$$\beta_{hj} = \gamma_{h0} + U_{hj} \ .$$

Substitution yields

$$Y_{ij} = \gamma_{00} + \sum_{h=1}^{p} \gamma_{h0} \, x_{hij} + U_{0j} + \sum_{h=1}^{p} U_{hj} \, x_{hij} + R_{ij} \ . \tag{6.5}$$

This model has a number of level-one variables with fixed and random effects, but it will usually not be necessary to include all random effects.

For the precise specification of the level-one model, the following steps are useful.

1. Select in any case the variables on which the research is focused. In addition, select relevant available level-one variables on the basis of subject matter knowledge. Also include plausible interactions between level-one variables.

2. Select, among these variables, those for which, on the basis of subject matter knowledge, a group-dependent effect (random slope!) is plausible.
   If one does not have a clue, one could select the variables that are expected to have the strongest fixed effects.

3. Estimate the model with the fixed effects of step 1 and the random effects of step 2.

4. Test the significance of the random slopes, and exclude the non-significant slopes from the model.

5. Test the significance of the regression coefficients, and exclude the non-significant coefficients from the model. This can also be a moment to consider the inclusion of interaction effects between level-one variables.

6. For a check, one could test whether the variables for which a group-dependent effect (i.e., a random slope) was not thought plausible in step 2, indeed have a non-significant random slope. (Keep in mind that, normally, including a random slope implies inclusion of the fixed effect!) Be reluctant to include random slopes for interactions; these are often hard to interpret.

With respect to the random slopes, one may be restricted by the fact that data usually contain less information about random effects than about fixed effects. Including many random slopes can therefore lead to long iteration processes of the estimation algorithm. The algorithm may even fail to converge. For this reason it may be necessary to specify only a small number of random slopes.

After this process, one has arrived at a model with a number of level-one variables, some of which have a random in addition to their fixed effect. It is possible that the random intercept is the only remaining random effect.

This model is an interesting intermediate product, as it indicates the within-group regressions and their variability.

Modeling between-group variability

The next step is to try and explain these random effects by level-two variables. The random intercept variance can be explained by level-two variables, the random slopes by interactions of level-one with level-two variables, as was discussed in Section 5.2. It should be kept in mind that aggregates of level-one variables can be important level-two variables.

For deciding which main effects of level-two variables and which cross-level interactions to include, it is again advisable first to select those effects that are plausible on the basis of substantive knowledge, then to test these and include or omit them depending on their importance (statistical and substantive), and finally to check whether other (less plausible) effects also are significant.

This procedure has a built-in filter for cross-level interactions: an interaction between level-one variable X and level-two variable Z is considered only if X has a significant random slope. However, this 'filter' should not be employed as a strict rule. If there are theoretical reasons to consider the $X \times Z$ interaction, this interaction can be tested even if X does not have a significant random slope. The background to this is the fact that if there is $X \times Z$ interaction, the test for this interaction has a higher power to detect this than the test for a random slope.

It is possible that, if one carries out both tests, the test of the random slope is non-significant whereas the test of the $X \times Z$ interaction is indeed significant. This implies that either an error of the first kind is made by the test on $X \times Z$ interaction (this is the case if there is no interaction), or an error of the second kind is made by the test of the random slope (this is the case if there is interaction). Assuming that the significance level is 0.05 and one focuses on the test of this interaction effect, the probability of the first event is less than 0.05 whereas the probability of an error of the second kind can be quite high, especially since the test of the random slope does not always have high power for testing the specific alternative hypothesis of an $X \times Z$ interaction effect. Therefore, provided that the $X \times Z$ interaction effect was hypothesized before looking at the data, the significant result of the test of this effect is what counts, and not the lack of significance for the random slope.

6.4.2 Joint consideration of level-one and level-two variables

The procedure of first building a level-one model and subsequently extending it with level-two variables is neat but not always the most efficient, or the most relevant. If there are level-two variables or cross-level interactions that are known to be important, why not include them in the model right from the start? For example, it could be expected that a certain level-one variable has a within-group regression differing from its between-group regression. In such a case, one may wish to include the group mean of this variable right from the start.

In this approach, the same steps are followed as in the preceding section, but without the distinction between level-one and level-two variables. This leads to the following steps.

1. Select relevant available level-one and level-two variables on the basis of subject matter knowledge. Also include plausible interactions. Don't forget group means of level-one variables to account for the possibility of difference between within-group and between-group regressions. Also don't forget cross-level interactions.

2. Select among the level 1 variables those for which, on the basis of subject matter knowledge, a group-dependent effect (random slope!) is plausible. (A possibility would be, again, to select those variables that are expected to have the strongest fixed effects.)

3. Estimate the model with the fixed effects of step 1 and the random effects of step 2.

4. Test the significance of the random slopes, and exclude the non-significant slopes from the model.

5. Test the significance of the regression coefficients, and exclude the non-significant coefficients from the model. This can also be a moment to consider the inclusion of more interaction effects.

6. Check if other effects, thought less plausible at the start of model building, indeed are not significant. If they are significant, include them in the model.

In an extreme instance of step 1, one may wish to include all available variables and a large number of interactions in the fixed part. Similarly, one might wish to give all level-one variables random effects in step 2. Whether this is practically possible will depend, among others, on the number of level-one variables. Such an implementation of these steps leads to a backward model-fitting process, where one starts with a large model and reduces it by stepwise excluding non-significant effects. The advantage is that masking effects (where a variable is excluded early in the model building process because of non-significance, whereas it would have reached significance if one had controlled for another variable) do not occur. The disadvantage is that it may be a very time-consuming procedure.

6.4.3 Concluding remarks about model specification

This section has suggested a general approach to specification of multilevel models rather than laying out a step-by-step procedure. This is in accordance with our view of model specification being a process with two steering wheels and without foolproof procedures. This implies that, given one data set, a researcher (let alone two researchers...) may come up with more than one model, each seeming in itself a satisfactory result of a model specification process. In our view, this reflects the basic indeterminacy that is inherent to model fitting on the basis of empirical data. It is well possible that several different models correspond to a given data set, and that there are no compelling arguments to choose between them. In such cases, it is better to accept this indeterminacy and leave it to be resolved by further research than to make an unwarranted choice between the different models.

This treatment of model specification may seem rather inductive, or data-driven. If one is in the fortunate situation of having *a priori* hypotheses to be tested (usually about regression coefficients), it is useful to distinguish between those parameters on which hypothesis tests are focused and the other parts of the model, required to have a well-fitting model and (consequently) a valid test of these hypotheses. An inductive approach is adequate for the latter part of the model, while the tested parameters evidently are to be included in the model anyway.

Another aspect of model specification is the checking of assumptions. Independence assumptions should be checked in the course of specifying the random part of the model. Distributional assumptions, specifically, the assumption of normal distributions for the various random effects, should be checked by residual analysis. Checks of assumptions are treated in Chapter 9.

7 How Much Does the Model Explain?

7.1 Explained variance

The concept of 'explained variance' is well-known in multiple regression analysis: it gives an answer to the question, how much of the variability of the dependent variable is accounted for by the linear regression on the explanatory variables. The usual measure for the ***explained proportion of variance*** is the squared multiple correlation coefficient, R^2. For the hierarchical linear model, however, the concept of 'explained proportion of variance' is somewhat problematic. In this section, we follow the approach of Snijders and Bosker (1994) to explain the difficulties and give a suitable multilevel version of R^2.

One way to approach this concept is to transfer its customary treatment, well-known from multiple linear regression, straightforwardly to the hierarchical random effects model: treat proportional reductions in the estimated variance components, σ^2 and τ_0^2 in the random-intercept model for two levels, as analogues of R^2 values. Since there are several variance components in the hierarchical linear model, this approach leads to several R^2 values, one for each variance component. However, this definition of R^2 now and then leads to unpleasant surprises: it sometimes happens that adding explanatory variables ***increases*** rather than decreases some of the variance components. Even negative values of R^2 are possible. Negative values of R^2 clearly are undesirable and are not in accordance with its intuitive interpretation.

In the discussion of R^2-type measures, it should be kept in mind that these measures depend on the distribution of the explanatory variables. This implies that these variables, denoted in this section by X, are supposed to be drawn at random from the population at level one and the population at level two, and not determined by the experimental design or the researcher's whims. In order to stress the random nature of the X-variable, the values of X are denoted X_{ij}, instead of by x_{ij} as in earlier chapters.

7.1.1 *Negative values of R^2?*

As an example, we consider data from a study by Vermeulen and Bosker (1992) on the effects of part-time teaching in primary schools. The dependent variable Y is an arithmetic test score; the sample consists of 718 grade 3 pupils in 42 schools. An intelligence test score X is used as predictor variable. Group sizes range from 1 to 33 with an average of 20. In the following, it sometimes is desirable to present an example for balanced data (i.e., with equal group sizes). The balanced data presented below are the data restricted to 33 schools with 10 pupils in each school, by deleting schools with less than 10 pupils from the sample and randomly sampling 10 pupils from each school from the remaining schools. For demonstration purposes, three models were fitted: the empty Model A; Model B, with a group mean, $\overline{X}_{.j}$, as predictor variable; and Model C, with a within-group deviation score, $(X_{ij} - \overline{X}_{.j})$, as predictor variable. Table 7.1 presents the results of the analyses both for the balanced and for the entire data set. The residual variance at level one is denoted σ^2; the residual variance at level two is denoted τ_0^2. From Table 7.1 we see that in the balanced as

Table 7.1 Estimated residual variance parameters $\hat{\sigma}^2$ and $\hat{\tau}_0^2$ for models with within-group and between-group predictor variables.

	$\hat{\sigma}^2$	$\hat{\tau}_0^2$
I. Balanced design		
A. $Y_{ij} = \beta_0 + U_{0j} + E_{ij}$	8.694	2.271
B. $Y_{ij} = \beta_0 + \beta_1 \overline{X}_{.j} + U_{0j} + E_{ij}$	8.694	0.819
C. $Y_{ij} = \beta_0 + \beta_2 (X_{ij} - \overline{X}_{.j}) + U_{0j} + E_{ij}$	6.973	2.443
II. Unbalanced design		
A. $Y_{ij} = \beta_0 + U_{0j} + E_{ij}$	7.653	2.798
B. $Y_{ij} = \beta_0 + \beta_1 \overline{X}_{.j} + U_{0j} + E_{ij}$	7.685	2.038
C. $Y_{ij} = \beta_0 + \beta_2 (X_{ij} - \overline{X}_{.j}) + U_{0j} + E_{ij}$	6.668	2.891

well as in the unbalanced case, $\hat{\tau}_0^2$ increases as a within-group deviation variable is added as an explanatory variable to the model. Furthermore, for the balanced case, $\hat{\sigma}^2$ is not affected by adding a group-level variable to the model. In the unbalanced case, $\hat{\sigma}^2$ increases slightly when adding the group variable. When R^2 is defined as the proportional reduction in residual variance, as discussed above, then R^2 on the group level is negative for Model C, while for the entire data set R^2 on the pupil level is negative for Model B. Estimating σ^2 and τ_0^2 using the REML method results in slightly different parameter estimates. The pattern, however, remains the same. It is argued below that defining R^2 as the proportional reduction in residual variance parameters $\hat{\sigma}^2$ and $\hat{\tau}_0^2$, respectively, is not the best way to define a measure analogous to R^2 in the linear regression model; and that the problems mentioned can be solved by using other definitions, leading to measures denoted below by R_1^2 and R_2^2.

7.1.2 Definitions of proportions of explained variance in two-level models
In multiple linear regression, the customary R^2 parameter can be introduced in several ways: e.g., as the maximal squared correlation coefficient between the dependent variable and some linear combination of the predictor variables, or as the proportional reduction in the residual variance parameter due to the joint predictor variables. A very appealing principle to define measures of modeled (or explained) variation is the principle of *proportional reduction of prediction error*. This is one of the definitions of R^2 in multiple linear regression, and can be described as follows. A population of values is given for the explanatory and the dependent variables, $(X_{1i}, ..., X_{qi}, Y_i)$, with a known joint probability distribution; β is the value for the vector v for which the expected squared error

$$\mathcal{E}(Y_i - \sum_{h=0}^{q} v_h X_{hi})^2$$

is minimal. (This is the definition of the ordinary least squares ('OLS') estimation criterion.) (In this equation, v_0 is defined as the intercept and $X_{0ij} = 1$ for all i, j.) If, for a certain case i, the values of $X_{1i}, ..., X_{qi}$ are unknown, then the best predictor for Y_i is its expectation $\mathcal{E}(Y)$, with mean squared prediction error $\text{var}(Y_i)$; if the values $X_{1i}, ..., X_{qi}$ are given, then the linear predictor of Y_i with minimum squared error is the regression value $\sum_h \beta_h X_{hi}$. The difference between the observed value Y_i and the predicted value $\sum_h \beta_h X_{hi}$ is the prediction error. Accordingly, the mean squared prediction error is defined as

$$\mathcal{E}(Y_i - \sum_h \beta_h X_{hi})^2 .$$

The proportional reduction of the mean squared error of prediction is the same as the proportional reduction in the unexplained variance, due to the use of the variables X_1 to X_q. In a formula, it can be expressed by

$$R^2 = \frac{\text{var}(Y_i) - \text{var}(Y_i - \sum_h \beta_h X_{hi})}{\text{var}(Y_i)}$$

$$= 1 - \frac{\text{var}(Y_i - \sum_h \beta_h X_{hi})}{\text{var}(Y_i)} ;$$

this formula expresses one of the equivalent ways to define R^2.

The same principle can be used to define 'explained proportion of variance' in the hierarchical linear model. For this model, however, there are several options with respect to what one wishes to predict. Let us consider a two-level model with dependent variable Y. In such a model, one can choose between predicting an individual value Y_{ij} at the lowest level, or a group mean $\overline{Y}_{.j}$. On the basis of this distinction, two concepts of explained proportion of variance in a two-level model can be defined. The first, and most important, is the *proportional reduction of error for predicting an individual outcome*. The second is the *proportional reduction of error for predicting a group mean*. To elaborate these concepts more specifically, first consider a two-level random effects model with a random intercept and some predictor variables with fixed effects but no other random effects:

$$Y_{ij} = \gamma_0 + \sum_{h=1}^{q} \gamma_h X_{hij} + U_{0j} + R_{ij} . \tag{7.1}$$

Since we wish to discuss the definition of 'explained proportion of variance' as a population parameter, we assume temporarily that the vector γ of regression coefficients is known.

Level one

For the level-one explained proportion of variance, we consider the prediction of Y_{ij} for a randomly drawn level-one unit i within a randomly drawn level-two unit j. If the values of the predictors X_{ij} are unknown, then the best predictor for Y_{ij} is its expectation; the associated mean squared prediction error is $\text{var}(Y_{ij})$. If the value of the predictor vector X_{ij} for the given unit is known, then the best linear predictor for Y_{ij} is the regression value $\sum_{h=0}^{q} \gamma_h X_{hij}$ (where X_{h0j} is defined as 1 for all h, j.) The associated mean squared prediction error is

$$\text{var}(Y_{ij} - \sum_h \gamma_h X_{hij}) = \sigma^2 + \tau_0^2 .$$

The level-one explained proportion of variance is defined as the proportional reduction in mean squared prediction error:

$$R_1^2 = 1 - \frac{\text{var}(Y_{ij} - \sum_h \gamma_h X_{hij})}{\text{var}(Y_{ij})} . \tag{7.2}$$

Now let us proceed from the population to the data. The most straightforward way to estimate R_1^2 is to consider $\hat{\sigma}^2 + \hat{\tau}_0^2$ for the empty model,

$$Y_{ij} = \gamma_0 + U_{0j} + E_{ij} , \tag{7.3}$$

as well as for the fitted model (7.1), and compute 1 minus the ratio of these values. In other words, R_1^2 is just the proportional reduction in the value of $\hat{\sigma}^2 + \hat{\tau}_0^2$ due to including the X-variables in the model. For a sequence of nested models, the contributions to the estimated value of (7.2) due to adding new predictors can be considered to be the contribution of these predictors to the explained variance at level one.

To illustrate this, we once again use the data from the first (balanced) example, and estimate the proportional reduction of prediction error for a model where within-group and between-groups regression coefficients may be different.

Table 7.2 Estimating the level-one explained variance (balanced data).

	$\hat{\sigma}^2$	$\hat{\tau}_0^2$
A. $Y_{ij} = \beta_0 + U_{0j} + E_{ij}$	8.694	2.271
D. $Y_{ij} = \beta_0 + \beta_1(X_{ij} - \overline{X}_{.j}) + \beta_2 \overline{X}_{.j} + U_{0j} + E_{ij}$	6.973	0.991

From Table 7.2 we see that $\hat{\sigma}^2 + \hat{\tau}_0^2$ for model (A) amounts to 10.965, and for model (D) to 7.964. R_1^2 is thus estimated to be $1 - (7.964/10.965) = 0.274$.

Level two

The level-two explained proportion of variance can be defined as the proportional reduction in mean squared prediction error, for the prediction of $\overline{Y}_{.j}$ for a randomly drawn level-two unit j. If the values of the predictors X_{hij} for the level-one units i within level-two unit j are completely unknown, then the best predictor for $\overline{Y}_{.j}$ is its expectation; the associated mean squared prediction error is $\text{var}(\overline{Y}_{.j})$. If the values of all predictors X_{hij} for all i in this particular group j are known, then the best linear predictor for $\overline{Y}_{.j}$ is the regression value $\sum_h \gamma_h \overline{X}_{h.j}$; the associated mean square prediction error is

$$\text{var}(\overline{Y}_{.j} - \sum_h \gamma_h \overline{X}_{h.j}) = \frac{\sigma^2}{n} + \tau_0^2 \ ,$$

where n is the number of level-one units on which the average is based. The level-two explained proportion of variance is now defined as the proportional reduction in mean squared prediction error for $\overline{Y}_{.j}$:

$$R_2^2 \ = 1 - \frac{\text{var}(\overline{Y}_{.j} - \sum_h \gamma_h \overline{X}_{h.j})}{\text{var}(\overline{Y}_{.j})} \ . \tag{7.4}$$

To estimate the level-two explained proportion of variance, we follow a similar approach as above: R_2^2 is estimated as the proportional reduction in the value of $\hat{\sigma}^2/n + \hat{\tau}_0^2$, where n is a representative value for the group size.

In the example given earlier, let us use for n a usual group size of $n = 30$. For model (a) the value of $\hat{\sigma}^2/n + \hat{\tau}_0^2$ is $8.694/30 + 2.271 = 2.561$, whereas for model (b) this amounts to $6.973/30 + 0.991 = 1.223$. R_2^2 is thus estimated at $1 - (1.223/2.561) = 0.52$.

It is natural that the mean squared error for predicting a group mean should depend on the group size. Often one can use for n a value which is deemed *a priori* to be 'representative'. For example, if a normal class size is considered to be 30, and even if because of missing data the values of n_j in the data set are on average less than 30, it is advisable to use the representative value $n = 30$. In the case of varying group sizes, if it is unclear how a representative group size should be chosen, one possibility is to use the harmonic mean, defined by $N/\{\sum_j (1/n_j)\}$.

It is quite common that the data within the groups are based on a sample and not the entire groups in the population are observed, so that the group sizes in the data set do not reflect the group sizes in the population. In such a case, since it is more relevant to predict population group averages than sample group averages, it is advisable to let n reflect the group sizes in the population rather than the sample group sizes. If group sizes are very large, e.g., when the level-two units are defined by municipalities or other regions and the level-one units by their inhabitants, this means that, practically speaking, R_2^2 is the proportional reduction in the intercept variance.

Population values of R_1^2 and R_2^2 are non-negative
What happens to R_1^2 and R_2^2, when predictor variables are added to the multilevel model? Is it possible that adding predictor variables leads to smaller values of R_1^2 or R_2^2? Can we even be sure at all that these quantities are positive?

It turns out that a distinction must be made between the population parameters R_1^2 and R_2^2 and their estimates from data. *Population* values of R_1^2 and R_2^2 in correctly specified models, with a constant group size n, become smaller when predictor variables are deleted, provided that the variables U_{0j} and E_{ij} on one hand are uncorrelated with all the X_{ij} variables on the other hand (the usual model assumption).

For *estimates* of R_1^2 and R_2^2, however, the situation is different: these estimates sometimes do increase when predictor variables are deleted. When it is observed that an estimated value for R_1^2 or R_2^2 becomes smaller by the addition of a predictor variable, or larger by the deletion of a predictor variable, there are two possibilities: either this is a chance fluctuation, or the larger model is misspecified. It will depend on how large the change in R_1^2 or R_2^2 is, and on the subject-matter insight of the researcher, whether the researcher will deem that either the first or the second possibility is more likely. In this sense changes in R_1^2 or R_2^2 in the 'wrong' direction serve as a *diagnostic* for possible misspecification. This possibility of misspecification refers to the fixed part of the model, i.e., the specification of the explanatory variables having fixed regression coefficients, and not to the random part of the model. We return to this in Section 9.2.3.

7.1.3 *Explained variance in three-level models*

In three-level random intercept models (Section 4.8), the residual variance, or mean squared prediction variance, is the sum of the variance components at the three levels, $\sigma^2 + \tau_0^2 + \varphi_0^2$. Accordingly, the level-one explained proportion of variance can be defined here as the proportional reduction in the sum of these three variance parameters.

Example 7.1 *Variance in maths performance explained by IQ.*
In Example 4.8, Table 4.5 exhibits the results of the empty model (Model 1) and a model in which IQ has a fixed effect (Model 2). The total variance in the empty model is 7.816 + 1.746 + 2.124 = 11.686 while the total unexplained variance in Model 2 is 6.910 + 0.701 + 1.109 = 8.720. Hence the level-one explained proportion of variance is 1 - (8.720/11.686) = 0.25.

7.1.4 *Explained variance in models with random slopes*

The idea of using the proportional reduction in the prediction error for Y_{ij} and $\overline{Y}_{.j}$, respectively, as the definitions of explained variance at either level, can be extended to two-level models with one or more random regression coefficients. The formulae to calculate R_1^2 and R_2^2 can be found in Snijders and Bosker (1994). However, the estimated values for R_1^2 and R_2^2 usually change only very little when random regression coefficients are included in the model.

The formulae for estimating R_1^2 and R_2^2 in models with random intercepts only are very easy. Estimating R_1^2 and R_2^2 in models with random slopes is more tedious. The software package HLM (Bryk et al., 1996), however, provides the necessary estimates, since it not only produces estimates of the variance components but also of the observed residual between group variance

$$\text{var}(\overline{Y}_{.j} - \sum_h \gamma_h \overline{X}_{h.j}) \,. \tag{7.5}$$

Using this estimate, denoted in the HLM output as D_BAR, one can calculate the estimate of R_2^2 (as the proportional reduction in D_BAR). In HLM versions 2.20 and higher, the output includes estimates of τ_0^2 and the reliability of the intercepts, but no longer the D_BAR. However, this observed residual between-group variance can now be calculated as τ_0^2/reliability. Since the level-two variance is not constant in random slope models it is usually advisable here to center explanatory variables around their grand means, so that R_2^2 now refers to the explained variance at level two for the average level-two unit.

The simplest possibility to estimate R_1^2 and R_2^2 in models with random slopes is to re-estimate the models as random intercept models with the same fixed parts (omitting the random slopes), and use the resulting parameter estimates to calculate R_1^2 and R_2^2 in the usual (simple) way for random intercept models. This will usually yield values that are very close to the values for the random slopes model.

Example 7.2 *Explained variance for language scores.*
In Table 5.4, a model was presented for the data set on language scores in elementary schools used throughout Chapters 4 and 5.

When a random intercept model is fitted with the same fixed part, the estimated variance parameters are $\hat{\tau}_0^2 = 7.61$ for level two and $\hat{\sigma}^2 = 39.82$ for level one. For the empty model, Table 4.1 shows that the estimates are $\hat{\tau}_0^2 = 19.42$ and $\hat{\sigma}^2 = 64.57$. This implies that explained variance at level one is $1-(39.82+7.61)/(64.57+19.42) = 0.44$ and, using an average class size $n = 25$, explained variance at level two is $1-[(39.82/25)+7.61]/[(64.57/25)+19.42] = 0.58$. These explained variances are quite high, and can be attributed mainly to the explanation by IQ.

7.2 Components of variance[1]

The preceding section focused on the total amount of variance that can be explained by the explanatory variables. In these measures of explained variance, only the fixed effects contribute. It can also be theoretically illuminating to decompose the observed variance of Y into parts that correspond to the various constituents of the model. This is discussed in this section for a two-level model.

For the dependent variable Y, the level-one and level-two variances in the empty model (4.6) are denoted σ_E^2 and τ_E^2, respectively. The total

[1] This is a more advanced section which may be skipped by the reader.

variance of Y therefore is $\sigma_E^2 + \tau_E^2$, and the components of variance are the parts into which this quantity is split. The first split, obviously, is the split of $\sigma_E^2 + \tau_E^2$ into σ_E^2 and τ_E^2, and was extensively discussed in our treatment of the intraclass correlation coefficient.

To obtain formulae for a further decomposition, it is necessary to be more specific about the distribution of the explanatory variables. It is usual in single-level as well as in multilevel regression analysis to condition on the values of the explanatory variables, i.e., to consider those as given values. In this section, however, all explanatory variables are regarded as random variables with a given distribution.

7.2.1 Random intercept models

For the random intercept model, we distinguish the explanatory variables in level-one variables X and level-two variables Z. Deviating from the notation in other parts of this book, matrix notation is used, and X and Z denote vectors.

The explanatory variables $X_1, ..., X_p$ at level one are collected in the vector X with value X_{ij} for unit i in group j. It is assumed more specifically that X_{ij} can be decomposed into independent level-one and level-two parts,

$$X_{ij} = X_{ij}^W + X_j^B \tag{7.6}$$

(i.e., a kind of multivariate hierarchical linear model without a fixed part). The expectation is denoted $\mathcal{E}X_{ij} = \mu_X$, the level-one covariance matrix is

$$\text{cov}(X_{ij}^W) = \Sigma_X^W ,$$

and the level-two covariance matrix is

$$\text{cov}(X_j^B) = \Sigma_X^B .$$

This implies that the overall covariance matrix of X is the sum of these two,

$$\text{cov}(X_{ij}) = \Sigma_X^W + \Sigma_X^B = \Sigma_X .$$

Further, the covariance matrix of the group average for a group of size n is

$$\text{cov}(\overline{X}_{.j}) = \frac{1}{n} \Sigma_X^W + \Sigma_X^B .$$

It may be noted that this notation deviates slightly from the common split of X_{ij} into

$$X_{ij} = (X_{ij} - \overline{X}_{.j}) + \overline{X}_{.j} . \tag{7.7}$$

The split (7.6) is a population-based split, whereas the more usual split (7.7) is sample-based. In the notation used here, the covariance matrix of the within-group deviation variable is

$$\text{cov}(X_{ij} - \overline{X}_{.j}) = \frac{n-1}{n} \Sigma_X^W ,$$

while the covariance matrix of the group means is

$$\text{cov}(\overline{X}_{.j}) = \frac{1}{n} \Sigma_X^W + \Sigma_X^B .$$

For the discussion in this section, the present notation is more convenient.

The split (7.6) is not a completely innocuous assumption. The independence between X_{ij}^W and X_j^B implies that the covariance matrix of the group means is larger[2] than $1/(n-1)$ times the within-group covariance matrix of X.

The vector of explanatory variables $Z = (Z_1, ..., Z_q)$ at level two has value Z_j for group j. The vector of expectations of Z is denoted

$$\mathcal{E} Z_j = \mu_Z ,$$

and the covariance matrix is

$$\text{cov}(Z_j) = \Sigma_Z .$$

In the random intercept model (4.7), denote the vector of regression coefficients of the X's by

$$\gamma_X = (\gamma_{10}, ..., \gamma_{p0})' ,$$

and the vector of regression coefficients of the Z's by

$$\gamma_Z = (\gamma_{01}, ..., \gamma_{0q})' .$$

Taking into account the stochastic nature of the explanatory variables then leads to the following expression for the variance of Y:

$$\begin{aligned} \text{var}(Y_{ij}) &= \gamma'_X \, \Sigma_X \, \gamma_X \; + \; \gamma'_Z \, \Sigma_Z \, \gamma_Z \; + \; \tau_0^2 \; + \; \sigma^2 \\ &= \gamma'_X \, \Sigma_X^W \, \gamma_X \; + \; \gamma'_X \, \Sigma_X^B \, \gamma_X \; + \; \gamma'_Z \, \Sigma_Z \, \gamma_Z \\ &\qquad + \tau_0^2 \; + \; \sigma^2 \; . \end{aligned} \tag{7.8}$$

It may be illuminating to remark that, for the special case that all explanatory variables are uncorrelated, this expression is equal to

$$\text{var}(Y_{ij}) = \sum_{h=1}^{p} \gamma_{h0}^2 \, \text{var}(X_h) \; + \; \sum_{h=1}^{q} \gamma_{0h}^2 \, \text{var}(Z_h) \; + \; \tau_0^2 \; + \; \sigma^2 \; .$$

(This holds, e.g., if there is only one level-one and only one level-two explanatory variable.) This formula shows that, in this special case, the contribution of each explanatory variable to the variance of the dependent variable is given by the product of the regression coefficient and the variance of the explanatory variable.

The decomposition of X into independent level-one and level-two parts allows us to indicate precisely which parts of (7.8) correspond to the unconditional level-one variance σ_E^2 of Y, and which parts to the unconditional level-two variance τ_E^2:

$$\begin{aligned} \sigma_E^2 &= \gamma'_X \, \Sigma_X^W \, \gamma_X \; + \; \sigma^2 \\ \tau_E^2 &= \gamma'_X \, \Sigma_X^B \, \gamma_X \; + \; \gamma'_Z \, \Sigma_Z \, \gamma_Z \; + \; \tau_0^2 \; . \end{aligned}$$

This shows how the within-group variation of the level-one variables eats some part of the unconditional level-one variance; parts of the level-two variance are eaten by the variation of the level-two variables, and also by the between-group (composition) variation of the level-one variables. Recall, however, the definition of Σ_X^B, which implies that the between-group variation of X is taken net of the 'random' variation of the group mean, that may be expected given the within-group variation of X_{ij}.

[2]The word 'larger' is meant here in the sense of the ordering of positive definite symmetric matrices.

7.2.2 Random slope models

For the hierarchical linear model in its general specification given by (5.12), a decomposition of the variance is very complicated because of the presence of the cross-level interactions. Therefore the decomposition of the variance is discussed for random slopes models in the formulation (5.14), repeated here as

$$Y_{ij} = \gamma_0 + \sum_{h=1}^{q} \gamma_h \, x_{hij} + U_{0j} + \sum_{h=1}^{p} U_{hj} \, x_{hij} + R_{ij} \ , \tag{7.9}$$

without bothering about whether some of the x_{hij} are level-one or level-two variables, or products of a level-one and a level-two variable.

Recall that in this section the explanatory variables X are stochastic. The vector $X = (X_1, ..., X_q)$ of all explanatory variables has mean $\mu_{X(q)}$ and covariance matrix $\Sigma_{X(q)}$. The sub-vector $(X_1, ..., X_p)$ of variables that have random slopes, has mean $\mu_{X(p)}$ and covariance matrix $\Sigma_{X(p)}$. These covariance matrices could be split into within-group and between-group parts, but this is left up to the reader.

The covariance matrix of the random slopes $(U_{1j}, ..., U_{pj})$ is denoted T_{11} and the $p \times 1$ vector of the intercept-slope covariances is denoted T_{10}.

With these specifications, the variance of the dependent variable can be shown to be given by

$$\begin{aligned} \text{var}(Y_{ij}) = \gamma' \, \Sigma_{X(q)} \, \gamma + \tau_0^2 + \mu'_{X(q)} \, T_{10} + \mu'_{X(q)} \, T_{11} \, \mu_{X(q)} \\ + \text{trace} \left(T_{11} \, \Sigma_{X(p)} \right) + \sigma^2 \ . \end{aligned} \tag{7.10}$$

(A similar expression, but without taking the fixed effects into account, is given by Snijders and Bosker (1993) as formula (21).) A brief discussion of all terms in this expression is as follows.

1. The first term,

$$\gamma' \, \Sigma_{X(q)} \, \gamma \ ,$$

gives the contribution of the fixed effects and may be regarded as the 'explained part' of the variance. This term could be split into a level-one and a level-two part as in the preceding subsection.

2. The part

$$\tau_0^2 + \mu'_{X(q)} \, T_{10} + \mu'_{X(q)} \, T_{11} \, \mu_{X(q)} \tag{7.11}$$

should be seen as one piece. One could rescale all variables with random slopes to have a zero mean (cf. the discussion in Section 5.1.2); this would lead to $\mu_{X(q)} = 0$ and leave of this piece only the intercept variance τ_0^2. In other words, (7.11) is just the intercept variance after subtracting the mean from all variables with random slopes.

3. The part

$$\text{trace} \left(T_{11} \, \Sigma_{X(p)} \right)$$

is the contribution of the random slopes to the variance of Y. In the extreme case that all variables X_1 to X_q would be uncorrelated and have unit variances, this expression reduces to the sum of squared random slope variances. This term also could be split into a level-one and a level-two part.

4. Finally,

$$\sigma^2$$

is the residual level-one variability that can neither be explained on the basis of the fixed effects, nor on the basis of the latent group characteristics that are represented by the random intercept and slopes.

8 Heteroscedasticity

The hierarchical linear model is a quite flexible model, and it has some other features in addition to the possibility of representing a nested data structure. One of these features is the possibility of representing multilevel as well as single-level regression models where the residual variance is not constant.

In ordinary least squares regression analysis, one of the standard assumptions is ***homoscedasticity***: residual variance is constant, i.e., it does not depend on the explanatory variables. This assumption was made in the preceding chapters, e.g., for the residual variance at level one and for the intercept variance at level two. The techniques used in the hierarchical linear model allow to relax this assumption and replace it by the weaker assumption that variances depend linearly or quadratically on explanatory variables. Thus, an important special case of heteroscedastic models (i.e., models with heterogeneous variances) is obtained, viz., heteroscedasticity where the variance depends on given explanatory variables. This feature is implemented at this moment in the MLn/MLwiN program and in HLM version 5, and can also be obtained in SAS (cf. Chapter 15). This chapter treats a two-level model, but the techniques treated (and the software mentioned) can be used also for heteroscedastic single-level regression models.

8.1 Heteroscedasticity at level one

8.1.1 Linear variance functions

In a hierarchical linear model it sometimes makes sense to consider the possibility that the residual variance at level one depends on one of the predictor variables. An example is a situation where two measurement instruments have been used, each with a different precision, resulting in two different values for the measurement error variance which is a component of the level-one variance.

If the level-one residual variance depends linearly on some variable X_1, it can be expressed by

$$\text{level-one variance } = \sigma_0^2 + 2\,\sigma_{01}\, x_{1ij} \ , \qquad (8.1)$$

where the value of X_1 for a given unit is denoted by x_{1ij} while the random part at level one now has two parameters, σ_0^2 and σ_{01}. The reason for incorporating the factor 2 will become clear later, when also quadratic variance functions are considered.

For example, when X_1 is a dummy variable with values 0 and 1, the residual variance is σ_0^2 for the units with $X_1 = 0$ and $\sigma_0^2 + 2\,\sigma_{01}$ for the units with $X_1 = 1$. When the level-one variance depends on more than variable, their effects can be added to the variance function (8.1) by adding terms $2\,\sigma_{02}\,x_{2ij}$, etc.

Example 8.1 *Residual variance depending on gender.*
In the example used in Chapters 4 and 5, the residual variance might depend on the pupil's gender. To investigate this in a model that is not overly complicated, we take the model of Table 5.4, delete the effects of multi-grade classes, and add the effect of gender (a dummy variable which is 0 for boys and 1 for girls).

Table 8.1 presents estimates for two models: one with constant residual variances, and one with residual variances depending on gender. Thus, Model 1 is a homoscedastic model and Model 2 a gender-dependent heteroscedastic model.

Table 8.1 Estimates for homoscedastic and heteroscedastic models.

	Model 1		Model 2	
Fixed Effect	Coefficient	S.E.	Coefficient	S.E.
Intercept	39.53	0.31	39.53	0.31
IQ	2.268	0.081	2.264	0.081
SES	0.152	0.014	0.151	0.014
Gender	2.64	0.26	2.64	0.26
$\overline{\text{IQ}}$	1.02	0.32	1.01	0.32
Random Effect	Parameter	S.E.	Parameter	S.E.
Level-two random effects:				
Intercept variance	8.27	1.33	8.24	1.33
IQ slope variance	0.169	0.088	0.171	0.088
Intercept - IQ slope covariance	−0.76	0.25	−0.78	0.26
Level-one variance parameters:				
σ_0^2 constant term	37.56	1.17	38.72	1.67
σ_{01} gender effect			−1.21	1.17
Deviance	15005.5		15004.4	

According to formula (8.1), the residual variance in Model 2 is 38.72 for boys and $38.72 - 2 \times 1.21 = 36.30$ for girls. The residual variance estimated in the homoscedastic Model 1 is very close to the average of these two figures. This is natural, since about half of the pupils are girls and half are boys. The difference between the two variances is, however, not significant: the deviance test yields $\chi^2 = 15005.5 - 15004.4 = 1.1$, $d.f. = 1$, $p > 0.2$.

The fixed effect of gender is quite significant ($t = 2.64/0.26 = 10.2, p < 0.0001$). Controlling for IQ and SES, girls score on average 2.64 higher than boys.

Analogous to the dependence due to the multilevel nesting structure as discussed in Chapter 2, heteroscedasticity has two faces: it can be a nuisance and it can be interesting. It can be a nuisance because the failure to take it into account may lead to a misspecified model and, hence, incorrect parameter estimates and standard errors. On the other hand, it can also be an interesting phenomenon in itself. When high values on some variable X_1 are associated with a higher residual variance, this means that for the units who score high on X_1 there is, within the context of the model being considered, more uncertainty about their value on the dependent variable Y. Thus, it may be interesting to look for explanatory variables that differentiate especially between units who score high on X_1. Sometimes a non-linear function of X_1, or an interaction involving X_1, could play such a role.

Example 8.2 *Heteroscedasticity related to IQ.*
Continuing the previous example, it is now investigated if residual variance depends on IQ. The corresponding parameter estimates are presented as Model 3 in Table 8.2.

Table 8.2 Heteroscedastic models depending on IQ.

	Model 3		Model 4	
Fixed Effect	Coefficient	S.E.	Coefficient	S.E.
Intercept	39.61	0.31	39.73	0.31
IQ	2.223	0.077	3.236	0.157
IQ^2_-			0.246	0.046
IQ^2_+			−0.306	0.039
SES	0.146	0.014	0.144	0.014
Gender	2.51	0.26	2.35	0.25
$\overline{IQ}$	1.02	0.32	1.21	0.31
Random Effect	Parameter	S.E.	Parameter	S.E.
Level-two random effects:				
Intercept variance	8.06	1.31	7.19	1.17
IQ slope variance	0.133	0.078	0.0	0.0
Intercept - IQ slope covariance	−0.51	0.24	0.0	0.0
Level-one variance parameters:				
σ_0^2 constant term	37.83	1.20	37.88	1.18
σ_{01} IQ effect	−2.01	0.26	−2.37	0.22
Deviance	14960.0		14908.1	

Comparing the deviance to Model 1 shows that there is a quite significant heteroscedasticity associated with IQ: $\chi^2 = 15005.5 - 14960.0 = 45.5$, $d.f. = 1$, $p < 0.0001$. The level-one variance function is (cf. (8.1))

$$37.83 - 4.02\,\mathrm{IQ}\ .$$

This shows that language scores of the less intelligent pupils are more variable than language scores of the more intelligent. The standard deviation of IQ is 2.07 and the mean is 0. Thus, the range of the level-one variance, when IQ is in the range of the mean ± twice the standard deviation, is between 29.79 and 45.87. This is an appreciable variation around the average value of 37.56 estimated in the homoscedastic Model 1.

Prompted by the IQ-dependent heteroscedasticity, the data were explored for effects that might differentiate between the pupils with lower IQ scores. Non-linear effects of IQ and some interactions involving IQ were tried. It appeared that a non-linear effect of IQ is discernible, represented better by a so-called spline function[1] than by a polynomial function of IQ. Specifically, the coefficient of the square of IQ turned out to be different for negative than for positive IQ values (recall that IQ was standardized to have an average of 0). This is represented in Model 4 of Table 8.2 by the variables

$$\begin{aligned} \mathrm{IQ}^2_- &= \begin{cases} \mathrm{IQ}^2 & \text{if IQ } < 0 \\ 0 & \text{if IQ } \geq 0\,, \end{cases} \\ \mathrm{IQ}^2_+ &= \begin{cases} 0 & \text{if IQ } < 0 \\ \mathrm{IQ}^2 & \text{if IQ } \geq 0\,. \end{cases} \end{aligned} \tag{8.2}$$

Adding these two variables to the fixed part gives a quite significant decrease of the deviance (51.9 for two degrees of freedom) and completely takes away the random slope of IQ. The IQ-related heteroscedasticity, however, becomes even stronger. The total fixed effect of IQ now is given by

$$\text{effect of IQ} = \begin{cases} 2.223\ \mathrm{IQ} + 0.246\ \mathrm{IQ}^2 & \text{if IQ } < 0 \\ 2.223\ \mathrm{IQ} - 0.306\ \mathrm{IQ}^2 & \text{if IQ } \geq 0\,. \end{cases} \tag{8.3}$$

The graph of this effect is shown in Figure 8.1. It is an increasing function which flattens out for low and for high values of IQ in a way that cannot be well represented by a quadratic or cubic function.

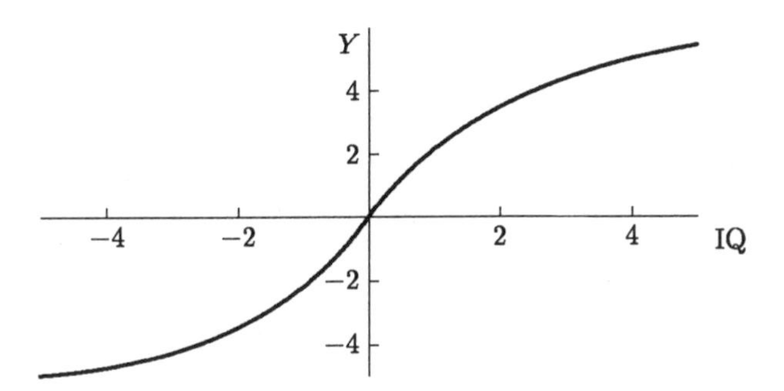

Figure 8.1 Effect of IQ on language test.

This is an interesting turn of this modeling exercise, and a nuisance only because it indicates once again that school performance is a quite complicated subject. Our interpretation of this data set toward the closing of Chapter 5 was that there is a random slope of IQ, i.e., schools differ in the effect of IQ. Now it turns out that the data are clearly better described by a model in which IQ has a non-linear effect, the effect of IQ being stronger in the middle range than toward its extreme values; in which the effect of IQ does not vary across schools; and in which the combined effects of IQ, SES, gender, and the school-average of IQ predict the language scores for high-IQ much better than for low-IQ pupils.

[1] Spline functions (introduced a bit more extensively in Section 12.2.2 and treated more fully, e.g., in Seber and Wild, 1989, Section 9.5) are a more flexible class of functions than polynomials. They are polynomials of which the coefficients may be different on several intervals.

8.1.2 *Quadratic variance functions*

The formal representation of level-one heteroscedasticity is based on including random effects at level one, as spelled out in Goldstein (1995, see the remarks on a complex random part at level one). In Section 5.1.1 it was remarked already that random slopes, i.e., random effects at level two, lead to heteroscedasticity. This also holds for random effects at level one. Consider a two-level model and suppose that the level-one random part is

$$\text{random part at level one} = R_{0ij} + R_{1ij}\, x_{1ij} \; . \tag{8.4}$$

Denote the variances of R_{0ij} and R_{1ij} by σ_0^2 and σ_1^2, respectively, and their covariance by σ_{01}. The rules for calculating with variances and covariances imply that

$$\text{var}(R_{0ij} + R_{1ij}\, x_{ij}) = \sigma_0^2 + 2\,\sigma_{01}\, x_{1ij} + \sigma_1^2\, x_{1ij}^2 \; . \tag{8.5}$$

Formula (8.4) is just a formal representation, used in MLn/MLwiN to specify this heteroscedastic model. For the interpretation one should rather look at (8.5). This formula can be used without the interpretation that σ_0^2 and σ_1^2 are variances and σ_{01} a covariance; these parameters might be any numbers. The formula only implies that the residual variance is a quadratic function of x_{1ij}. In the previous section, the case was used where $\sigma_1^2 = 0$, producing the linear function (8.1). If a quadratic function is desired, all three parameters are estimated from the data.

Example 8.3 *Educational level attained by pupils.*
This example is about a cohort of pupils entering secondary school in 1989, studied by Dekkers, Bosker, and Driessen (1998). The question is, how well the educational level attained in 1995 can be predicted from individual characteristics and school achievement at the end of primary school. The data is about 13,007 pupils in 369 secondary schools. The dependent variable is an educational attainment variable, being defined as 12 minus the minimum number of additional years of schooling it would take theoretically for this pupil in 1995 to gain a certificate giving access to university. The range is 4 to 13 (e.g., the value of 13 means that the pupil already is a first year university student). Explanatory variables are teacher's rating at the end of primary school (an advice on the most suitable type of secondary school, range 1 to 4), achievement on three standardized tests, so-called CITO tests, on language, arithmetic, and information processing (all with a mean value around 11 and a standard deviation between 4 and 5), socio-economic status (a discrete ordered scale with values 1 to 6), gender (0 for boys, 1 for girls), and minority status (based on the parents' country of birth, 0 for the Netherlands and other industrialized countries, 1 for other countries).

Table 8.3 presents the results of a random intercept model as Model 1. Note that standard errors are quite small due to the large number of pupils in this data set. The explained proportion of variance at level 1 is $R_1^2 = 0.39$. Model 2 shows the results for a model where residual variance depends quadratically on SES. It follows from (8.5) that residual variance here is given by

$$\text{residual variance} = 3.475 - 0.632\,\text{SES} + 0.056\,\text{SES}^2 \; .$$

The deviance difference between Models 1 and 2 ($\chi^2 = 97.3$, $d.f. = 2$, $p < 0.0001$) indicates that the dependence of residual variance on SES is quite significant. The variance function decreases curvilinearly from the value 2.91 for SES = 1 to a minimum value of 1.75 for SES = 6. This implies that when educational attainment is predicted by the variables in this model, the uncertainty in the prediction is highest for low-SES pupils. It is reassuring that the estimates and standard errors of other effects are not appreciably different between Models 1 and 2.

The specification of this random part was checked in the following way. First the models with only a linear or only a quadratic variance term were estimated separately. This showed that both variance terms are significant. Further, it might be possible that the SES-dependence of the residual variance is a random slope in disguise. Therefore a model with a random slope (i.e., a random effect at level two) for SES also was fitted. This showed that the random slope was barely significant and not very large, and did not take away the heteroscedasticity effect.

Table 8.3 Heteroscedastic model depending quadratically on SES.

	Model 1		Model 2	
Fixed Effect	Coefficient	S.E.	Coefficient	S.E.
Intercept	6.021	0.066	6.024	0.067
Teacher's rating	0.490	0.023	0.487	0.023
CITO Arithmetic	0.0567	0.0042	0.0578	0.0042
CITO Information	0.0651	0.0049	0.0648	0.0049
CITO Language	0.0361	0.0049	0.0365	0.0048
SES	0.168	0.013	0.166	0.013
Gender	0.322	0.028	0.318	0.027
Random Effect	Parameter	S.E.	Parameter	S.E.
Level-two random effect:				
Intercept variance	0.128	0.014	0.124	0.014
Level-one variance parameters:				
σ_0^2 constant term	1.982	0.025	3.475	0.247
σ_{01} linear SES effect			−0.316	0.064
σ_1^2 quadratic SES effect			0.056	0.016
Deviance	46205.1		46107.8	

The SES-dependent heteroscedasticity led to the consideration of non-linear effects of SES and interaction effects involving SES. Since the SES variable assumes values 1 through 6, five dummy variables were used contrasting the respective SES values to the reference category SES = 3. In order to limit the number of variables, the interactions of SES were defined as interactions with the numerical SES variable rather than with the categorical variable represented by these dummies. For the same reason, for the interaction of SES with the CITO tests only the average of the three CITO tests (range 1 to 20, mean 11.7) was considered. Product interactions were considered of SES with gender, with the average CITO score, and with minority status. As factors in the product, SES and CITO tests were centered approximately by using (SES − 3) and (CITO average − 12). Gender and minority status, being 0–1 variables, did not need to be centered.

This implies that although SES is represented by dummies (i.e., as a categorical variable) in the main effect, it is used as a numerical variable in the interaction effects. (The main effect of SES as a numerical variable is implicitly also included, because it can be represented also as an effect of the dummy variables. Therefore the numerical SES variable does not need to be added to the model.)

Minority status does not have a significant main effect when added to Model 2, but the main effect was added to facilitate interpretation of the interaction effect. The results are in Table 8.4.

Table 8.4 Heteroscedastic model with interaction effects.

	Model 3	
Fixed Effect	Coefficient	S.E.
Intercept	6.594	0.069
Teacher's rating	0.484	0.023
CITO Arithmetic	0.0564	0.0042
CITO Information	0.0632	0.0050
CITO Language	0.0355	0.0049
SES = 1	−0.827	0.155
SES = 2	−0.390	0.050
SES = 4	0.133	0.034
SES = 5	0.389	0.047
SES = 6	0.500	0.070
Gender	0.384	0.034
Minority	−0.005	0.064
SES × Gender	−0.074	0.023
SES × CITO	0.0056	0.0035
SES × Minority	−0.219	0.050
Random Effect	Parameter	S.E.
Level-two random effect:		
Intercept variance	0.119	0.014
Level-one variance parameters:		
σ_0^2 constant term	3.422	0.246
σ_{01} linear SES effect	−0.302	0.064
σ_1^2 quadratic SES effect	0.053	0.016
Deviance	46063.0	

Model 3 as a whole is a strong improvement over Model 2: the deviance difference is $\chi^2 = 44.8$ for $d.f. = 7$ ($p < 0.0001$). For the evaluation of the non-linear effect of SES, note that since SES = 3 is the reference category, the parameter value for SES = 3 should be taken as 0.0. This demonstrates that the effect of SES is non-linear but it is indeed an increasing function of SES. The differences between the SES values 1, 2, and 3 are larger than those between the values 3, 4, 5, and 6. The interaction effects of SES with gender and with minority status are significant. The main effect of minority status corresponds with its effect for SES = 3, since the product variable was defined using (SES − 3), and is practically nil. Thus, it turns out that, when the other included variables (including the main effect of SES!) are controlled for, pupils with parents from a non-industrialized country attain a higher educational level than those with parents from the Netherlands or another industrialized country when they come from a low-SES family, but a lower level if they come from a high-SES family.

In Model 3, residual variance is

$$\text{residual variance } = 3.422 - 0.604\,\text{SES} + 0.053\,\text{SES}^2.$$

This decreases from 2.87 for SES = 1 to 1.71 for SES = 6. Thus, with the inclusion of interactions and a non-linear SES effect, residual variance has hardly decreased.

Model 3 was obtained, on the spur of heteroscedasticity, in a data-driven rather than theory-driven way. Therefore one may question the validity of the tests for the newly included effects: are these not the result of chance capitalization? The interaction effect of SES with minority status has such a high t-value, 4.4, that its significance is beyond doubt, even when the data-driven selection is taken into account. For the interaction of SES with gender this is less clear. For convincing hypothesis tests, it would have been preferable to use cross-validation: split the data into two subsets and use one subset for the model selection and the other for the test of the effects. Since the two subsets should be independent, it would be best to select half the schools at random and use all pupils in these schools for one subset, and the other schools and their pupils for the other.

More generally, the residual variance may depend on more than one variable; in terms of representation (8.4), several variables may have random effects at level one. These can be level-two as well as level-one variables. If the random part at level one is given by

$$\text{random part at level one } = R_{0ij} + R_{1ij}\,x_{1ij} + \ldots + R_{pij}\,x_{pij}\ ,$$

while the variances and covariances of the R_{hij} are denoted σ_h^2 and σ_{hk}, then the variance function is

$$\text{residual variance } = \sum_{h=0}^{p} \sigma_h^2\, x_{hj}^2 + 2 \sum_{h=0}^{p-1} \sum_{k=h+1}^{p} \sigma_{hk}\, x_{hij}\, x_{kij}\ . \tag{8.6}$$

This complex level-one variance function can be used for any values for the parameters σ_h^2 and σ_{hk}, provided that the residual variance is positive. The simplest case is to include only σ_0^2 and the 'covariance' parameters σ_{0h}, leading to the linear variance function

$$\text{residual variance } = \sigma_0^2 + 2\,\sigma_{01}\,x_{1ij} + 2\,\sigma_{02}\,x_{2ij} + \ldots + 2\,\sigma_{0p}\,x_{pij}\ .$$

Correlates of diversity

It can be important to investigate the factors that are associated with outcome variability. For example, Raudenbush and Bryk (1987) (see also Bryk and Raudenbush, 1992, p. 169–172) investigated the effects of school policy and organization on mathematics achievement of pupils. They did this by considering within-school dispersion as a dependent variable. The preceding section offers an alternative approach which remains closer to the hierarchical linear model. In this approach, relevant level-two variables are considered as potentially being associated with level-one heteroscedasticity.

Example 8.4 *School composition and outcome variability.*

Continuing the preceding example on educational attainment predicted from data available at the end of primary school, it is now investigated whether composition of the school with respect to socio-economic status is associated with diversity in later educational attainment. It turns out that socio-economic status has an intraclass correlation of 0.25, which is quite high. Therefore the average socio-economic status of schools could be an important factor in the within-school processes associated with average outcomes but also with outcome diversity.

To investigate this, the school average of SES was added to Model 3 of Table 8.4 both as a fixed effect and as a linear effect on the level-one variance. The non-significant SES-by-CITO interaction was deleted from the model. The school average of SES ranges from 1.4 to 5.5, has a mean of 3.7, and a standard deviation of 0.59. This variable is denoted by SA-SES. Its fixed effect is 0.417 (S.E. 0.109, $t = 3.8$). We further present only the random part of the resulting model in Table 8.5.

Table 8.5 Heteroscedastic model depending on average SES.

	Model 4	
Random Effect	Parameter	S.E.
Level-two random effect:		
Intercept variance	0.101	0.013
Level-one variance parameters:		
σ_0^2 constant term	5.203	0.282
σ_{01} linear SES effect	−0.331	0.063
σ_1^2 quadratic SES effect	0.078	0.022
σ_{02} linear SA-SES effect	−0.265	0.022
Deviance	45893.0	

To test the effect of SA-SES on the level-one variance, the model was estimated also without this effect. This yielded a deviance of 46029.6, so the test statistic is $\chi^2 = 46029.6 - 45893.0 = 136.6$ with $d.f. = 1$, which is very significant. The quadratic effect of SA-SES was estimated both as a main effect and for level-one heteroscedasticity, but neither were significant.

How important is the effect of SA-SES on the level-one variance? The standard deviation of SA-SES is 0.59, so four times the standard deviation (the difference between the few percent highest-SA-SES and the few percent lowest-SA-SES schools) leads to a difference in the residual variance of $4 \times 0.59 \times 0.265 = 0.63$. For an average residual level-one variance of 2.0 (see Model 1 in Table 8.3), this is an appreciable difference.

This 'random effect at level 1' of SA-SES might be explained by interactions between SA-SES and pupil-level variables. The interactions of SA-SES with gender and with minority status were considered. Adding these to Model 4 yielded interaction effects of −0.219 (S.E. 0.096, $t = -2.28$) for SA-SES by minority status and −0.225 (S.E. 0.050, $t = 4.50$) for SA-SES by gender. This implies that, although a high school average for SES leads to higher educational attainment on average (the main effect of 0.417 reported above), this effect is weaker for minority pupils and for girls. These interactions did, however, not lead to a noticeably lower effect of SA-SES on the residual level-one variability.

8.2 Heteroscedasticity at level two

For the intercept variance and the random slope variance in the hierarchical linear model it was assumed in preceding chapters that they are constant across groups. This is a homoscedasticity assumption at level two. If there are theoretical or empirical reasons to drop this assumption, it could be replaced by the weaker assumption that these variances depend on some level-two variable Z. For example, if Z is a dummy variable with values 0 and 1, distinguishing two types of groups, the assumption would be that the intercept and slope variances depend on the group. In this section we only discuss the case of level-two variances depending on a single variable Z; this discussion can be extended to variances depending on more than one level-two variable.

Consider a random intercept model in which one assumes that the intercept variance depends linearly or quadratically on a variable Z. The intercept variance then can be expressed by

$$\text{intercept variance} \ = \tau_0^2 \ + \ 2\,\tau_{01}\,z_j \ + \ \tau_1^2\,z_j^2 \ , \tag{8.7}$$

for parameters τ_0^2, τ_{01}, and τ_1^2. For example, in organizational research, when the level-two units are organizations, it is possible that small-sized organizations are (because of greater specialization or other factors) more different from one another than large-sized organizations. Then Z could be some measure for the size of the organization, and (8.7) would indicate that it depends on Z whether differences between organizations tend to be small or large; where 'differences' refer to the intercepts in the multilevel model. The expression (8.7) is a quadratic function of z_j, so that it can represent a curvilinear dependence of the intercept variance on Z. If a linear function is used (i.e., $\tau_1^2 = 0$), the intercept variance is either increasing or decreasing over the whole range of Z.

Analogous to (8.4), this variance function can be obtained by using the 'random part'

$$\text{random part at level two} \ = U_{0j} \ + \ U_{1j}\,z_j \ . \tag{8.8}$$

Thus, strange as it may sound, the level-two variable Z formally gets a random slope at level two. (Note that in Section 5.1.1 it was discussed that random slopes for level-one variables also create some kind of heteroscedasticity, viz., heteroscedasticity of the observations Y.)

The parameters τ_0^2, τ_1^2, and τ_{01} are, like in the preceding section, not to be interpreted themselves as variances and a corresponding covariance. The interpretation is by means of the variance function (8.8). Therefore it is not required that $\tau_{01}^2 \leq \tau_0^2 \times \tau_1^2$. Stated differently: 'correlations' defined formally by $\tau_{01}/(\tau_0\tau_1)$ may be larger than 1 or smaller than -1, even infinite, because the idea of a correlation does not make sense here. An example of this is provided by the linear variance function for which $\tau_1^2 = 0$ and only the parameters τ_0^2 and τ_{01} are used.

In a similar way, random slope variances can be made to depend on suitable level-two variables.

9 Assumptions of the Hierarchical Linear Model

As all statistical models, the hierarchical linear model is based on a number of assumptions. If these assumptions are not satisfied, the procedures for estimating and testing coefficients can be invalid. The assumptions are about the linear dependence of the dependent variable, Y, on the explanatory variables and the random effects; the independence of the residuals, at level one as well as at the higher level or levels; the specification of the variables having random slopes (which implies a certain variance and correlation structure for the observations); and the normal distributions for the residuals. It is advisable, when analysing multilevel data, to devote some energy to checks of the assumptions. This chapter is devoted to such checks.[1]

Why are model checks important at all? The main dangers of model misspecification are the general misrepresentation of the relations in the data (e.g., if the effect of X on Y is curvilinear increasing for X below average and decreasing again for X above average, but only the linear effect of X is tested, then one might obtain a non-significant effect and conclude mistakenly that X has no effect on Y) and the invalidity of hypothesis tests (e.g., if the random part of the model is grossly misspecified, the standard errors and therefore the hypothesis tests for fixed effects can be completely off the mark). Constructing the model on the basis of a good insight in the studied phenomenon is the main basis for a good specification. In addition, there are some helpful statistical tools such as those described in this chapter.

9.1 Assumptions of the hierarchical linear model

The hierarchical linear model was introduced in Chapter 5. In this chapter we only consider the two-level model. The basic definition of the model for observation Y_{ij} on level-one unit i within level-two unit j, for $i = 1, ..., n_j$,

[1] Some of the diagnostic procedures discussed below are not yet widely implemented in multilevel software. MLn macros to compute the diagnostics mentioned in Sections 9.4.1 to 9.6.2 can be downloaded from the website, http://stat.gamma.rug.nl/snijders/multilevel.htm .

is given in formula (5.14) as

$$Y_{ij} = \gamma_0 + \sum_{h=1}^{r} \gamma_h \, x_{hij} + U_{0j} + \sum_{h=1}^{p} U_{hj} \, x_{hij} + R_{ij} \; . \tag{9.1}$$

As before, we shall refer to level-two units also as 'groups'. The assumptions were formulated as follows after equation (5.12). The vectors $(U_{0j}, ..., U_{pj})$ of level-two random coefficients, or level-two residuals, are independent between groups. These level-two random coefficients are independent of the level-one residuals R_{ij} and all residuals have population means 0, given the values of all explanatory variables. Further, the level-one residuals R_{ij} have a normal distribution with constant variance σ^2. The level-two random effects $(U_{0j}, \ldots, U_{pj})$ have a multivariate normal distribution with a constant covariance matrix. The property of constant variance is also called homoscedasticity, non-constant variance being referred to as heteroscedasticity.

To check these assumptions, the following questions may be posed.

1. Does the fixed part contain the right variables (now X_1 to X_r)?
2. Does the random part contain the right variables (now X_1 to X_p)?
3. Are the level-one residuals normally distributed?
4. Do the level-one residuals have constant variance?
5. Are the level-two random coefficients normally distributed?
6. Do the level-two random coefficients have a constant covariance matrix?

These questions are answered in various ways in this chapter. The answers are necessarily incomplete, because a completely convincing argument that a given specification is correct, is impossible to give. For complicated models it can be sensible, if there are enough data, to employ cross-validation (e.g., Mosteller and Tukey, 1977). Cross-validation means that the data are split into two independent halves, one half being used for the search for a satisfactory model specification and the other half for testing of effects. This has the advantage that testing and model specification are separated, so that tests do not lose their validity because of capitalization on chance. For a two-level model, two independent halves are obtained by randomly distributing the level-two units into two subsets.

9.2 Following the logic of the hierarchical linear model

The hierarchical linear model itself is an extension of the linear regression model and relaxes one of the crucial assumptions of this model, the independence of the residuals. The following model checks follow immediately from the logic of the hierarchical linear model, and were mentioned accordingly in Section 6.4.

9.2.1 *Include contextual effects*

In the spirit of Chapters 4 and 5, for every level-one explanatory variable one should consider the possibility that the within-group regression is different from the between-group regression (see Section 4.5), and the possibility that X has a random slope. A difference between the within- and between-group regression coefficients of a variable X is modeled by including the group mean of X also in the fixed part of the model. This group mean is a meaningful contextual variable, as follows from Chapters 2 and 3.

Also for interaction variables it can be checked whether there are contextual effects. This is done by applying the principles of Section 4.5 to, e.g., the product variable $X_{ij} Z_j$ rather than the single variable X_{ij}. The cross-level interaction variable $X_{ij} Z_j$ can be split into

$$X_{ij} Z_j = (X_{ij} - \bar{X}_{.j}) Z_j + \bar{X}_{.j} Z_j ,$$

and it is possible that the two variables $(X_{ij} - \bar{X}_{.j}) Z_j$ and $\bar{X}_{.j} Z_j$ have different regression coefficients. This is tested by checking whether the level-two interaction variable $\bar{X}_{.j} Z_j$ has an additional fixed effect when $X_{ij} Z_j$ belongs already to the fixed part of the model. If there is such an additional effect, one has the choice between including in the model either $(X_{ij} - \bar{X}_{.j} Z_j)$ and $\bar{X}_{.j} Z_j$, or $X_{ij} Z_j$ and $\bar{X}_{.j} Z_j$; these two options will yield the same fit (cf. Section 4.5).

The assumption that the population mean of U_{hj} is 0, conditionally given all the explanatory variables, implies that these random intercepts and slopes are uncorrelated with all explanatory variables. If this assumption is incorrect, the problem can be remedied by adding relevant explanatory variables to the model. A non-zero correlation between U_{hj} and a level-two variable Z_j is remedied by including the product variable $X_{hij} Z_j$, as discussed in the preceding paragraph. A non-zero correlation between U_{hj} and a level-one variable X_{kij} is remedied by including the product variable $X_{hij} \bar{X}_{k.j}$. Since for $h = 0$ it holds that $X_{hij} = 1$, applying this procedure to the random intercept U_{0j} leads to including the main effect variables Z_j and $\bar{X}_{k.j}$, respectively.

Sometimes it makes sense also to include other contextual variables, e.g., the standard deviation within group j of a relevant level-one variable. For example, heterogeneity of a school class, as measured by the standard deviation of the pupils' intelligence or of their prior achievements, may be a complicating factor in the teaching process and thus have a negative effect on the pupils' achievements.

9.2.2 *Check whether variables have random effects*

It is possible that the effect of explanatory variables differs from group to group, i.e., that some variables have random slopes. In the analysis of covariance (ANCOVA), this is known as heterogeneity of regression. It is usual in ANCOVA to check whether regressions are homogeneous. Similarly, in the hierarchical linear model it is advisable to check for each level-one variable in the fixed part whether it has a random slope.

Depending on the amount of data and the complexity of the model, estimating random slopes for each explanatory variable may be a time-consuming affair. A faster method to test random slopes was proposed by Berkhof and Snijders (1998). These tests do not require the complete ML or REML estimates. Suppose that some model, denoted by M_0, has been estimated, and it is to be tested whether a random slope for some variable should be added to the model. Thus, M_0 is to be tested as the null hypothesis. The method proceeds as follows.

Recall from Section 4.6 that the estimation algorithms for the hierarchical linear model are iterative procedures. If the IGLS, RIGLS, or Fisher scoring algorithm is used, it is possible to base a test of the random slope already on the results of the first step of the algorithm, so that it is not necessary to carry out the full estimation process with all iteration steps. To test whether some variable has a random slope, first estimate the parameters for the model M_0 and then add this random slope to model M_0 (it is sufficient to add only the variance parameter, without the corresponding covariance parameters). Now make just one step of the (R)IGLS or Fisher scoring algorithm, starting from the parameter estimates obtained when fitting M_0. Denote the provisional slope variance estimate after this single step by $\tilde{\tau}^2$ with associated standard error (also after the single step) $\text{S.E.}(\tilde{\tau}^2)$. Then the t-ratio,

$$\frac{\tilde{\tau}^2}{\text{S.E.}(\tilde{\tau}^2)} ,$$

can be tested in the standard normal distribution. (Note that, as is explained in Berkhof and Snijders (1998), this test of a t-ratio for a variance parameter can be applied only to the result of the first step of the estimation algorithm and not to the final ML estimate! The main reason is that the standard error of the usual estimator, obtained on convergence of the algorithm, is to a certain extent proportional to the estimated variance component, in contrast to the standard error of the single step estimate.)

Another question that is relevant when a random slope is specified for a variable X of which the between-group regression coefficient differs from the within-group coefficient, is whether the original variable X_{ij} or the within-group deviation variable $X_{ij} - \bar{X}_{.j}$ should get the random slope. This was discussed in Section 5.3.1.

9.2.3 Explained variance

In the discussion in Section 7.1 of the concept of explained variance for multilevel models, it was mentioned that the fractions of explained variance at level one and two, R_1^2 and R_2^2, respectively, can be used as misspecification diagnostics. If the fraction of explained variance, as defined in Section 7.1, decreases when a fixed effect is added to the model, this can be a sign of misspecification. A small decrease, however, can be a result of chance fluctuations. For reasonably large data sets, a decrease by a magnitude of 0.05 or more should be taken as a warning of possible misspecification.

9.3 Specification of the fixed part

Whether the fixed part contains the right variables is an equivocal question in almost all research in the social and behavioral sciences. For many dependent variables there is not one single correct explanation, but different points of view may give complementary and valid representations. The performance of a pupil at the final moment of compulsory education can be studied as a function of the characteristics of the child's family; as a function of the child's achievements in primary school; as a function of the child's personality; or as a function of the mental processes during the final examination. All these points of view are valid, taken alone and also combined. Therefore, what is the right set of variables in the fixed part depends in the first place on the domain in which the explanation is being considered.

In the second place, once the data collection is over one can only use variables based on the available data. If some variable, say X_q, is omitted from the fixed part but other variables are included that are correlated with X_q, then the effects of the included variables will take up some of the effect of the omitted variable. This is well-known in regression analysis. It implies that (unless the variables are experimentally manipulated) one should be careful with interpretation and very reluctant in making causal interpretations. For the hierarchical linear model the consequences of omitting variables from the fixed part of the model are elaborated in Bryk and Raudenbush (1992, Chapter 9).

In a multilevel design, one should be aware in any case of the possibility that supposed level-one effects are in reality, completely or partially, higher level effects of aggregated variables. This was discussed in Section 9.2.1.

Transformation of explanatory variables

One important option for improving the specification of the fixed part of the hierarchical linear model is the transformation of explanatory variables. Aggregation to group means or to group standard deviations is one kind of transformation. Calculating products or other interaction variables is another kind. The importance given in multilevel modeling to cross-level interactions should not diminish the attention paid to the possibility of within-level interactions!

A third kind of transformation is the non-linear transformation of variables. As examples of non-linear effects of an explanatory variable X on Y, one may think of an effect which is always positive but levels off toward low or high values of X; or a U-shaped effect, expressed by a function that first decreases and then increases again.

The simplest way to investigate the possible non-linearity of the effect of a variable X is to conjecture some non-linear transformation of X, e.g., X^2 or (for positive variables) $\ln(X)$ or $1/X$, add this effect to the fixed part of the model – retaining the original linear effect of X – and test the significance of the effect of this non-linear transformed variable. Instead of adding just one non-linear transformed variable, added flexibility is obtained by adding several transformations of the same variable X, because the linear combinations of these transformations will be able to represent quite a variety of shapes of the effect of X on Y.

We found it quite convenient to use quadratic splines for this purpose. Spline functions are discussed in Section 12.2.2; a more extensive discussion is given in Seber and Wild (1989, Section 9.5). Examples of splines in the fixed part of a multilevel model are given in the figures on pages 113 and 217. A quadratic spline transformation of some variable X can be chosen as follows. Choose a value x_0 in the middle of the range of X (e.g., the grand mean of X) as a reference point for the square of X and choose a few other values $x_1, ..., x_K$ within the range of X. Usually, K is a low number, such as 1, 2, or 3. Higher values of K yield more flexibility to approximate many shapes of functions of X. Calculate, for each $k = 1, ..., K$, the 'half-squares' $f_k(X)$, defined as 0 left of x_k and $(X - x_k)^2$ right of x_k:

$$f_k(X) = \begin{cases} 0 & (X \leq x_k) \\ (X - x_k)^2 & (X > x_k) \end{cases}$$

Use in the fixed part the linear effect X, the quadratic effect $(X - x_0)^2$, and the effects of the 'half squares' $f_1(X)$, ..., $f_K(X)$. Together these functions can represent a wide variety of smooth functions of X, as is evident from the figures on pages 113 and 217. If some of the $f_k(X)$ have non-significant effects they can be left out of the model, and by trial and error the choice of the so-called nodes x_k may be improved. Such an explorative procedure was used to obtain the functions displayed on the mentioned pages.

9.4 Specification of the random part

The specification of the random part was discussed already above and in Section 6.4. If certain variables are mistakenly omitted from the random part, the tests of their fixed coefficients may also be unreliable. Therefore it is advisable to check the randomness of slopes of all variables of main interest, and not only those for which a random slope is theoretically expected.

The random part specification is directly linked to the structure of the covariance matrix of the observations. In Section 5.1.1 we already saw that a random slope implies a heteroscedastic specification of the variances of the observations and of the covariance between level-one units in the same group (level-two unit). When different specifications of the random part yield a similar structure for the covariance matrix, it will be empirically difficult or impossible to distinguish between them. But also a misspecification of the fixed part of the model can lead to a misspecification of the random part; sometimes an incorrect fixed part shows up in an unnecessarily complex random part. For example, if an explanatory variable X with a reasonably high intraclass correlation has in reality a curvilinear (e.g., quadratic) effect without a random component, whereas it is specified as having a linear fixed and random effect, the excluded curvilinear fixed effect may show up in the shape of a significant random effect. The latter effect then will disappear when the correctly specified curvilinear effect is added to the fixed part of the model. This was observed in Example 8.2. The random slope of IQ disappeared in this example when a curvilinear effect of IQ was considered.

9.4.1 Testing for heteroscedasticity

In the hierarchical linear model with random slopes, the ***observations*** are heteroscedastic because their variances depend on the explanatory variables, as expressed by equation (5.5). However, the ***residuals*** R_{ij} and U_{hj} are assumed to be homoscedastic, i.e., to have constant variances.

In Chapter 8 it was explained that the hierarchical linear model can also represent models in which the level-one residuals have variances depending linearly or quadratically on an explanatory variable, say, X. Such a model can be specified by the technical device of giving this variable X a random slope at level one. Similarly, giving a level-two variable Z a random slope at level two leads to models for which the level-two random intercept variance depends on Z. Also random slope variances can be made to depend on some variable Z. Neglecting such types of heteroscedasticity may lead to incorrect hypotheses tests for variables which are associated to the variables responsible for this heteroscedasticity (X and Z in this paragraph). Checking for this type of heteroscedasticity is straightforward using the methods of Chapter 8. However, this requires that variables are available that are thought to be possibly associated with residual variances.

A different method, described in Bryk and Raudenbush (1992, Chapter 9), can be used to detect heteroscedasticity in the form of between-group differences in the level-one residual variance, without a specific connection to some explanatory variable. It is based on the estimated least-squares residuals within each group, further called the OLS residuals. This method only is applicable if many (or all) groups are considerably larger than the number of explanatory variables. Only level-one explanatory variables are considered, the level-two variables are disregarded. What follows applies to the groups for which $n_j - r - 1$ is not too small, say, 10 or more, where r is the number of level-one explanatory variables. For each of these groups separately an ordinary least squares regression is carried out with the level-one variables as explanatory variables. Denote by s_j^2 the resulting estimated residual variance for group j and by $df_j = n_j - r - 1$ the corresponding number of degrees of freedom. The weighted average of the logarithms,

$$ls_{\text{tot}} = \frac{\sum_j df_j \ln(s_j^2)}{\sum_j df_j} , \tag{9.2}$$

must be calculated. If the hierarchical linear model is well-specified this weighted average ls_{tot} will be close to the logarithm of the maximum likelihood estimate of σ^2.

From the group-dependent residual variance s_j^2 a standardized residual dispersion measure can be calculated using the formula

$$d_j = \sqrt{\frac{df_j}{2}} \left\{ \ln(s_j^2) - ls_{\text{tot}} \right\} . \tag{9.3}$$

If the level-one model is well specified and the population level-one residual variance is the same in all groups, then the distribution of the values d_j is close to the standard normal distribution. The sum of squares,

$$H = \sum_j d_j^2 \,, \tag{9.4}$$

can be used to test the constancy of the level-one residual variances. Its null distribution is chi-squared with $N - 1$ degrees of freedom, where N is the number of groups included in the summation.

If the within-groups degrees of freedom df_j are less than 10 for many or all groups, the null distribution of H is not chi-squared. Since the null distribution depends only on the values of df_j and not on any of the unknown parameters, it is feasible to obtain this null distribution by straightforward computer simulation. This can be carried out as follows: generate independent random variables V_j according to chi-squared distributions with df_j degrees of freedom, calculate $s_j^2 = V_j / df_j$, and apply equations (9.2), (9.3) and (9.4). The resulting value H is one random draw from the correct null distribution. Repeating this, say, 1,000 times, gives a random sample from the null distribution with which one can compare the observed value from the real data set.

If this test yields a significant result, one can inspect the individual d_j values to investigate the pattern of heteroscedasticity. For example, it is possible that the heteroscedasticity is due to a few unusual level-two units for which d_j has a large absolute value.

Example 9.1 *Level-one heteroscedasticity.*
The example of pupils' language performance used, e.g., in Chapters 4 and 5 is considered again. We investigate whether there is evidence of level-one heteroscedasticity where the explanatory variables at level one are IQ, SES, and gender (this is the same level-one model as in Table 8.1).

Those groups were used for which the residual degrees of freedom are at least 10. There were 86 such groups. The sum of squared standardized residual dispersions defined in (9.4) is $H = 79.165$, a chi-squared value with $d.f. = 85$, $p = 0.66$. Hence this test does not give evidence of heteroscedasticity.

All d_j values were smaller in absolute value than 2.3. This is quite reasonable for a sample of 86 standard normal deviates, and therefore confirms the conclusion that there is no evidence that some groups have a larger within-group residual variance than others.

An advantage of this test is that it is based only on the specification of the within-groups regression model. The level-two variables and the level-two random effects play no role at all, so what is checked here is purely the level-one specification. However, the mentioned null distributions of the d_j and of H do depend on the normality of the level-one residuals. A more heavy-tailed distribution for these residuals in itself will also lead to higher values of H, even if the residuals do have constant variance. Therefore, if H leads to a significant result, one should investigate the possible pattern of heteroscedasticity by inspecting the d_j values, but one should also inspect the distribution of the OLS within-group residuals for normality.

9.4.2 What to do in case of heteroscedasticity

If there is evidence for heteroscedasticity, it may be possible to find variables accounting for the different values of the level-one residual variance. These could be level-one as well as level-two variables. Sometimes such variables can be proposed on the basis of theoretical considerations. In addition, plots of d_j versus relevant level-two variables, or plots of squared unstandardized residuals (see Section 9.5) can be informative for suggesting such variables. When there is a conjecture that the non-constant residual variance is associated with a certain variable, one can apply the methods of Chapter 8 to test whether this is indeed the case, and fit a heteroscedastic model.

In some cases, a better approach to deal with heteroscedasticity is to apply a non-linear transformation to the dependent variable, e.g., a square root or logarithmic transformation. This can be useful, e.g., when the dependent variable is highly skewed. How to choose transformations of the dependent variable in single-level models is discussed in Atkinson (1985). The use of the Box-Cox transformation family for multilevel models is discussed by Hodges (1998, p. 506). When there is heteroscedasticity and the dependent variable has a small number of categories, another option is to use the multilevel ordered logit model for multiple ordered categories (Section 14.4) or to dichotomize the variable and apply multilevel logistic regression (Section 14.2).

9.5 Inspection of level-one residuals

A plethora of methods have been developed for the inspection of residuals in ordinary least squares regression; see, e.g., Atkinson (1985) and Cook and Weisberg (1982, 1994). Inspection of residuals can be used, e.g., to find outlying cases that have an undue high influence on the results of the statistical analysis, to check the specification of the fixed part of the model, to suggest transformations of the dependent or the explanatory variables, or to point to heteroscedasticity.

These methods may be applied to the hierarchical linear model, but some changes are necessary because of the more complex nature of the hierarchical linear model and the fact that there are several types of residuals. For example, in a random intercept model there is a level-one and also a level-two residual. Various methods of residual inspection for multilevel models were proposed by Hilden-Minton (1995). He noted that a problem in residual analysis for multilevel models is that the observations depend on the level-one and level-two residual jointly, whereas for model checking it is desirable to consider these residuals separately. It turns out that level-one residuals can be estimated so that they are unconfounded by the level-two residuals, but the other way around is impossible.

Analysis of level-one residuals can be based on the ordinary least squares (OLS) regressions calculated within each group separately. This was proposed by Hilden-Minton (1995). We used the within-group OLS regression already in Section 9.4.1, where we remarked that an advantage of this approach is its being based only on the specification of the level-one model. Thus, it is possible to inspect estimated level-one residuals without confounding by level-two residuals or by level-two misspecification.

The estimated residuals are obtained by carrying out OLS regressions in each group separately, using only the level-one variables. It is advisable to omit the residuals for groups with low within-group residual degrees of freedom, because those residuals are quite unstable.

The variance of the OLS residuals depends on the explanatory variables. The OLS residuals can be standardized by dividing them by their standard deviation. This yields the ***standardized OLS residuals***, cf. Atkinson (1985) or other texts on regression diagnostics. This provides us with two sets of level-one residuals: the raw OLS residuals and the standardized OLS residuals. The advantage of the standardized residuals is their constant variance, the disadvantage is that the standardization may distort the (possibly non-linear) relations with explanatory variables.

Various possibilities for employing residuals to inspect model fit were discussed by Hilden-Minton (1995). For basic model checking we propose to use the level-one within-group OLS residuals in the following ways:

1. Plot the unstandardized OLS residuals against level-one explanatory variables to check for possible non-linear fixed effects.

   If there are in total a few hundred or more level-one units a raw plot may exhibit just random scatter, in which case it is advisable to make a smoothed plot. The residual plot can be smoothed as follows.

   If the explanatory variable has a limited number of categories, the mean residual can be calculated for each category and plotted together with vertical bars extending to twice the standard error of the mean within this category. (This standard error can be calculated in the usual way, without taking account of the non-constant variance and the mutual dependence of the residuals.)

   For continuous explanatory variables a convenient way of smoothing is the following construction of a simple moving average. Denote by $x_1, ..., x_M$ the ordered values of the explanatory variable under consideration and by $r_1, ..., r_M$ the corresponding values of the residuals. This means that x_i and r_i are values of the explanatory variable and the within-group residual for the same level-one unit, reordered so that $x_1 \leq x_2 \leq ... \leq x_M$. (Units with identical x-values are ordered arbitrarily.) Since groups with a small number of units may have been omitted, the number M may be less than the original total sample size. The smoothed residuals now are defined as averages of $2K$ consecutive values,

$$\bar{r}_i = \frac{1}{2K} \sum_{h=-K+1}^{K} r_{i+h} ,$$

for $i = K$ to $M - K$. The value of K will depend on data and sample size; e.g., if the total number of residuals M is at least 1,500, one could take moving averages of $2K = 100$ values.

One may add to the plot horizontal lines plotted at r equal to plus or minus twice the standard error of the mean of $2K$ values, i.e., $r = \pm 2\sqrt{\bar{\sigma}^2/(2K)}$, where $\bar{\sigma}^2$ is the variance of the OLS residuals. This indicates roughly that values $\bar{r}_i$ outside this band, i.e., $|\bar{r}_i| > 2\sqrt{\bar{\sigma}^2/(2K)}$, may be considered to be relatively large.

2. Make a normal probability plot of the standardized OLS residuals to check the assumption of a normal distribution. This is done by plotting the values $(z_i, \tilde{r}_i)$ where $\tilde{r}_i$ is the standardized OLS residual and z_i the corresponding normal score (i.e., the expected value from the standard normal distribution according to the rank of $\tilde{r}_i$). Especially when this shows that the residual distribution has longer tails than the normal distribution, there is a danger of parameter estimates being unduly influenced by outlying level-one units.

When a data exploration is carried out along these lines, this may suggest model improvements which then can be tested. One should realize that these tests are suggested by the data; if the tests use the same data that were used to suggest them, which is usual, this will lead to chance capitalization and inflated probabilities of errors of the first kind. The resulting improvements are convincing only if they are 'very significant' (e.g., $p < 0.01$ or $p < 0.001$). If the data set is big enough, it is preferable to employ cross-validation, cf. p. 117.

Example 9.2 *Level-one residual inspection.*
We continue Exampe 9.1, in which the data set also used in Chapters 4 and 5 is considered again, the explanatory variables at level one being IQ, SES, and gender. Within-group OLS residuals were calculated for all groups with at least 10 within-group residual degrees of freedom.

The variables IQ and SES both have a limited number of categories. For each category, the mean residual was calculated and the standard error of this mean was calculated in the usual way. The mean residuals are plotted in Figure 9.1 for the categories containing 10 or more pupils.

The vertical lines indicate the intervals bounded by the mean residual plus or minus twice the standard error of the mean. For IQ the mean residuals exhibit a clearer pattern than for SES. The left-hand figure suggests a non-linear function of IQ which has a local minimum for IQ between −2 and 0

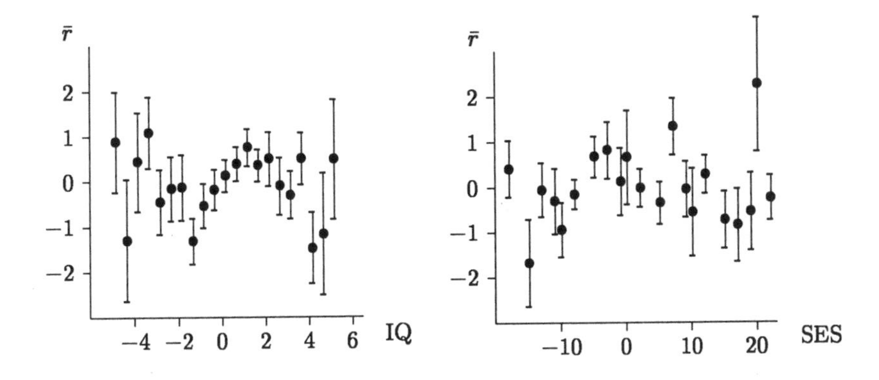

Figure 9.1 Mean level-one OLS residuals (with bars extending to twice the standard error of the mean) as function of IQ (left) and SES (right).

and a local maximum for IQ somewhere around 2. There are few pupils with IQ values less than −3 or greater than +4, so in this IQ range the error bars are very wide and not very informative. Thus the figures point toward a non-linear effect for IQ, having a local minimum for a negative IQ value and a local maximum for a positive IQ value. Examples of such functions are a third-degree polynomial and a quadratic spline with two nodes (cf. Section 12.2.2). The first option was explored by adding IQ^2 and IQ^3 to the fixed part. The second option was explored by adding IQ_-^2 and IQ_+^2, as defined in (8.2), to the fixed part. The second option gave a much better model improvement and therefore was selected. When IQ_-^2 and IQ_+^2 are added to Model 1 of Table 8.1, the deviance goes down by 45.8 ($d.f. = 2$, $p < 0.00001$). This is strongly significant, so this non-linear effect of IQ is convincing even though it was not hypothesized beforehand but suggested by the data.

The mean residuals for the resulting model are graphed as functions of IQ and SES in Figure 9.2.

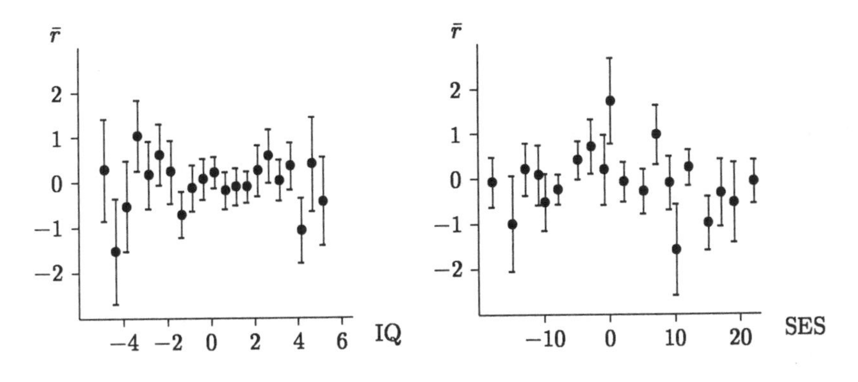

Figure 9.2 Mean level-one OLS residuals (with bars extending to twice the standard error of the mean) as function of IQ (left) and SES (right), for model with non-linear effect of IQ.

These plots do not exhibit a remaining non-linear effect for IQ. For SES one could be tempted to see a curvilinear pattern, but explorations with a third power of SES and with some quadratic splines did not produce any convincing non-linear effects.

A normal probability plot of the residuals for the model that includes the non-linear effect of IQ is given in Figure 9.3. The distribution looks quite normal except for the very low values, where the residuals are somewhat more strongly negative (i.e., larger in absolute value) than expected. However, this deviation from normality is rather small.

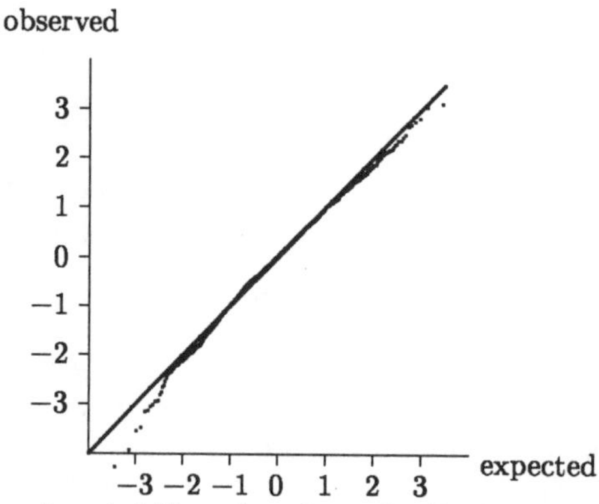

Figure 9.3 Normal probability plot of standardized level-one OLS residuals.

9.6 Residuals and influence at level two

Estimated level-two residuals always are confounded with the estimated level-one residuals. Therefore one should check the specification of the level-one model before moving on to checking the specification at level two.

9.6.1 Empirical Bayes residuals

The empirical Bayes estimates (also called posterior means) of the level-two random effects, treated in Section 4.7, are the 'estimates' of the level-two random variables U_{hj} $(h = 0, ..., p)$ in the model specification 9.1. These can be used as estimated level-two residuals. They can be standardized by dividing by their standard deviation. Similarly to the level-one residuals, one may plot the unstandardized level-two residuals as a function of relevant level-two variables, and make normal probability plots of the standardized level-two residuals. Smoothing the plots will be less necessary because of the usually much smaller number of level-two units. If the plots contain outliers, one may inspect the corresponding level-two units to check whether anything unusual can be found. Checking empirical Bayes residuals is discussed more extensively in Langford and Lewis (1998, Section 2.4).

Example 9.3 *Level-two residual inspection.*
The example above is continued now with the inspection of the level-two residuals. The model is specified with fixed effects of IQ (including the two non-linear functions of IQ) and the school mean of IQ, SES, and gender; and with a random effect of IQ at level 2. (This is like Model 4 of Table 8.2, except that the level-one heteroscedasticity is left out.) Level-two empirical Bayes residuals are calculated for the intercept and for the slope of IQ. Figure 9.4 shows the unstandardized residuals as a function of class size, a level-two variable that is not included in the model.

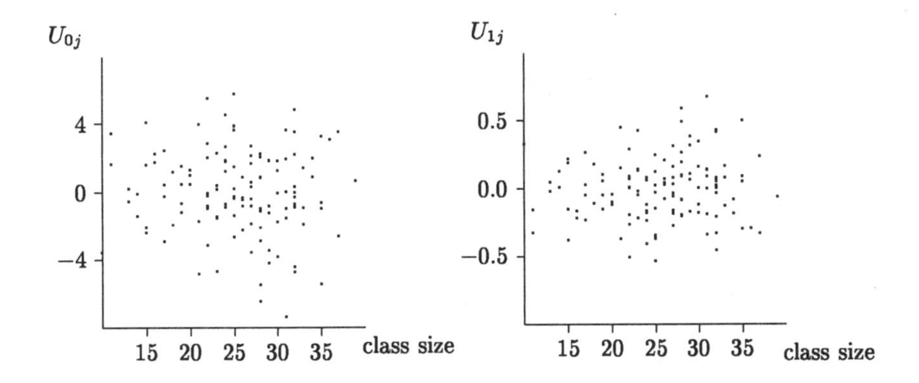

Figure 9.4 Level-two residuals as function of class size; left residuals for the intercept, right residuals for the IQ slope.

This figure does not show any consistent pattern. This supports the exclusion of this variable, and its interaction with IQ, from the model.

In Figure 9.5 the normal probability plots of the standardized residuals are shown. These normal plots support the approximate normality of the level-two residuals.

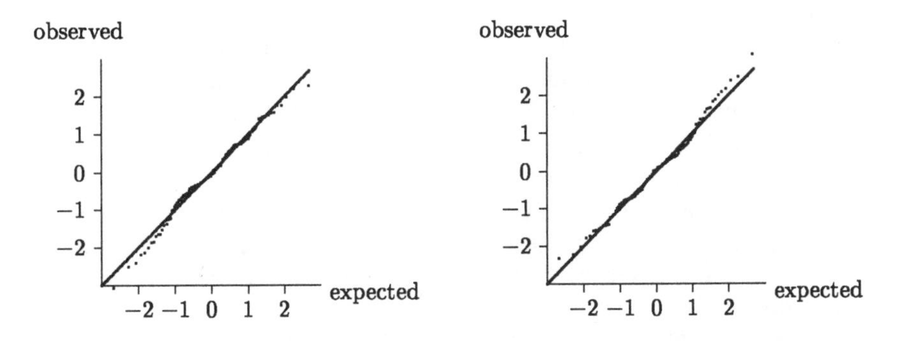

Figure 9.5 Normal probability plot of standardized level-two residuals.

9.6.2 *Influence of level-two units*

The diagnostic value of residuals can be supplemented by so-called *influence diagnostics*, which give an indication of the effect of certain parts of the data on the obtained parameter estimates. In this section we present diagnostics to investigate the influence of level-two units. These diagnostics show how strongly the parameter estimates are affected if unit j is deleted from the data set. This section is based to a large extent on Lesaffre and Verbeke (1998), although we do not precisely follow all their recommendations. Other influence measures for multilevel models may be found in Langford and Lewis (1998, Section 2.3).

For the definition of the diagnostics we shall need matrix notation, but readers who do not know this notation can base their understanding on the verbal explanations.

Denote by $\gamma = (\gamma_0, ..., \gamma_r)$ the vector of all parameters of the fixed part, consisting of the general intercept and all regression coefficients. Further denote by $\hat{\gamma}$ the vector of estimates produced when using all data and by $\hat{\Sigma}_{\mathrm{F}}$ the covariance matrix of this vector of estimates (so the standard errors of the elements of γ are the square roots of the diagonal elements of $\hat{\Sigma}_{\mathrm{F}}$). Denote the parameter estimate obtained if level-two unit j is deleted from the data by $\hat{\gamma}_{(-j)}$. Then unit j has a large influence if $\hat{\gamma}_{(-j)}$ differs much from $\hat{\gamma}$. This difference can be measured on the basis of the covariance matrix $\hat{\Sigma}_{\mathrm{F}}$, because this matrix indicates the uncertainty that exists anyway in the elements of $\hat{\gamma}$. Unit j has a large impact on the parameter estimates if, for one or more of the individual regression coefficients $\hat{\gamma}_h$, the difference between $\hat{\gamma}_h$ and $\hat{\gamma}_{(-j)h}$ is not much smaller, or even larger, than the standard error of $\hat{\gamma}_h$. Given that the vector γ has a total of $r + 1$ elements, a standardized measure of the difference between the estimated fixed effects for the entire data set and the estimates for the data set excluding unit j is given by

$$C_j^{0\mathrm{F}} = \frac{1}{r+1} \left(\hat{\gamma} - \hat{\gamma}_{(-j)}\right)' \, \hat{\Sigma}_{\mathrm{F}}^{-1} \, \left(\hat{\gamma} - \hat{\gamma}_{(-j)}\right) \; .$$

This can be interpreted as the average squared deviation between the estimates with and those without unit j, where the deviations are measured proportional to the standard errors (and account is taken of the correlations between the parameter estimates). For example, if there is only one fixed parameter and the value of $C_j^{0\mathrm{F}}$ for some unit is 0.5, then leaving out this unit changes the parameter estimate by a deviation equal to $\sqrt{0.5} = 0.7$ times its standard error, which is quite appreciable.

This influence diagnostic is a direct analogue of Cook's distance (treated, e.g., in Cook and Weisberg, 1982, and Atkinson, 1985) for linear regression analysis. If the random part at level two is empty, then $C_j^{0\mathrm{F}}$ is equal to Cook's distance.

It may be quite time-consuming to calculate the estimates $\hat{\gamma}_{(-j)}$ for each group j. Since the focus here is on the diagnostic value of the influence statistic rather than on the precise estimation for the data set without group j, an approximation can be used instead. It was proposed by Pregibon (1981), in a different context, to substitute the one-step estimator for $\hat{\gamma}_{(-j)}$. This is the estimator which starts from the estimate obtained for the full data set and then makes a single step of the IGLS, RIGLS, or Fisher scoring algorithm. Denoting this one-step estimator by $\tilde{\gamma}_{(-j)}$, we obtain the influence diagnostic

$$C_j^{\mathrm{F}} = \frac{1}{r+1} \left(\hat{\gamma} - \tilde{\gamma}_{(-j)}\right)' \, \hat{\Sigma}_{\mathrm{F}}^{-1} \, \left(\hat{\gamma} - \tilde{\gamma}_{(-j)}\right) \; . \tag{9.5}$$

A similar influence diagnostic can be defined for the influence of group j on the estimates of the random part. Denote by φ the vector of all parameters of the random part and by q the number of elements of this vector. If the covariance matrix of the random effects is unrestricted, then φ contains the variances and covariances of the random slopes as well as the level-one residual variance. Given that there are p random slopes, this is a total of $q = (p+1)(p+2)/2 + 1$ parameters. Now define $\hat{\varphi}$ as the estimate based on the complete data set, $\hat{\Sigma}_{\mathrm{R}}$ as the covariance matrix of this estimate (having on its diagonal the squared standard errors of the elements of $\hat{\varphi}$), and $\tilde{\varphi}_{(-j)}$ as the one-step estimate when deleting level-two unit j. Then the diagnostic for the influence of unit j on the parameters of the random part is defined by

$$C_j^{\mathrm{R}} = \frac{1}{q} \left(\hat{\varphi} - \tilde{\varphi}_{(-j)}\right)' \, \hat{\Sigma}_{\mathrm{R}}^{-1} \, \left(\hat{\varphi} - \tilde{\varphi}_{(-j)}\right) \; . \tag{9.6}$$

One can also consider the combined influence of group j on the parameters of the fixed and those of the random part of the model. Since the estimates $\hat{\gamma}$ and $\hat{\varphi}$ are approximately uncorrelated (Longford, 1987), such a combined influence diagnostic can be defined simply as the weighted average of the two previously defined diagnostics,

$$C_j = \frac{1}{r+q+1} \left((r+1)\, C_j^{\mathrm{F}} \, + \, q\, C_j^{\mathrm{R}}\right) \; . \tag{9.7}$$

Lesaffre and Verbeke (1998) propose local influence diagnostics which are closely related to those proposed here.[2]

Fit and leverage

Whether a group has a large influence on the parameter estimates for the whole data set depends on two things. The first is the *leverage* of the group, i.e., the potential of this group to influence the parameters because of its size n_j and the values of the explanatory variables in this group. Groups with a large size and with strongly dispersed values of the explanatory variables have a high leverage. The second is the extent to which this group fits the model as defined by (or estimated from) the other groups, which is closely related to the values of the residuals in this group. A poorly fitting group that has low leverage, e.g., because of its small size, will not strongly affect

[2]Lesaffre and Verbeke's diagnostics are very close to the diagnostics (9.5), (9.6), and (9.7), multiplied by twice the number of parameters, but using different approximations. We have chosen for the diagnostics proposed here because their values are on the same scale as the well-known Cook's distance for the linear regression model, and because the one-step estimates can be conveniently calculated when one uses software based on the Fisher scoring or (R)IGLS algorithms.

the parameter estimates. Vice versa, a group with high leverage will not strongly affect the parameter estimates if it has residuals very close to 0.

If the model fits well and the explanatory variables are approximately randomly distributed across the groups, then the expected value of the diagnostics (9.5), (9.6), and (9.7) is roughly proportional to the group size n_j. A plot of these diagnostics as a function of n_j will reveal whether some of the groups influence the fixed parameter estimates more strongly than should be expected on the basis of group size.

The fit of the level-two units to the model can be measured by the ***standardized multivariate residual*** for each unit. This measure was also proposed by Lesaffre and Verbeke (1998), who call it the 'squared length of the residual'. It is defined as follows. The predicted value for observation Y_{ij} on the basis of the fixed part is given by

$$\hat{Y}_{ij} = \hat{\gamma}_0 + \sum_{h=1}^{r} \hat{\gamma}_h \, x_{hij} \; .$$

The multivariate residual for group j is the vector of deviations between the observations and these predicted values,

$$D_j = \begin{pmatrix} Y_{1j} - \hat{Y}_{1j} \\ Y_{2j} - \hat{Y}_{2j} \\ \cdot \\ \cdot \\ Y_{n_j\,j} - \hat{Y}_{n_j\,j} \end{pmatrix} \; . \tag{9.8}$$

This multivariate residual can be standardized on the basis of the covariance matrix of the vector of all observations in group j. For the hierarchical linear model with one random slope, the elements of this covariance matrix are given by (5.5) and (5.6). Denote, for a general model specification, the covariance matrix of the observations in group j by

$$\Sigma(Y_j) = \text{Cov}(Y_j) \; .$$

This covariance matrix is a function of the parameters of the random part of the model. Substituting estimated parameters yields the estimated covariance matrix, $\hat{\Sigma}(Y_j)$. Now the standardized multivariate residual is defined by

$$S_j^2 = D_j' \left(\hat{\Sigma}(Y_j)\right)^{-1} D_j \; . \tag{9.9}$$

This definition can also be applied to models with heteroscedasticity at level one (see Chapter 8).

The standardized residual can be interpreted as the sum of squares of the multivariate residual for group j after transforming this multivariate residual to a vector of uncorrelated elements each with unit variance. If the model is correctly specified and the number of groups is large enough for the parameter estimates to be quite precise, then S_j^2 has a chi-squared distribution with n_j degrees of freedom. This distribution can be used to test the value of S_j^2 and thus investigate the fit of this group to the hierarchical linear model as defined by the total data set. (This proposal was made also by Waternaux, Laird, and Ware, 1989.) Some caution should be exercised with this test, however, because it is to be expected that among all the groups there may occur some significant results by chance even for a well-specified model. For example, if there are 200 groups and the model is correctly specified, it may be expected that S_j^2 has a significantly large value at the 0.05 significance level for ten groups, and at the 0.01 level for two groups. One can apply here the Bonferroni correction for multiple testing, which implies that the model specification is doubtful if the smallest p-value of the standardized multivariate residuals is less than $0.05/N$, where N is the total number of groups.

Values of C_j are assessed only in comparison with the other groups, while values of S_j^2 can be assessed on the basis of the chi-squared distribution. When checking the model specification, one should worry mainly about groups j for which the influence diagnostic C_j is large compared to the other groups, and simultaneously there is a poor fit as measured by the standardized multivariate residual S_j^2. For such groups one should investigate whether there is anything particular. For example, there may be data errors or the group may not really belong to the investigated population. This investigation may, of course, also lead to a model improvement.

Example 9.4 *Influence of level-two units.*
We continue checking the model for the language test score which was also investigated in Example 9.3. Recall that this is Model 4 of Table 8.2, but without the level-one heteroscedasticity. The twenty largest influence diagnostics (9.7) are presented in Table 9.1 together with the p-values of the standardized multivariate residuals (9.9).

Table 9.1 Twenty largest influence diagnostics.

School	n_j	C_j	p_j
218	24	0.089	0.752
107	17	0.066	0.017
170	26	0.055	0.173
256	10	0.031	0.227
67	26	0.026	0.238
18	24	0.026	0.006
1	25	0.022	0.211
176	23	0.021	0.313
52	21	0.021	0.088
121	10	0.021	0.279
15	8	0.020	0.00032
40	35	0.020	0.113
55	30	0.019	0.482
101	23	0.019	0.111
228	21	0.018	0.764
130	13	0.017	0.838
108	9	0.015	0.00020
147	22	0.014	0.724
47	8	0.013	0.755
182	9	0.013	0.599

The school with the largest influence has a good fit ($p = 0.752$). Schools 15 and 108 fit very poorly, while schools 18 and 107 have a moderately poor fit. The Bonferroni-corrected bound for the smallest p-value is $0.05/131 = 0.00038$. The two poorest fitting schools have p-values smaller than 0.00038, which suggests that the fit is unsatisfactory. This led to the exploration of a more detailed model. As intelligence has the most important effect on the language test, it was tried to represent this effect in a more precise way.

Table 9.2 Model with a more detailed effect of IQ.

Fixed Effect	Coefficient	S.E.
Intercept	39.54	0.31
IQ	3.011	0.178
IQ_-^2	0.255	0.042
IQ_+^2	−0.276	0.050
performal IQ	0.697	0.065
SES	0.140	0.014
Gender	2.93	0.26
IQ × SES	−0.023	0.006
IQ × gender	−0.262	0.125
$\overline{\text{IQ}}$	1.04	0.30
Random Effect	**Parameter**	**S.E.**
Level-two random effects:		
Intercept variance	7.34	1.19
IQ slope variance	0.076	0.068
Intercept - IQ slope covariance	−0.76	0.22
Level-one variance:		
σ_0^2 constant term	35.19	1.20
Deviance	14831.4	

An important variable in the data set, not used in the earlier examples, is performal IQ. (The IQ variable used up to now is a measure of verbal IQ, a natural first IQ dimension for explaining language achievement. Performal IQ is an important second IQ dimension.) Performal IQ in this data set has mean 0 and standard deviation 2.20. Exploration of some interactions led to considering also the interaction of verbal IQ with gender and with SES. Including these fixed effects in the model led to the parameter estimates in Table 9.2. These estimates show that performal IQ has an effect additional to the verbal IQ, and that IQ has a weaker effect for pupils whose parents have a higher socio-economic status and also for girls. The twenty largest influence diagnostics for this model are presented in Table 9.3.

The three largest influence diagnostics have become markedly smaller compared to Table 9.1. However, the fit of school 108 has further deteriorated, and is significantly bad even when taking into consideration (by using the Bonferroni correction) that this is the poorest fit in a collection of 131 schools. This poor fit is not too alarming, however, as the sample size for this school is small ($n_j = 9$) and its influence diagnostic is not very large ($C_j = 0.018$) compared to C_j values of other schools. (The same applies, by the way, to the diagnostics presented in Table 9.1.)

A look at the data for school 108 revealed that the predicted values $\hat{Y}_{ij}$ for this school are quite homogeneous and close to the average, but two of

Table 9.3 Twenty largest influence diagnostics for extended model.

School	n_j	C_j	p_j
170	26	0.059	0.094
107	17	0.056	0.032
101	23	0.032	0.038
67	26	0.031	0.273
52	21	0.030	0.099
1	25	0.026	0.299
121	10	0.024	0.116
218	24	0.023	0.812
18	24	0.021	0.008
40	35	0.021	0.124
15	8	0.021	0.0016
55	30	0.019	0.388
256	10	0.019	0.359
228	21	0.018	0.770
108	9	0.018	0.000032
24	24	0.017	0.215
76	16	0.017	0.478
125	26	0.014	0.699
182	9	0.013	0.642
20	23	0.013	0.590

the nine outcome values Y_{ij} are very low and have, correspondingly, very low residuals, −21.8 and −28.6. When these two pupils are deleted from the data set, the parameter estimates hardly change, but the lowest p-value for the standardized multivariate residuals becomes 0.0014, which is not significant when the Bonferroni correction is applied. Concluding, the investigation of the influence statistics led to an improved fit due to the inclusion of performal IQ and the interactions of IQ with SES and with gender, and to the detection of two deviant cases in the data set. These two cases out of a total of 2,287, however, did not have any noticeable influence on the parameter estimates.

9.7 More general distributional assumptions

If it is suspected that the normality and homoscedasticity assumptions of the hierarchical linear model are not satisfied, one could employ methods based on less restrictive model assumptions. For example, one could assume that the residuals have a distribution with more heavy tails than the normal distribution, or that residual variances vary randomly between level-two units. This is an area of active research. E.g., Verbeke and Lesaffre (1997) propose a 'sandwich' modification to make the standard errors of the normal theory estimators applicable also to non-normally distributed random effects; Richardson (1997) derives robust estimators; Seltzer (1993), Seltzer, Wong, and Bryk (1996), and Kasim and Raudenbush (1998) propose estimators based on Gibbs sampling. Several multilevel computer programs provide robust ('sandwich') standard errors, which can be used if the random effects are not normally distributed and the sample size is fairly large.

10 Designing Multilevel Studies

Up to now it was assumed that the researcher wishes to test interesting theories on hierarchically structured systems (or phenomena that can be thought of as having a hierarchical structure, such as repeated data) on available data. Or that multilevel data exist, and that one wishes to explore the structure of the data. This, of course, is the other way around. Normally a theory (or a practical problem that has to be investigated) will direct the design of the study and the data to be collected. This chapter focuses on one aspect of this research design, namely, the sample sizes. Sample size questions in multilevel studies were treated also by Snijders and Bosker (1993), Mok (1995), and Cohen (1998). Another aspect, the allocation of treatments to subjects or groups, and the gain in precision obtained by including a covariate, is discussed in Raudenbush (1997). Hedeker, Gibbons, and Waternaux (1999) present methods for sample size determination for longitudinal data analysed by multilevel methods.

This chapter presents some methods to choose sample sizes that will yield a high power for testing, or (equivalently) small standard errors for estimating, certain parameters in two-level designs, given financial and practical constraints. A problem in the practical application of these methods is that sample sizes which are optimal, e.g., for testing some cross-level interaction effect, are not necessarily optimal, e.g., for estimating the intraclass correlation. The fact that optimality depends on one's objectives, however, is a general problem of life that cannot be solved by this textbook. If one wishes to design a good multilevel study it is advisable to determine first the primary objective of the study, express this objective in a tested or estimated parameter, and then choose sample sizes for which this parameter can be estimated with a small standard error, given financial, statistical, and other practical constraints. Sometimes it is possible to check, in addition, whether also for some other parameters (corresponding to secondary objectives), these sample sizes yield acceptably low standard errors.

A relevant general remark is that the sample size at the highest level is usually the most restrictive element in the design. For example, a two-level design with 10 groups, i.e., a macro-level sample size of 10, is at least as uncomfortable as a single-level design with a sample size of 10. Requirements on the sample size at the highest level, for a hierarchical linear model with q explanatory variables at this level, are at least as stringent as requirements on the sample size in a single level design with q explanatory variables.

10.1 Some introductory notes on power

When a researcher is designing a multi-stage sampling scheme, e.g., to assess the effects of schools on the achievement of students, or to test the hypothesis that citizens in impoverished neighborhoods are more often victims of crime than other citizens, important decisions must be made with respect to the sample sizes at the various levels. For the two-level design in the first example the question may be phrased as follows: should one investigate many schools with few students per school or few schools with many students per school? Or, for the second example: should we sample many neighborhoods with only few citizens per neighborhood or many citizens per neighborhood and only few neighborhoods? In both cases we assume, of course, that there are budgetary constraints for the research to be conducted. To phrase this question more generally and more precisely: how should researchers choose sample sizes at the macro- and micro-level in order to ensure a desired level of power given a relevant (hypothesized) effect size and a chosen significance level? The average micro-level sample size per macro-level unit will be denoted by n and the macro-level sample size by N. In practice the sizes of the macro-level units will usually be variable (if it were only for unintentionally missing data), but for calculations of desired sample sizes it normally is adequate to use approximations based on the assumptions of constant 'group' sizes.

A general introduction to power analysis can be found in the standard work by Cohen (1988), or, for a quick introduction, Cohen's (1992) power primer. The basic idea is that we would like to find support for a research hypothesis (H_1) stating that a certain effect exists, and therefore we test a null hypothesis about the absence of this effect (H_0) using a sample from the population of interest. The significance level α represents the risk of mistakenly rejecting H_0. This mistake is known as a Type I error. Vice versa, β denotes the risk of disappointingly not rejecting H_0, in the case that the effect does exist in the population. This mistake is known as a Type II error. The statistical power of a significance test is the probability of rejecting H_0 given the effect size in the population, the significance level α, and the sample size and study design. Power is therefore given by $1 - \beta$. As a rule of thumb, Cohen suggests that power is moderate when it is 0.50 and high when it is at least 0.80. Power increases as α increases, and also as the sample size and/or the effect size increase. The effect size can be conceived as the researcher's idea about 'the degree to which the null hypothesis is believed to be false' (Cohen, 1992, p. 156).

We suppose that the effect size is expressed by some parameter γ that can be estimated with a certain standard error, denoted by $S.E.(\hat{\gamma})$. Bear in mind that the size of the standard error is a monotone decreasing function of the sample size: the larger the sample size the smaller the standard error! In most single-level designs, the standard error of estimation is inversely proportional (or roughly so) to the square root of sample size.

The relation between effect size, power, significance level, and sample size can be presented in one formula. This formula is an approximation that is valid for practical use when the test in question is a one-sided t-test for γ with a reasonably large number of degrees of freedom (say, $d.f. \geq 10$). Recall that the test statistic for the t-test can be expressed by the ratio $t = \hat{\gamma}/S.E.(\hat{\gamma})$. The formula is

$$\frac{\text{effect size}}{\text{standard error}} \approx (z_{1-\alpha} + z_{1-\beta}) = (z_{1-\alpha} - z_{\beta}) \ , \tag{10.1}$$

where $z_{1-\alpha}$, $z_{1-\beta}$, and z_β are the z-scores (values from the standard normal distribution) associated with the indicated cumulative probability values. If, for instance, α is chosen at 0.05 and $1 - \beta$ at 0.80 (so that $\beta = 0.20$), and an effect size of 0.50 is what we expect, then we can derive that we are searching for a minimum sample size that satisfies:

$$\text{standard error} \leq \frac{0.50}{1.64 + 0.84} = 0.20 \ .$$

Formula (10.1) contains four 'unknowns'. This means that if three of these are given, then we can compute the fourth. In most applications that we have in mind, the significance level α is given and the effect size is hypothetically considered (or guessed) to have a given value; either the standard error is also known and the power $1 - \beta$ is calculated, or the intended power is known and the standard error calculated. Given this standard error, we can then try to calculate the required sample size.

For many types of design one can choose the sample size necessary to achieve a certain level of power on the basis of Cohen's work. For nested designs, however, there are two kinds of sample size: the sample size of the micro-units within each macro-unit n and the sample size of the macro-units N, with $N \times n$ being the total sample size for the micro-units.

10.2 Estimating a population mean

The most simple case of a multilevel study occurs when one wishes to estimate a population mean for a certain variable of interest (e.g., income, age, literacy), and one is willing to use the fact that the respondents are regionally clustered. This makes sense, since if one is interviewing persons it is a lot cheaper to sample, let us say, 100 regions and then interviewing 10 persons per region, than to randomly sample 1,000 persons that may live scattered all over a country. In educational assessment studies it is of course more cost-efficient to sample a number of schools and then to take a subsample of students within each school, than to sample students completely at random (ignoring their being clustered in schools).

Cochran (1977, Chapter 9) provides formulae to calculate desired sample sizes in case of such two-stage sampling. On p. 242, he defines the ***design effect*** for a two-stage sample, which is the factor by which the variance of an estimate (which is the square of the standard error of this estimate) is increased because of using a two-stage sample rather than a simple random sample with the same total sample size. Since estimation variance for a simple random sample is inversely proportional to total sample size, the design effect is also the factor by which total sample size needs to be increased to achieve the same estimation variance (or, equivalently, the same standard error) as a simple random sample of the given size. In Section 3.4 we saw that the design effect is given by

$$\text{design effect} = 1 + (n-1)\,\rho_{\mathrm{I}}\,, \tag{3.17}$$

where n is the average sample size in the second stage of the sample and ρ_{I} is the intraclass correlation. Now we can use (3.17) to calculate required sample sizes for two-stage samples.

Example 10.1 *Assessing mathematics achievement.*
Consider an international assessment study on mathematics achievement in secondary schools. The mathematics achievement variable has a unit variance, and within each country the mean should be estimated with a standard error of 0.02. If a simple random sample of size n is considered, it can readily be deduced from the well-known formula

$$\text{standard error} = \frac{\text{standard deviation}}{\sqrt{n}}$$

that the sample size should be $n = 1/(0.02^2) = 2{,}500$.

What will happen to the standard error if a two-stage sampling scheme is employed (first schools then students), in case the intraclass correlation is 0.20, and assuming that there are no direct extra budgetary consequences of sampling schools (this might be the case where one is estimating costs, when the standard errors are imposed by some international board)?

The design effect now is $1 + (30-1) \times 0.20 = 6.8$. So instead of having a simple random sample size of 2,500 one should use a two-stage sample size of $2{,}500 \times 6.8 = 17{,}000$ in order to achieve the same precision.

Further suppose that one also wishes to test whether the average maths achievement in the country exceeds some predetermined value. The power in this design, given this sample size, depends on the effect size and α. The effect size here is the difference between the actual average achievement and the tested predetermined value. Let us assume that α is set at 0.01 and that one wishes to know the effect size for which the power is as large as $\beta = 0.80$. According to (10.1), this will be the case when effect size is at least $0.02 \times (2.33 + 0.84) = 0.063$.

10.3 Measurement of subjects

In Section 3.5 the situation was discussed where the macro-unit may be, e.g., an individual subject, a school, a firm, etc., and the mean $\mu + U_j$ is regarded as the 'true score' of unit j which is to be measured. The level-one deviation R_{ij} then is regarded as measurement error. This means that, whereas in the previous section the question was how to measure the overall population mean, now we are interested in measuring the mean of a level-two unit, which we shall refer to as a *subject.* The observations Y_{ij} then can be regarded as *parallel test items* for measuring subject j. Suppose we wish to use an unbiased estimate, so that the posterior mean of Section 4.7 is excluded (because it is biased toward the population mean, see p. 59). The true score for subject j therefore will be measured by the observed mean, $\overline{Y}_{.j}$.

In equation (3.21) the reliability of an estimate defined as the mean of n_j measurements was given by

$$\lambda_j = \frac{n_j\, \rho_I}{1 + (n_j - 1)\, \rho_I} \, . \tag{3.21}$$

Now suppose that it is desired to have a reliability of at least a given value λ_0. The equation for the reliability implies that this requires to have at least

$$n_{\min} = \frac{\lambda_0\, (1 - \rho_I)}{(1 - \lambda_0\,)\, \rho_I} \tag{10.2}$$

measurements for each subject.

Example 10.2 *Hormone measurement.*
Consider individual persons as subjects, for whom the concentration of some hormone is to be measured from saliva samples. Suppose it is known that the intraclass correlation of these measurements is 0.40, so that the reliability of an individual measurement is 0.40 (cf. equation (3.19)). When an aggregate measurement based on several saliva samples is to be made with a reliability of $\lambda_0 = 0.80$, this requires at least $n_{\min} = (0.80 \times 0.60)/(0.20 \times 0.40) = 6$ measurements per subject.

10.4 Estimating association between variables

When one is interested in the effect of an explanatory variable on some dependent variable in a two-level hierarchical linear model, two situations are possible: both variables are micro-level variables, or the dependent variable is a micro-level variable whereas the explanatory variable is defined at the macro level. Generally, in both cases, standard errors are inversely proportional to $\sqrt{N}$, the square root of the sampled number of macro-units, which implies that taking as many macro-units as possible (while n, the sample size within each sampled macro-unit, is constant) reduces the standard errors in the usual way, which in turn has a positive effect on the power of the statistical tests. Now let us suppose that the researcher is stuck to a given total sample size $M = N \times n$, but that he or she is free in choosing either N large with n small or N small with n large.

Snijders and Bosker (1993) developed formulae for standard errors of fixed effects in hierarchical linear models. These formulae can be of help in making precisely this decision. Their software program PinT (see Chapter 15) can help researchers to settle these problems. This program calculates standard errors for estimating fixed effects in two-level designs, as a function of sample sizes. The greatest practical difficulty in using this program is that the user has to specify the means, variances, and covariances of all explanatory variables as well as the variances and covariances of all random effects. This specification has to be based on prior knowledge and/or reasonable guesswork. If one is uncertain about these input values, one may try several combinations of them to see how sensitive the resulting optimal sample sizes are for the input values within the likely range.

We refrain from presenting the exact formulae, but give two examples that can help in gaining some notion about how to decide in such a situation. More information can be found in the PinT manual and in Snijders and Bosker (1993).

Example 10.3 *A design to study level-one associations.*
Suppose one wants to assess the association between income (INCOME) and total number of years spent in school (YEARS) as part of a larger national survey using face-to-face interviews. Since interviewers have to travel to the respondents it seems worthwhile to reduce travelling costs and to take a two-stage sample: randomly select neighborhoods and within each neighborhood select a number of respondents.

First we have to make some educated guess on the possible structure of the data. The dependent variable is INCOME, while YEARS is the explanatory variable. For convenience let us assume that both variables are standard normal scores (z-scores, with mean 0 and variance 1). (Maybe monetary income has to be logarithmically transformed to obtain an approximately normal distribution.) Suppose that for INCOME the intraclass correlation is 0.50 and for YEARS it is 0.30. Now let us furthermore guess that the correlation between both variables is 0.447 (which implies that YEARS accounts for 20 percent of the variation in INCOME). The residual intraclass correlation for INCOME may be 0.40 (since the between-neighborhood regression of INCOME on YEARS generally will be larger than the within-neighborhood regression: dwellings of educated, rich people tend to cluster together). The standard errors of the effect of YEARS on INCOME for various combinations of the sample sizes n and N, restricted so that the total sample size $M = n \times N$ is at most 3,000, are given in Table 10.1.

Since N and n are integer numbers and M may not exceed 3,000, the total sample size fluctuates slightly. Nevertheless, it is quite clear that in this example it is optimal to take $n = 1$ respondent per neighborhood while maximizing the number of neighborhoods in the sample. If we test some target value and assume an effect size of 0.05 (the degree to which the results may be 'off target') and set α at 0.01, for $n = 1$ the power $1 - \beta$ would be 0.92, as follows from (apply (10.1))

$$0.01348 = \frac{0.05}{2.33 + z_{1-\beta}} \Leftrightarrow z_{1-\beta} = 1.38 \Leftrightarrow 1 - \beta = 0.92 \, .$$

Using $n = 1$ implies a single-level design, in which it is impossible to estimate the intraclass correlation. If one wishes to estimate the within-neighborhood variance and the intraclass correlation, one needs at least 2 respondents per neighborhood.

In the example given the researchers clearly had an almost unlimited budget. Of course it is far more expensive to travel to 3,000 neighborhoods and interview one person per neighborhood, than to go to 100 neighborhoods and interview 30 persons per neighborhood. It is therefore common to impose budgetary constraints on the sampling design. One can express the costs of sampling an extra macro-unit (neighborhood, school, hospital) in terms of micro-unit costs.

Table 10.1 Standard errors as a function of n and N.

$M = N \times n$	N	n	S.E.
3000	3000	1	0.01348
3000	1500	2	0.01390
3000	1000	3	0.01414
3000	750	4	0.01431
3000	600	5	0.01442
3000	500	6	0.01451
2996	428	7	0.01459
3000	375	8	0.01463
2997	333	9	0.01468
3000	300	10	0.01471
3000	250	12	0.01477
2996	214	14	0.01482
2992	187	16	0.01487
2988	166	18	0.01491
3000	150	20	0.01490
2992	136	22	0.01494
3000	125	24	0.01493
2990	115	26	0.01497
2996	107	28	0.01497
3000	100	30	0.01497

To sketch the idea in the context of educational research: assume that observation costs are composed of salary, travel costs, and the material required. Assume that the costs of contacting one school and the travel for the visit are \$150. Further assume that the salary and material costs for testing one student are \$5. For example, investigating 25 students at one school costs a total of $\$150 + 25 \times \$5 = \$275$. More generally, this means that the cost function can be taken as $\$150\,N + \$5\,Nn = \$5\,N(n + 30)$.

In many practical cases, for some value of c (in the example, $c = 30$) the cost function is proportional to $N(n + c)$. The number c is the ratio indicating how much more costly it is to sample one extra macro-unit (without changing the overall total number of micro-units sampled), than it is to sample one extra micro-unit within an already sampled macro-unit.

Usually for mail or telephone surveys using a two-stage sampling design there is no efficiency gain in using a two-stage sample as compared to using a simple random sample, so that $c = 0$. But for research studies in which face-to-face interviewing or supervised testing is required, efficiency gains can be made by using a two-stage sampling design.

So we are searching for sample sizes at micro and macro level that satisfy the inequality

$$N(n + c) \leq K \,, \tag{10.3}$$

in which K is the available budget, expressed in monetary units equal to the cost of sampling one additional micro-level unit. In the example given, K would be the budget in dollars divided by 5.

Example 10.4 *Setting up a class size experiment.*
Let us now turn to an example in which a macro-level variable is supposed to have an effect on a micro-level variable, and apply the idea just unfolded on cost restrictions.

A case might be an experiment with class size, in which a class size reduction of 6 pupils per class is compared to a usual class size of 26 in its effect on achievement of young pupils. Let us assume that achievement is measured in z-scores, i.e., their variance is 1. If there are additional covariates in the design to control for differences between the experimental and control group we assume that the residual variance, given these covariates, is equal to 1.

Let us say that the experiment is set up to detect an effect of 0.20 (or larger) in the population of interest. Transforming the effect size d into a correlation coefficient r using the formula (Rosenthal, 1991, p. 20)

$$d = \frac{2r}{\sqrt{1 - r^2}}$$

results in $r = 0.10$. Stated otherwise, the experimental factor will account for $r^2 = 1$ percent of the variation in the achievement score. The astonishing thing about this example is, that most researchers would be inclined to test all pupils within a class and would take the class size of 23 (being the average of 20 and 26) as given. There is no need at all to do so! If the total sample size $M = N \times n$ would be predetermined, it would be optimal (from a statistical point of view) to take as many classes as possible with only one pupil per class. It is only because of budgetary constraints that a two-stage sample may be preferred. Let us take $c = 23$ (taking an extra school is 23 times the price of taking one extra student within an already sampled school). Running the software program PinT with a budget constraint $N(n + 23) \leq 1{,}000$ leads to the results of Table 10.2.

The optimum for the standard error associated with the effect of the treatment appears to be when we take a sample of 9 pupils per school and 31 schools. In that case the standard error of interest reaches its minimum of 0.09485. It will be clear from the outset that this standard error is too large to reach a satisfactory power for this effect size of 0.20. Even with α as high as 0.10, the power will be only 0.42.

The first solution to this problem would be to increase the budget. Applying formula (10.1) shows that to reach a power of 0.80 with $\alpha = 0.10$, the standard error should be made twice as small. This implies a multiplication of the number of schools by $2^2 = 4$, which amounts to $4 \times 31 = 124$ schools in each of which 9 pupils are sampled. To make it even worse: politicians do not like to spend money on interventions that may not work, and for that reason it may be required to have α as low as 0.01. Applying formula (10.1) shows that the original standard error of 0.09485 should be brought down to one third of its value, implying a required sample size of $3^2 \times 31 = 279$ schools.

If, instead of the optimum value $n = 9$, one would use the earlier mentioned value of $n = 23$, the standard error would become $0.10346/0.09485 = 1.09$ times as large, which could be offset by sampling $1.09^2 = 1.19$ as many schools. To reach a power of 0.80 with $\alpha = 0.01$ and $n = 23$, one would need to sample $3^2 \times 1.19 \times 31 = 332$ schools.

The example illustrates that when one is interested in effects of macro-level variables on micro-level outcomes it usually is sensible (in view of

Table 10.2 Standard errors in case of budget constraints.

$M = N \times n$	N	n	S.E.
41	41	1	0.15539
80	40	2	0.12145
114	38	3	0.10962
148	37	4	0.10267
175	35	5	0.10000
204	34	6	0.09752
231	33	7	0.09602
256	32	8	0.09520
279	31	9	0.09485
300	30	10	0.09487
319	29	11	0.09518
336	28	12	0.09574
351	27	13	0.09652
378	27	14	0.09567
390	26	15	0.09674
400	25	16	0.09798
425	25	17	0.09738
432	24	18	0.09884
437	23	19	0.10046
460	23	20	0.10000
462	22	21	0.10182
484	22	22	0.10144
483	21	23	0.10346

reducing the standard error of interest) to sample many macro-units with relatively few micro-units. The example given may be regarded as a cluster randomized trial: whole classes of students are randomly assigned to either the experimental (a class size of 20) or the control (a class size of 26) condition. Raudenbush (1997) shows how much the standard error for the experimental effect can be decreased (and thus the power increased) if a covariate is added that is strongly related to the outcome variable. In this case this would imply, in practice, to administer a pretest or intelligence test.

10.4.1 Cross-level interaction effects

Especially interesting in social research are cross-level interaction effects. If one is interested in such an effect, how can one improve the power of the test to be employed? Should one increase the sample size of the macro-units or should one increase the sample size of the micro-units? If only the total sample size is fixed (i.e., $c = 0$ in (10.3)), then it usually is optimal to take the sample size for the macro-units as large as possible. This is illustrated in the following example, taken from Snijders and Bosker (1993).

Example 10.5 *Design for the estimation of a cross-level effect.*
Suppose we want to assess the effect of some school policy measure to enhance the achievement of students from low socio-economic status families, while taking into account aptitude differences between students. More specif-

ically we are interested in the question whether this policy measure reduces differences between students from low and high socio-economic status families. In this case we are interested in the effect of the cross-product variable POLICY × SES on achievement, POLICY being a school-level variable and SES being a student-level variable. Assume that we control for the students' IQ and that SES has a random slope. The postulated model is

$$\begin{aligned} Y_{ij} = \; & \gamma_{00} + \gamma_{10}\,\mathrm{SES}_{ij} + \gamma_{20}\,\mathrm{IQ}_{ij} + \gamma_{01}\,\mathrm{POLICY}_j \\ & + \gamma_{11}\,\mathrm{SES}_{ij} \times \mathrm{POLICY}_j + U_{0ij} + U_{1ij}\,\mathrm{SES}_{ij} + R_{ij}\,, \end{aligned}$$

where Y_{ij} is the achievement of student i in school j.

The primary objective of the study is to test whether γ_{11} is 0. It is assumed that all observed variables are scaled to have mean 0 and unit variance. With respect to the explanatory variables it is assumed that SES is measured as a within-school deviation variable (i.e., all school means are 0), the intraclass correlation of IQ is 0.20, the correlation between IQ and SES is 0.30, and that the covariance between the policy variable and the school mean for IQ is −0.13 (corresponding to a correlation of −0.3). With respect to the random effects it is assumed that the random intercept variance is 0.09, the slope variance for SES is 0.0075, and the intercept-slope covariance is −0.01. Residual variance is assumed to be 0.5. Some further motivation for these values is given in Snijders and Bosker (1993).

Given these parameter values, the PinT program can calculate the standard errors for the fixed coefficients γ. Figure 10.1 presents the standard errors of the regression coefficient associated with the cross-level effect as a function of n (the micro-level sample size), subject to the constraint that total sample size M is not more than 1,000. N is taken as M/n rounded to the nearest lower integer. As n increases, N will decrease (or, sometimes, remain constant).

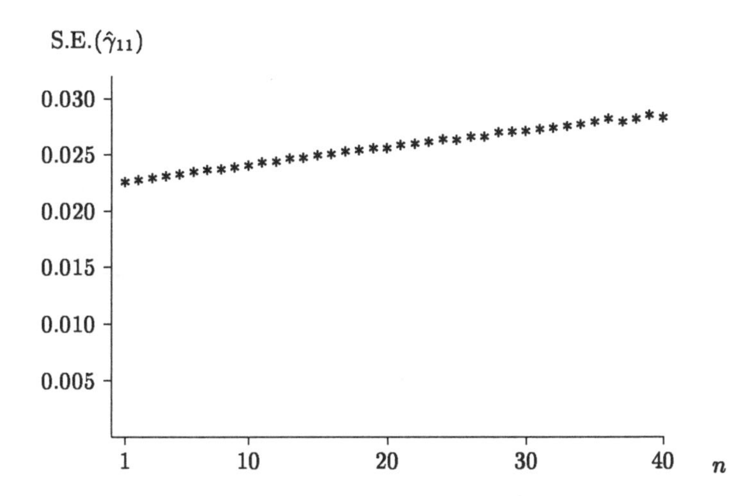

Figure 10.1 Standard error of a cross-level effect (for $N \times n \leq 1{,}000$).

The graph clearly illustrates that it is optimal to take n as small as possible, and (by implication) N as large as possible, since the clustering of the data will then affect the standard errors the least. This may appear to be somewhat counter-intuitive: in the two-step formulation of the model, at the school level we regress the SES slope on the POLICY variable, which suggests that we need a reliable estimate for the SES-slope for each school, and that 'thus' n should be large. In this case, however, one should realize that the cross-product variable POLICY $\times$ SES is simply a student-level variable, and that as long as we have variation in this variable, the effect can be estimated. Even when n is as small as 1! Although to estimate intercept and slope variances we need n to be at least 3.

This conclusion can be generalized to studies into the development over time of a certain characteristic (cf. Chapter 12). A series of cross-sectional measurements for people of different ages, for example, can be used to make assessments on growth differences between the sexes.

We again assumed in the previous example that, given the total sample size, it is not extra costly to sample many schools. This assumption is not very realistic, and therefore we will look at the behavior of the relevant standard error once again but now assuming that the price of sampling an extra school is $c = 5$ times the price of sampling a pupil, and the total budget in (10.3) still is $K = 1{,}000$. Running the software program PinT once again, we obtain the results presented in Figure 10.2.

Because of the budget restriction the total sample size M decreases and the number of groups N increases as the group size n decreases. Balancing between N large (with n small) and N small (with n large), we see that in this case it is optimal to take n somewhere between 15 and 20 (implying that N is between 50 and 40, and M between 750 and 800). More precisely, the minimum standard error is 0.0286, obtained for n equal to 17 and 18. If one is willing to accept a standard error that is 5 percent higher than this minimum, then n can be as low as 9.

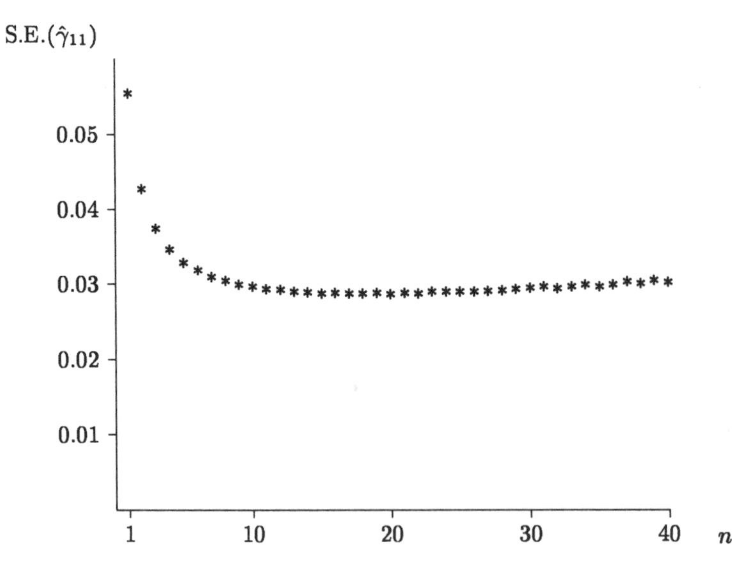

Figure 10.2 Standard error of a cross-level effect (for a budget restriction with $c = 5$, $K = 1{,}000$).

10.5 Exploring the variance structure

In all examples given so far, we presented cases in which either we wanted to assess the mean value of a variable or the association between two variables. Another series of interesting questions may have to do with unspecified effects: does it matter which school a child attends? do neighborhoods differ in their religious composition? do children differ in their growth patterns? is the effect of social status on income different in different regions? In all these instances we are interested in the magnitude of the variance components, and the power of the tests on the variance components is affected by n and N in a different manner than the power of the tests on the fixed effects.

10.5.1 The intraclass correlation

These example questions about unspecified effects can be answered in first instance by estimating or testing the corresponding intraclass correlation coefficient. Note that in Sections 10.2 and 10.3, the minimum required sample sizes were expressed in terms of the intraclass correlation coefficient, so that also to determine those required sample sizes it is necessary to estimate this coefficient.

In Section 3.13 the standard error for estimating the intraclass correlation was given as

$$\text{S.E.}(\hat{\rho}_{\text{I}}) = (1 - \rho_{\text{I}})(1 + (n-1)\rho_{\text{I}})\sqrt{\frac{2}{n(n-1)(N-1)}} \,. \tag{3.13}$$

To investigate whether it is better to take many small groups or few large ones, a graph of this formula can be drawn as a function of n, the group size, assuming either a constant total sample size or the budget constraint (10.3). For simplicity, let us forget about the fact that N must be an integer number, and employ the sample size constraint $M = N \times n$ and the budget constraint $N(n + c) = K$ as equations for computing N as a function of n.

For a fixed total sample size M, the substitution $N = M/n$ implies that the factor with the square root sign in (3.13) must be replaced by

$$\sqrt{\frac{2}{(n-1)(M-n)}} \,.$$

The optimal group size is the value of n for which the resulting expression is minimal. This optimum for a fixed total sample size depends on ρ_{I}. If ρ_{I} is precisely equal to 0, the standard error decreases with n and the optimum is achieved for $N = 2$ groups each of size $n = M/2$. For ρ_{I} close to 1, on the other hand, the standard error increases with n and it is best to have $N = M/2$ groups each of size 2. It is unpleasant that the optimum values can be so totally different, depending on the value of ρ_{I} which is unknown to begin with! However, in most cases the researcher will have some prior knowledge about the likely values of ρ_{I}. In most social science research, the intraclass correlation ranges between 0.0 and 0.4, and often narrower bounds can be indicated. For such a range of values, the researcher can determine a group size that gives a relatively small standard error for all ρ_I values within this range.

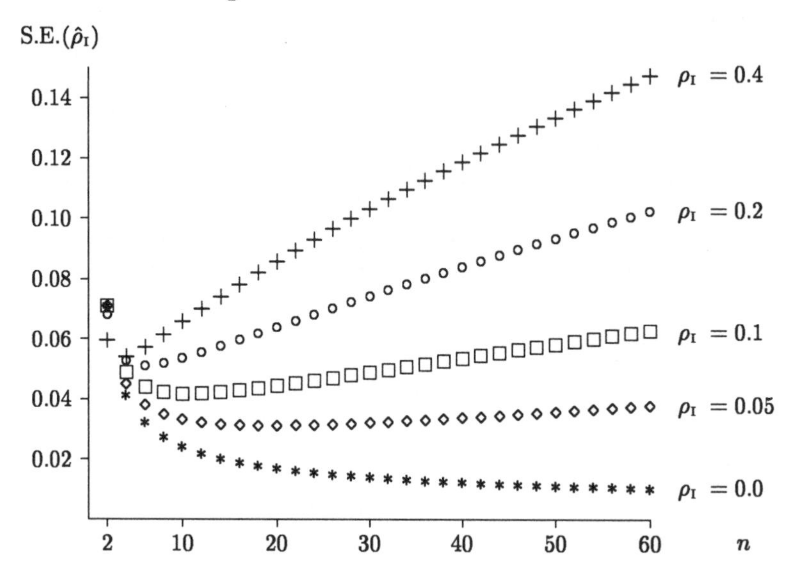

Figure 10.3 Standard error for estimating the intraclass correlation coefficient for $M = 400$.

For a fixed total sample size $M = 400$, the standard error is graphed in Figure 10.3 for several values of ρ_I. This figure shows indeed that the optimum group size depends strongly on the value of ρ_I. For example, if $\rho_I = 0.4$, group sizes over 20 are unattractive whereas for $\rho_I = 0.0$ this is the case for group sizes less than 5. This suggests that if prior knowledge implies that the intraclass correlation might be as small as 0.0 or as large as 0.4, and if the focus of the research is on estimating ρ_I while the total sample size is fixed at 400, one should preferably use group sizes between 5 and 20.

Example 10.6 *A good compromise for a fixed total sample size.*
Suppose that, like in Figure 10.3, the total sample size must be close to $M = 400$, while it is believed to be likely that the intraclass correlation ρ_I is between 0.05 and 0.2. Calculating the standard errors for these two values of ρ_I shows that for $\rho_I = 0.05$ the smallest standard error is equal to 0.0308, achieved for $n = 19$; for $\rho_I = 0.2$ the minimum is 0.0510, achieved for $n = 6$. This implies that it is best to use an intermediate group size, not too high because that would lead to a loss in standard error for low values of ρ_I, and not too low because that would entail a disadvantage for high ρ_I.

Inspecting the standard errors for intermediate group sizes (this can be done visually from Figure 10.3) shows that group sizes between 9 and 12 are a good compromise in the sense that, for each value $0.05 < \rho_I < 0.2$, the standard error for $9 \leq n \leq 12$ is not more than 10 percent higher than the minimum possible standard error for this value of ρ_I.

If the optimum group size is to be found for the budget constraint (10.3), the value $N = K/(n+c)$ must be substituted in expression (3.13) for the standard error, and the standard error calculated or graphed for a suitable range of n and some representative values of ρ_I.

Example 10.7 *A good compromise for a budget constraint.*
Suppose that the budget constraint must be satisfied with $c = 10$, so each additional group costs ten times as much as an additional observation in an already sampled group, and with a total available budget of $K = 600$. The extreme possible situations then are $N = 50$ groups of size $n = 2$, and on the other hand $N = 2$ groups of size $n = 290$. Suppose that practical constraints imply that the group size cannot be more than 20. The results obtained when substituting $N = 600/(n+10)$ in (3.13) and calculating the resulting expression are shown in Figure 10.4.

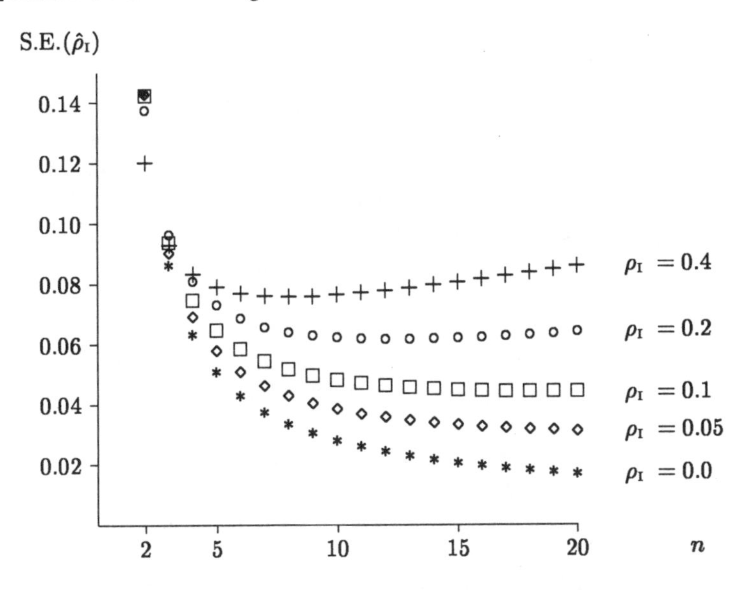

Figure 10.4 Standard error for estimating the intraclass correlation coefficient for a budget constraint with $c = 10$, $K = 600$.

The figure shows that for larger values of ρ_I, the optimum group size has an intermediate value (e.g., $n = 8$ for $\rho_I = 0.4$), but the standard error does not depend very strongly on the group size as long as the group size is at least 4. For the smaller values of ρ_I, it is best to take the group size as large as possible within the allowed range. Group sizes $n = 17$ and 18 give a good compromise between what would be best for ρ_I as small as 0.0 and what would be best for $\rho_I = 0.4$, but smaller group sizes may also be reasonable.

Now suppose, to modify the example, that from the point of view of estimating fixed effects it is desirable to use smaller group sizes, and that it is likely that ρ_I is at least 0.05. On the basis of this figure, one could then decide to sacrifice some of the standard error for low values of ρ_I, and use a group size as low as 8 or 10.

10.5.2 Variance parameters

For a two-level random intercept model, approximations to the estimation variances of the level-one variance σ^2 and the intercept variance τ_0^2 are given by Longford (1993a, p. 58). For the standard errors, these formulae are

$$\text{S.E.}(\hat{\sigma}^2) \approx \sigma^2 \sqrt{\frac{2}{N\,(n-1)}}$$

and

$$\text{S.E.}(\hat{\tau}_0^2) \approx \sigma^2 \, \frac{2}{N n} \sqrt{\frac{1}{n-1} + \frac{2\,\tau_0^2}{\sigma^2} + \frac{n\,\tau_0^4}{\sigma^4}}\,.$$

If one prefers standard errors for the estimated standard deviations, according to formula (5.15) these formulae can be transformed by

$$\text{S.E.}(\hat{\sigma}) \approx \frac{\text{S.E.}(\hat{\sigma}^2)}{2\,\sigma}\,, \tag{10.4}$$

and similarly for S.E.($\hat{\tau}_0$).

The standard error of $\hat{\sigma}^2$ is minimized by using large level-one sample sizes n, but in practice the standard error is, for a given reasonable budget constraint, so little sensitive to the value of n that one does not really care about obtaining precisely the minimal possible standard error for this parameter (this is also concluded by Cohen, 1998). To obtain a good design for estimating the variance parameters, the intercept variance is more important.

If optimal sample sizes are to be determined to estimate the intercept variance, the same approach can be applied as in the preceding section. One can plot these standard errors against n, with $M = N \times n$ fixed to a given value, for various values of the intraclass correlation coefficient $\rho_{\mathrm{I}} = \tau_0^2/(\sigma^2 + \tau_0^2)$. If there is a budget constraint, one can substitute $N = K/(n+c)$ into Longford's approximation expressions and plot the standard errors while keeping $N \times (n+c)$ fixed. An example is given by Cohen (1998, Section 5.1).

Standard errors of random slope variances are quite complicated functions of the sample sizes, the variation of the explanatory variables within and between the groups, and the variances themselves. Further research is needed for elaborating guidelines on how to choose sample sizes appropriately for the estimation of variance parameters in random slope models. If precise estimation of such parameters is required, it seems advisable to use large sample sizes (30 or higher) at either level and to make sure that the explanatory variable of which one wishes to estimate the random slope variance has enough dispersion within the level-two units. This may be achieved, e.g., by some kind of stratification.

For complicated models for which no formulae are available for the standard errors, Monte Carlo simulation can be an alternative way to get an impression about standard errors. Simulation analysis of multilevel models is possible in MLn/MLwiN and in MLA (see Chapter 15).

11 Crossed Random Coefficients

Random effects are not always nested. Crossed random effects, i.e., unexplained variation that is accounted for by crossed factors, arise in many domains of study. Although multilevel analysis focuses on nested random coefficients, differently structured random coefficients can also be incorporated in the multilevel model and in some multilevel software. Some examples of crossed random effects are as follows.

- In a study of pupils within schools, neighborhoods could also have an effect. Schools may be physically nested within neighborhoods, but they usually attract pupils from several neighborhoods. Schools and neighborhoods then are crossed factors.

- In a study of pupils within classes within schools, classes may have been taught by several teachers. In this case, classes and teachers are crossed factors, jointly nested within schools.

- In generalizability studies (e.g., Shavelson and Webb, 1991), one studies the joint effects of several random factors, and the extent to which the performance of subjects can be generalized over the domain defined by these factors. For example, in a study of the results of a course in communication skills, the students are taught by one teacher, examined in one or several role plays with an actor, and rated by one or several raters. Teacher, version of the role play, actor, and rater all have crossed effects.

After the initial exclusive focus of multilevel analysis on nested random effects, the later multilevel literature also contains theoretical results and applications of crossed random effects, e.g., Goldstein (1987, 1995), Raudenbush (1993), Rasbash and Goldstein (1994), and Raudenbush and Chan (1994). This chapter indicates how crossed effects can be modeled in the multilevel framework.

11.1 A two-level model with a crossed random factor

As the basic example of how to incorporate a crossed random factor in a multilevel model, we consider the case of a two-level study, e.g., of pupils (indicated by i) nested within schools (indicated by j), with a crossed random factor, e.g., the neighborhood in which the pupil lives. The neighborhoods are indexed by the letter f, running from 1 to F (the total number of neighborhoods in the data). The neighborhood of pupil i in school j is indicated by $f(i,j)$. The hierarchical linear model for an outcome variable Y_{ij} without the neighborhood effect reads, cf. (5.14),

$$Y_{ij} = \gamma_0 + \sum_{h=1}^{q} \gamma_h \, x_{hij} + U_{0j} + \sum_{h=1}^{p} U_{hj} \, x_{hij} + R_{ij} \, .$$

The random effect of neighborhood f, indicated by W_f, just is added to this model:

$$Y_{ij} = \gamma_0 + \sum_{h=1}^{q} \gamma_h \, x_{hij} + U_{0j} + \sum_{h=1}^{p} U_{hj} \, x_{hij} + W_{f(i,j)} + R_{ij} . \quad (11.1)$$

For the neighborhood effects W_f, the usual assumption is made that W_1 to W_F are mutually independent random variables, also independent of the other random effects (U and R), all W_f having the normal distribution with mean 0 and variance τ_W^2. The interpretation is just as that of other random effects: part of the variability in the dependent variable is accounted for by neighborhoods, and these are regarded as a random sample from a population of neighborhoods (the population may be hypothetical). The next subsection indicates how to solve the statistical complications caused by the fact that the structure is not neatly nested.

The analysis by multilevel software of random effects models that combine nested and crossed effects is technically and computationally more demanding than the usual multilevel models. Therefore, it is advisable to make preliminary analyses focusing on first one, then the other crossed factor; such analyses can be done employing the usual two-level model. An important question is, to what extent the two crossed factors are associated. In a study of schools and neighborhoods, these two factors will be associated, and when only one of them is incorporated as a random effect in the model it will draw to itself part of the other (temporarily neglected) random effect. In a generalizability study, on the other hand, usually the various crossed factors are not associated and they will have (nearly) independent effects.

11.1.1 Random slopes of dummy variables

Computer programs for the analysis of random coefficient models that are not specifically geared toward multilevel analysis, such as SAS Proc Mixed (see Chapter 15), can incorporate crossed effects in a straightforward way. Handling crossed random effects in multilevel software, however, requires some extra tricks.

To incorporate the random effect $W_{f(i,j)}$ in (11.1) in the multilevel formulation, this term is written differently. Dummy variables b_f ($f = 1, \ldots, F$) are defined, indicating the neighborhood of each pupil:

$$b_{fij} = \begin{cases} 1 & \text{if pupil } i \text{ in school } j \text{ lives in neighborhood } f \\ 0 & \text{if this pupil lives in a different neighborhood.} \end{cases}$$

The random neighborhood effect is rewritten as

$$W_{f(i,j)} = \sum_{f=1}^{F} W_f \, b_{fij} \; . \qquad (11.2)$$

This complicated-looking notation is chosen because it does away with the notation $f(i,j)$, which cannot be handled directly in the multilevel framework. Equation (11.2) reduces the crossed random effect to random slopes of dummy variables b_1 to b_F. However, these random slopes are not nested in the units at level one or two. The trick now is the creation of an extra level, the third, consisting of only one unit (encompassing therefore the whole data set). This may well be called a dummy level or pseudo level. The dummy variables b_f are given uncorrelated random slopes at this third level, and the slope variances must satisfy the restriction

$$\text{var}(W_1) = \text{var}(W_2) = \ldots = \text{var}(W_F) \; . \qquad (11.3)$$

These ingredients are sufficient to model the crossed random effects within the framework of the hierarchical linear model.

It is clear which requirements are made to multilevel software for modeling random effects in this way. An extra level with only one unit must be allowed, and it must be possible to make equality restrictions between random slopes. This is possible, e.g., in MLn/MLwiN.[1]

A practical requirement is that the software can accommodate the required number of F random slopes. In practice, this will set limits to the possibility of incorporating random effects. In model (11.1), the roles of schools j and neighborhoods f could be interchanged: pupils then would be nested within neighborhoods, and these would be crossed with schools. Both representations are equivalent. Which one to choose, is a matter of convenience. It will usually be most efficient for multilevel software to choose the factor with the larger number of units as the nesting factor, and the factor with the fewer units as the crossed factor.

This method can be extended to situations where pupils may have lived part of the time in one, and part of the time in another neighborhood. (Or, abstractly: where level-one units can be classified as belonging partially to more than one category of the crossed random factor.) This is done by defining the variable b_f not by 0–1 values as above, but by positive values summing to 1, indicating the fraction of time in which they belonged to neighborhood f, see Hill and Goldstein (1998).

Example 11.1 *Sustained primary school effects.*
We continue with the Belgian dataset of Opdenakker and Van Damme (1997) which was used also in the examples in Sections 4.8 and 5.5.

The dataset also contains information on the specific primary school attended by the students before they went to secondary education. In total 376 different primary schools were attended by the 3,792 students, whereas only 57 secondary schools were involved. In order to find out whether the effect of the primary school attended has a sustained effect on the mathematics

[1] This program also contains the special command SETX for defining a model with crossed random effects.

score of the students at the end of the second grade of secondary education, a cross-classified multi-level model analysis is performed on the data. In this case we use the nesting structure defined by students (level-one units) within primary schools (level-two units) and create a third 'pseudo level' at which we have 57 random dummies representing the secondary schools. Since in this chapter we focus on crossed random coefficients we ignore the fact that students are grouped in classes (although we know that this is an important feature of the data!). Table 11.1 contains the results of the analysis.

Table 11.1 Two cross-classified models for maths achievement.

	Model 1		Model 2	
Fixed Effect	Coeff.	S.E.	Coeff.	S.E.
γ_0 Intercept	7.99	0.23	8.62	0.14
γ_1 IQ			0.054	0.005
γ_2 Pretest			0.170	0.008
γ_3 Motivation			0.037	0.008
γ_4 Father's education			0.057	0.015
γ_5 Gender			0.237	0.112
Random Effect	Var. Comp.	S.E.	Var. Comp.	S.E.
Crossed random effect:				
$\tau_W^2 = \text{var}(W_f)$ secondary school	2.590	0.567	0.727	0.179
Level-two random effect:				
$\tau_0^2 = \text{var}(U_{0j})$ primary school	0.179	0.076	0.174	0.059
Level-one variance:				
$\sigma^2 = \text{var}(R_{ij})$	9.102	0.217	6.370	0.152
Deviance	19334.4		17958.7	

From the results in Table 11.1, Model 1, it can be seen that the average mathematics score is 7.99 and the total variance is 11.871. A substantial part of this variance, 22 percent, is associated with the secondary schools, the crossed random factor. Only 2 percent of the variance resides at the level of the primary schools. At the age of fourteen, differences in mathematics abilities therefore have much more to do with the secondary schools than with the primary schools attended before the students entered secondary education.

Of course we could argue that it is superfluous to add a random effect (the primary school the students went to) to the model if variables on manifest abilities and background factors are available. Table 11.1, Model 2 presents the results of an analysis in which the available information on previous math performance, achievement motivation, IQ, gender, and father's educational level is also contained in the model. In this model the effect of the primary school attended may be interpreted as the unmeasured quality of the education provided by the primary school and the effect of the secondary school may be interpreted as the value added by this school.

Having added the student predictor variables to the model, the total unexplained variance decreases to 7.271. These variables thus account for 39 percent of the total variation in mathematics ability. Interestingly, the variance at the primary school level does not change much, whereas the variance at the pseudo level of the secondary schools decreases by 72 percent. This indicates that the secondary schools in this Belgian sample are highly selective, some schools having the more advanced and privileged groups of students (high IQ, high pretest-score, high SES etc.), whereas others serve more disadvantaged groups. Still, however, the secondary schools differ markedly in their value added, since 10 percent of the residual variance still resides at the secondary school level. The fact that the variance associated with secondary schools is so much less in Model 2 than in Model 1 leads to the conclusion that some secondary schools 'pick' the better students from the primary schools, whereas other secondary schools have to deal with the other students. For the students it seems not to matter much for their mathematics ability which specific primary school they have attended: the unmeasured effect is still rather small.

11.2 Crossed random effects in three-level models

A crossed random effect in a three-level model can occur in two ways. An example of the first is pupils nested in classes nested in schools, with neighborhoods as a crossed effect. In this case *the extra factor is crossed with the level-three units:* neighborhoods are crossed with schools. The random effects of neighborhoods is modeled just like in the two-level case: an extra 'dummy' level is created, but now it is the fourth level; dummy variables are defined that indicate, for each neighborhood, whether the pupil lives in this neighborhood; these dummy variables get random slopes at level 4, with the restriction of equal slope variances.

The second kind of random effect in a three-level model is exemplified by pupils nested in schools nested in towns, again with neighborhoods as crossed effects. Here *the extra factor is crossed with the level-two units and nested in the level-three units:* neighborhoods are crossed with schools and nested in towns. If pupils are indicated by i, schools by j, and towns by k, the effect of neighborhood f in town k should be denoted W_{fk}. The usual assumptions are made again (normal distributions with mean 0 and a common variance τ_W^2, independence). Again, dummy variables b_f need to be created. They are defined by

$$b_{fijk} = \begin{cases} 1 & \text{if pupil } i \text{ in school } j \text{ in town } k \\ & \text{lives in neighborhood } f \\ 0 & \text{if this pupil lives in a different neighborhood.} \end{cases} \tag{11.4}$$

How many of these dummy variables are there? Denote by F_k the number of neighborhoods in town k, and their maximum by

$$F_{\max} = \max_k F_k \ . \tag{11.5}$$

The number of dummy variables then is $F_{\max}$.

Since neighborhoods are nested within towns, there is no need here to create an extra level to accommodate the crossed random effect. The dummy variables b_f, for $f = 1, \ldots, F_{\max}$, must have uncorrelated random slopes at level three, with slope variances constrained to be equal. Each single town is modeled just like the whole data set of Section 11.1. In other words, just as the dataset in the present section is a juxtaposition of towns, the model is the juxtaposition of the models described in Section 11.1.

Since the number of random slopes in this model is not the total number of neighborhoods in the data set, but the maximum of the number of neighborhoods in each town, this model economizes on the number of random slopes, compared to Section 11.1. Suppose now that one has a data set of pupils in schools and neighborhoods, these being nested in towns, but one is not interested in town effects or there are no random town effects. Then it still can be advisable to use towns as a third level, because this leads to a much smaller number of random slopes and thereby to a model that is much more efficient given the constraints of existing software. It might even be impossible to implement the large number of random slopes, unless one uses the towns as a nesting factor.

11.3 Correlated random coefficients of crossed factors

The principle of modeling crossed random effects, explained in Section 11.1, can be extended in many ways. This section treats three such variations with correlated random coefficients of crossed factors.

11.3.1 Random slopes in a crossed design

The neighborhoods in Section 11.1 could have random slopes in addition to their random effects. The notation of that section is continued here. If neighborhood f has a random intercept, and also a random slope for the variable X (denoted here as a level-one variable, but allowed also to be a level-two variable), the total contribution of neighborhood f to the random part is

$$W_{0f} + W_{1f}\, x_{ij} \; . \tag{11.6}$$

For the neighborhood of pupil i in school j, this is

$$W_{0,f(i,j)} + W_{1,f(i,j)}\, x_{ij} \; . \tag{11.7}$$

With the dummy variables b_f, this is rewritten as

$$\sum_{f=1}^{F} (W_{0f}\, b_{fij} + W_{1f}\, b_{fij}\, x_{ij}) \; . \tag{11.8}$$

This shows that, in addition to the dummy variables b_f, we now need the product variables $b_f X$ with values

$$b_{fij}\, x_{ij} \; . \tag{11.9}$$

To implement the random part (11.8), one gives random slopes to the variables b_f and $b_f\, X$. Their variances are denoted τ^2_{W0} and τ^2_{W1}. For different values of f, these random slopes are uncorrelated. The slopes of b_f and $b_f\, X$ for each single value f, however, must be assumed to be correlated. Their covariance is denoted τ_{W01}. The assumption (11.3) about equal slope variances now must be extended to

$$
\begin{aligned}
&\text{var}(W_{01}) = \text{var}(W_{02}) = \ldots = \text{var}(W_{0F}) , \\
&\text{var}(W_{11}) = \text{var}(W_{12}) = \ldots = \text{var}(W_{1F}) , \qquad (11.10) \\
&\text{cov}(W_{01}, W_{11}) = \text{cov}(W_{02}, W_{12}) = \ldots = \text{cov}(W_{0F}, W_{1F}) .
\end{aligned}
$$

The interpretation of the random slope of the crossed factor is similar to the interpretation in other random slope models.

11.3.2 Multiple roles

In some designs there are factors affecting the outcome via more than one 'role'. As an example, consider a course that is taught as well as examined by several persons. There are F persons acting as teachers and examiners; each of these persons teaches a group of students, and examines a subset of the students (but not those from his own group). The dependent variable is the student's grade, Y.

A positive effect of a person as a teacher means that this person has students with a good accomplishment. A positive effect of a person as an examiner means that this person gives high grades. It is likely that these effects are correlated, negatively or positively. The effects of the examiners could be modeled as the crossed random factor in Section 11.1, but this would yield uncorrelated effects of the teacher and examiner roles of a person. If we wish to model correlated effects, the nesting of students within teacher groups should not be used in the technical specification of the multilevel model. Rather, a two-level model should be used with the second level consisting of only one unit (a dummy level). Denote the student by i, his teacher by $f(i)$, and his examiner by $g(i)$. Note that both $f(i)$ and $g(i)$ have the same domain, viz., the set of teachers, indicated by $f = 1, \ldots, F$. The random effect of person f as a teacher is denoted W_{0f}, his random effect as an examiner is denoted W_{1f}. Different persons have uncorrelated effects, but a correlation is allowed between the two random effects W_{0f}, W_{1f} of the same person.

It will be no surprise that, in this case, two sets of dummy variables are needed. For the teaching roles there are dummy variables $b_1, \ldots, b_F$ with values defined by

$$
b_{fi} = \begin{cases} 1 & \text{if student } i \text{ was taught by person } f \\ 0 & \text{if this student was taught by another person.} \end{cases} \qquad (11.11)
$$

Dummy variables $c_1, \ldots, c_F$ for the examining roles are defined by

$$
c_{fi} = \begin{cases} 1 & \text{if student } i \text{ was examined by person } f \\ 0 & \text{if this student was examined by another person.} \end{cases} \qquad (11.12)
$$

The random contribution of teacher and examiner, represented by

$$
W_{0,f(i)} + W_{1,g(i)} , \qquad (11.13)
$$

can now be written as

$$
\sum_{f=1}^{F} (W_{0f}\, b_{fi} + W_{1f}\, c_{fi}) . \qquad (11.14)
$$

The dummy variables b_f and c_f have random slopes at level two, that are correlated for the same f and uncorrelated for different f. Restrictions must be made on the variances and covariances of these random slopes, namely exactly those given also in (11.10).

11.3.3 Social networks

An extension of the example of the preceding subsection allows modeling continuous dependent variables in social networks. For a given group of persons (or other actors) indicated by $1, \ldots, F$, a social network is understood to be the pattern of pairwise relationships between these persons. The most usual kind of relationship considered in social network analysis is binary (e.g., being a friend versus not being a friend), cf. Wasserman and Faust (1994). Here we consider a continuous measurement Y_{fg} that indicates a characteristic (e.g., the strength) of the directed relationship from actor f to actor g. For example, this could be the outcome variable when each person in a closed group expresses his positive feelings ('like') towards each of the other persons on a scale from 1 to 10. Usually, outcomes Y_{fg} are available only for $f \neq g$. The following model was discussed by Snijders and Kenny (1999).

Important effects that can be discerned here are:
reciprocity: if f likes g, then chances are higher that g also likes f;
popularity, indicated by a person receiving high scores from the others;
and *activity* or *outgoingness*, indicated by a person giving high scores to the others.
These effects are modeled for binary-valued relations in the so-called p_1 model (Holland and Leinhardt, 1981; Wasserman and Faust, 1994). In a random effects model for continuous variables, these effects are implied by the formula

$$Y_{fg} = \mu + A_f + B_g + U_{fg} + R_{fg} \, , \tag{11.15}$$

where μ is the population mean, A_f is the outgoingness effect, B_g the popularity effect, U_{fg} the reciprocity effect constrained by the requirement $U_{fg} = U_{gf}$, and R_{fg} the residual. In the terminology of social networks, f is called the *sender* and g the *receiver* of the directed relation Y_{fg}. All effects are assumed to be independent, except that $U_{fg} = U_{gf}$, and that the sender and receiver effects of the same person, A_f and B_f, are correlated.

Like in the preceding subsection, each actor has two effects in this formula: a sender effect and a receiver effect. Since a correlation must be allowed between these effects, it is impossible to use either the factor 'sender' or the factor 'receiver' as a nesting factor. Instead, we work again with dummy variables. The extra reciprocity effect can be modeled by using the *dyad*, i.e., the pair (Y_{fg}, Y_{gf}) of two-way relations between two persons, to define the second level. The dyads are numbered arbitrarily by j running from 1 to $F(F-1)/2$. Within each dyad, the two relations are arbitrarily numbered $i = 1, 2$. So every level-two unit contains only two level-one units. The data structure accordingly is redefined as a two-level structure with outcomes denoted by Z_{ij}.

Dummy variables s_f and r_f are defined, indicating the sender and receiver of each relationship: $s_{fij} = 1$ if person f is the sender of the relationship expressed by variable Z_{ij}, and 0 otherwise; $r_{fij} = 1$ if f is the receiver of relationship Z_{ij}, and 0 otherwise. Note that this implies that $s_{f1j} = r_{f2j}$ and $s_{f2j} = r_{f1j}$.

Recasting the reciprocity effect $U_{fg} = U_{gf}$ for dyad j in the notation U_j, and recasting the residual R_{fg} as R_{ij}, model (11.15) can be reformulated as

$$Z_{ij} = \mu + \sum_{f=1}^{F}(A_f\, s_{fij} + B_f\, r_{fij}) + U_j + R_{ij} \; . \tag{11.16}$$

In other words, the sender, or outgoingness, effect is the random slope of the dummy variables s_f; the receiver, or popularity effect, is the random slope of the dummy variables r_f; and the reciprocity effect is the random intercept at level two. It is again necessary to make restrictions on the variances and covariances of the random effects. Covariances between the effects for different persons f and f' are assumed to be zero. Further, the slopes for all persons have the same variances and covariances:

$$\begin{aligned} &\text{var}(A_1) = \text{var}(A_2) = \ldots = \text{var}(A_F)\;, \\ &\text{var}(B_1) = \text{var}(B_2) = \ldots = \text{var}(B_F)\;, \\ &\text{cov}(A_1, B_1) = \text{cov}(A_2, B_2) = \ldots = \text{cov}(A_F, B_F)\;. \end{aligned} \tag{11.17}$$

It is clear that, if variables associated to the sender, the receiver, or the (directed) pair (f, g) are available, these can be used as variables in the fixed part of (11.16), cf. Snijders and Kenny (1999). There may, of course, be more effects in the social network than the effects expressed in (11.16); e.g., transitivity effects ('a friend of my friend is my friend') or subgroup effects. If such effects are important and cannot be modeled by fixed effects of available covariates, the models of this section are not adequate.

Example 11.2 *Communication between high school teachers.*
In a study of relations between high school teachers, Heyl (1996) asked teachers in 12 schools about their communication on a number of topics. There were 195 responding teachers, nested within schools, and forming 1,544 dyads. Of the many dimensions of contacts and the various covariates investigated, we present here the example of communication about individual pupils and the effect of the teachers' gender on this communication. Frequency of communication between teachers was reported by the teachers on a 5-point scale, ranging from 1 = 'less than 4 times per year' to 5 = '(almost) daily'.

The model is like indicated above, but the schools are also a nesting factor. The schools define level three, and the random effects of senders and receivers are nested within schools. This means that there is no (fourth) dummy level, the number of dummy variables for senders and receivers is the maximum number of teachers in a single school (cf. formula (11.5)), and these dummy variables have random effects at level three. The parameter estimates of a model without explanatory variables (the 'empty model') are reported as Model 1 in Table 11.2.

Except for the school intercept variance, all variance components are large and many times larger than their standard errors. The largest variance component is of the senders: teachers differ strongly in the frequency of communi-

Table 11.2 Estimates for two social network models.

	Model 1		Model 2	
Fixed Effect	Coefficient	S.E.	Coefficient	S.E.
Intercept	2.32	0.08	2.20	0.17
Sender's gender (F)			−0.22	0.12
Receiver's gender (F)			−0.06	0.07
Similar gender (F-F)			0.74	0.16
Random Effect	Var. Comp.	S.E.	Var. Comp.	S.E.
Level-three random effects:				
School intercept	0.01	0.03	0.03	0.03
Sender variance	0.47	0.05	0.42	0.05
Receiver variance	0.18	0.02	0.13	0.02
Sender-receiver covariance	0.12	0.03	0.09	0.02
Level-two random effect:				
Reciprocity	0.19	0.02	0.17	0.01
Level-one variance:				
Residual	0.33	0.01	0.33	0.01
Deviance	7419.0		7237.7	

cation which they report. The reciprocity component also is important: when teacher f reports much communication with teacher g, then it is likely that, conversely, teacher g also reports much communication with f. The correlation between sender and receiver effects is $\rho(A_f, B_f) = 0.12/\sqrt{0.47 \times 0.18} = 0.41$, a considerable value. This means that teachers who report that they communicate much with others are reported about by others in a corresponding way.

The model with fixed effects of teacher's gender is reported as Model 2. The sender's gender, the receiver's gender, and the similarity of these two, all are potentially relevant explanatory variables. These are represented by dummy variables, where the male category is the reference, and where the similarity variable is 1 only if sender as well as receiver are female, and 0 otherwise. The parameter estimates indicate that only the similarity effect is significant: the combination of two female teachers leads to a considerably greater frequency of communication than all three other gender combinations. This gender effect explains part of the variances of the sender, the receiver, and the reciprocity effects.

Undirected relations

Social network data sometimes are symmetric by nature. Continuing the example above, this is the case if the liking between persons f and g is expressed by an outside observer who indicates the value of the mutual liking between the persons; or if, instead of liking, Y_{fg} expresses the objective amount of joint activity by the two persons. In this case the distinction between sender and receiver vanishes, together with the reciprocity effect. What remains of model (11.15) is

$$Y_{fg} = \mu + A_f + A_g + R_{fg} \ . \tag{11.18}$$

In this case it is superfluous to use a second level for the reciprocity effect. A one-level model is sufficient, like in the case of multiple roles. The pairs (f, g) for $f < g$ are arbitrarily renumbered as $i = 1, \ldots, F(F-1)/2$, and the observations Y_{fg} recast as Z_i. Define the dummy variables s_f for $f = 1, \ldots, F$ by

$$s_{fi} = \begin{cases} 1 & \text{if person } f \text{ is involved in pair } i \\ 0 & \text{if person } f \text{ is not involved in this pair.} \end{cases} \tag{11.19}$$

This definition leads to the reformulation of (11.18) as

$$Y_{fg} = \mu + \sum_{f=1}^{F} A_f \, s_{fi} + R_i \ . \tag{11.20}$$

The person effect A_f is the random slope of the dummy variable s_f, and the constraint is that these person effects are uncorrelated and have a common variance.

12 Longitudinal Data

The paradigmatic nesting structure used until now in this book has been the structure of individuals (level-one units) nested in groups (level-two units). The hierarchical linear model is very useful also for analysing repeated measures, or longitudinal data. Since the appearance of the path-breaking paper by Laird and Ware (1982), this is the main type of application of the hierarchical linear model in the biological and medical sciences. This chapter is devoted to the specific two-level models and modeling considerations that are relevant for data where the level-one units are measurement occasions and the level-two units individuals. In Section 12.1.3 a multivariate model is treated which can also be used for multivariate data that do not have a longitudinal nature.

Of the advantages of the hierarchical linear model approach to repeated measures, one deserves to be mentioned here. It is the flexibility to deal with unbalanced data structures, e.g., repeated measures data with fixed measurement occasions where the data for some (or all) individuals is incomplete, or longitudinal data where some or even all individuals are measured at different sets of time points.

The subject of repeated measures analysis is too vast to be treated in one chapter. For a more extensive treatment of this topic we refer to, e.g., Maxwell and Delaney (1990) or Crowder and Hand (1990). This chapter only explains the basic hierarchical linear model formulation of models for repeated measures. In economics this type of model is discussed mainly under the heading of panel data, e.g., Baltagi (1995), Chow (1984), and Hsiao (1995).

This chapter is about the two-level structure of measurements within individuals. When the individuals, in their turn, are nested in groups, the data have a three-level structure: longitudinal measurements nested within individuals nested within groups. Models for such data structures can be obtained by adding the group level as a third level to the models of this chapter. Such three-level models are not explicitly treated in this chapter. However, this three-level extension of the fully multivariate model of Section 12.1.3 is the same as the multivariate multilevel model of Chapter 13.

12.1 Fixed occasions

In fixed occasion designs, there is a fixed set $t = 1, ..., m$ of measurement occasions. This could be, e.g., in a study of some educational or therapeutic program, an intake test, pretest, mid-program test, post-test, and follow-up test ($m = 5$). Another example is a study of attitude change in early adulthood, with attitudes measured shortly after each birthday from the 18th to the 25th year of age. If data are complete, each individual has provided information on all these occasions. It is quite common, however, that data are incomplete. When they are, we assume that the absent data are missing at random, and the fact that they are missing does not itself provide relevant information about the studied phenomena.

The measurement occasions are denoted $t = 1, ..., m$, but each individual may have a smaller number of measurements because of missing data. Y_{ti} denotes the measurement for individual i at occasion t. It is allowed that, for any individual, some measurements are missing. Even individuals with only one measurement do not need to be deleted from the data set, although they contribute, of course, only little information.

Note that, differently from the other chapters, the level-one units now are the measurement occasions indexed by t, while the level-two units are the individuals indexed by i.

The models treated in this section differ primarily with respect to the random part. There are three kinds of motive for choosing between these models. The first motive is that if one works with a random part that is not an adequate description of the dependence between the m measurements (i.e., it does not satisfactorily represent the covariance matrix of these measurements), then the standard errors for estimated coefficients in the fixed part are not reliable. Hence, the tests for these coefficients also are unreliable. This motive points to the least restrictive model, i.e., the fully multivariate model, as the preferred one.

The second motive is the interpretation of the random part. Simpler models often allow a nicer and easier interpretation than more complicated models. This point of view favors the more restrictive models, e.g., the random intercept (compound symmetry) and random slope models.

The third motive is the possibility to get results at all and the desirability of standard errors that are not unnecessarily large (while not having a bias at the same time). If the amount of data is relatively small, the estimation algorithms may not converge for models that have a large number of parameters, such as the fully multivariate model. Or even if the algorithm does converge, having a bad 'data-to-parameter ratio' may lead to large standard errors. This suggests that, if one has a relatively small data set, one should not entertain models with too many parameters.

Concluding, one normally should use the simplest model for the random part (i.e., the model having the least parameters and being the most restrictive) that still yields a good fit to the data. The model for the random part can be selected with the aid of deviance tests (see Section 6.2).

12.1.1 *The compound symmetry model*

The classical model for repeated measures, called the compound symmetry model (e.g., Maxwell and Delaney, 1990), is the same as the random intercept model. It is sometimes referred to as the random effects model or the mixed model, but one should realize that it is only a simple specification of the many possibilities of the random effects model, or mixed model. If there are no explanatory variables except for the measurement occasions, i.e., the design is a pure ***within-subjects design***, the expected value for measurement occasion t can be denoted μ_t and this model can be expressed by

$$Y_{ti} = \mu_t + U_{0i} + R_{ti} \ . \tag{12.1}$$

The usual assumptions are made: the U_{0i} and R_{ti} are independent normally distributed random variables with expectations 0 and variances τ_0^2 for U_{0i} and σ^2 for R_{ti}.

To fit this model, note that the fixed part does not contain a constant term, but is based on m dummies for the m measurement occasions. This can be expressed by the following formula. Let d_{hti} be m dummy variables, defined for $h = 1, ..., m$ by

$$d_{hti} = \begin{cases} 1 & t = h \ , \\ 0 & t \neq h \ . \end{cases} \tag{12.2}$$

Then the fixed part μ_t in (12.1) can be written as

$$\mu_t = \sum_{h=1}^{m} \mu_h \, d_{hti}$$

and the compound symmetry model can be formulated as

$$Y_{ti} = \sum_{h=1}^{m} \mu_h \, d_{hti} + U_{0i} + R_{ti} \ , \tag{12.3}$$

the usual form of the fixed part of a hierarchical linear model.

Example 12.1 *Development of teachers' evaluations.*
Brekelmans and Créton (1993) made a study of the development over time of evaluations of teachers by their pupils. Starting from the first year of their teaching career, teachers were evaluated on their interpersonal behavior in the classroom. This happened repeatedly, at intervals of about one year. In this example, results are presented about the 'proximity' dimension, representing the degree of cooperation or closeness between a teacher and his or her students. The higher the proximity score of a teacher, the more cooperation is perceived by his or her students.

There are four measurement occasions: after 0, 1, 2, and 3 years of experience. Thus, the time variable t assumes the values 0 through 3. A total of 51 teachers was studied. The number of observations for the 4 moments decreased from 46 at $t = 0$ to 32 at $t = 3$. The non-response at various moments may be considered to be random.

First two models are considered with a random intercept only, i.e., with a compound symmetry structure for the covariance matrix. The first is the empty model of Chapter 4. In this model it is assumed that the 4 measurement occasions have the same population mean. The second model is model (12.1), which allows the means to vary freely over time. In this case, writing out (12.3) results in the model

$$Y_{ti} = \mu_1 \, d_{1ti} + \mu_2 \, d_{2ti} + \mu_3 \, d_{3ti} + \mu_4 \, d_{4ti} + U_{0i} + R_{ti} \ .$$

Applying definition (12.2) implies that the expected score for an individual, e.g., at time $t = 2$, is

$$\hat{Y}_{ti} = \mu_1 \times 0 + \mu_2 \times 1 + \mu_3 \times 0 + \mu_4 \times 0 = \mu_2 \ .$$

Table 12.1 Estimates for random intercept models.

	Model 1		Model 2	
Fixed Effect	Coefficient	S.E.	Coefficient	S.E.
μ_0 Mean at time 0	0.648	0.053	0.585	0.063
μ_1 Mean at time 1	0.648	0.053	0.718	0.067
μ_2 Mean at time 2	0.648	0.053	0.672	0.067
μ_3 Mean at time 3	0.648	0.053	0.639	0.070
Random Effect	Parameter	S.E.	Parameter	S.E.
Level-two (i.e., individual) variance:				
$\tau_0^2 = \text{var}(U_{0i})$	0.121	0.029	0.121	0.029
Level-one (i.e., occasion) variance:				
$\sigma^2 = \text{var}(R_{ti})$	0.074	0.010	0.072	0.010
Deviance	123.39		118.35	

The results are in Table 12.1. It is dangerous to trust results for the compound symmetry model before its assumptions have been tested, because standard errors of fixed effects may be incorrect if one uses a model with a random part that has an unsatisfactory fit. Later we will fit more complicated models and show how the assumption of compound symmetry can be tested.

The results suggest that individual (level two) variation is more important than random differences between occasions (level one variation). Further, the mean seems to increase from time $t = 0$ to time $t = 1$ and then to decrease again. However, the deviance test for the difference in mean between the four time points is not significant: $\chi^2 = 123.39 - 118.35 = 5.04$, $d.f. = 3$, $p > 0.10$.

Just like in our treatment of the random intercept model in Chapter 4, any number of relevant explanatory variables can be included in the fixed part. Often there are individual-dependent explanatory variables Z_k, $k = 1, ..., q$, e.g., traits or background characteristics. If these variables are categorical, they can be represented by dummy variables. Such individual-dependent variables are level-two variables in the multilevel approach, and they are called *between-subjects variables* in the terminology of repeated measures analysis. In addition, there often are one or more numerical variables describing the measurement occasion. Denote such a variable by $s(t)$; this can be, e.g., a measure for the time elapsing between the occasions, or the rank order of the occasion. Such a variable is called a *within-subjects variable.* In addition to the between-subjects and within-subjects variables having main effects, they may have interaction effects with each other. These within–between interactions are a kind of cross-level interactions in the multilevel terminology. Substantive interest in repeated measures analysis often focuses on these interactions.

When the main effect parameter of z_{ki} is denoted α_k and the interaction effect parameter between z_{ki} and $s(t)$ is denoted γ_k, the model is given by

$$Y_{ti} = \sum_{k=1}^{q} \alpha_k \, z_{ki} + \sum_{k=1}^{q} \gamma_k \, z_{ki} \, s(t) + \sum_{h=1}^{m} \mu_h \, d_{hti} + U_{0i} + R_{ti} \, . \quad (12.4)$$

The fixed part is an extension of (12.1) or the equivalent form (12.3), but the random part still is the same. Therefore this is still called a compound symmetry model. Inclusion of this kind of variables in the fixed part suggests, however, that the random part could also contain random slopes of the time variables such as $s(t)$. Therefore we defer giving an example of the fixed part (12.4) until after the treatment of such a random slope.

Classical analysis of variance methods are available to estimate and test parameters of the compound symmetry model if all data are complete (e.g., Maxwell and Delaney, 1990). The hierarchical linear model formulation of this model, and the algorithms and software available, permit the statistical evaluation of this model also for incomplete data without any additional complication.

Covariance matrix

In the fixed occasion design one can talk about the complete data vector

$$Y^{c} = \begin{pmatrix} Y_{1i} \\ \cdot \\ \cdot \\ \cdot \\ Y_{mi} \end{pmatrix} .$$

Even if there would be no subject at all with complete data, still the complete data vector would make sense from a conceptual point of view.

The compound symmetry model (12.1), (12.3) or (12.4) implies that for the complete data vector, all variances are equal and also all covariances are equal. The expression for the covariance matrix of the complete data vector, conditional on the explanatory variables, is the $m \times m$ matrix

$$\Sigma(Y^{c}) = \begin{pmatrix} \tau_0^2 + \sigma^2 & \tau_0^2 & \cdot\;\cdot\;\cdot & \tau_0^2 \\ \tau_0^2 & \tau_0^2 + \sigma^2 & \cdot\;\cdot\;\cdot & \tau_0^2 \\ \tau_0^2 & \tau_0^2 & \cdot\;\cdot\;\cdot & \tau_0^2 \\ \cdot & \cdot & \cdot\;\cdot\;\cdot & \cdot \\ \cdot & \cdot & \cdot\;\cdot\;\cdot & \cdot \\ \tau_0^2 & \tau_0^2 & \cdot\;\cdot\;\cdot & \tau_0^2 + \sigma^2 \end{pmatrix} ; \quad (12.5)$$

cf. p. 46. This matrix is referred to as the compound symmetry covariance matrix. In this covariance matrix, all residual variances are the same and all residual within-subject correlations are equal to

$$\rho_{\mathrm{I}} = \rho\{Y_{ti}, Y_{si}\} = \frac{\tau_0^2}{\tau_0^2 + \sigma^2} \, , \qquad (t \neq s) \quad (12.6)$$

the residual intraclass correlation which we encountered already in Chapter 4.

The compound symmetry model is a very restrictive model, and often an unlikely one. For example, if measurements are ordered in time, the correlation often is larger between nearby measurements than between measurements that are far apart. (Only for $m = 2$ this condition is not so very restrictive. In this case, formula (12.5) only means that the two measurements have the same variance and a positive correlation.)

Example 12.2 *Covariance matrix for teachers' evaluations.*
In Table 12.1 the estimates $\sigma^2 = 0.074$ and $\tau_0^2 = 0.121$ were obtained. This implies that the measurement variances are $0.121 + 0.074 = 0.195$ and the within-subjects correlations are $\rho_{\text{I}} = 0.121/0.195 = 0.621$. This number, however, is conditional on the validity of the compound symmetry model; we shall see below that the compound symmetry model does not fit very well to these data.

12.1.2 Random slopes

There are various ways in which the assumption of compound symmetry (which states that the covariance matrix has constant variances and also constant covariances, cf. (12.5)), can be relaxed. In the hierarchical linear model framework, the simplest way is to include one or more random slopes in the model. This makes sense if there is some meaningful dimension, such as time, underlying the measurement occasions. It will be assumed here that the index t used to denote the measurement occasions is a meaningful numerical variable, and that it is relevant to consider the regression of Y on t. This variable will be referred to as 'time'. It is easy to modify the notation so that some other numerical function of the measurement occasion t gets a random slope. It is assumed further that there is some meaningful reference value for t, denoted t_0. This could refer, e.g., to one of the time points, such as the first. The choice of t_0 affects only the parameter interpretation, not the fit of the model.

Since the focus now is on the random rather than on the fixed part, the precise formulation of the fixed part is left out of the formulae.

The model with a random intercept and a random slope for t is given by

$$Y_{ti} = \text{ fixed part } + U_{0i} + U_{1i}\,(t - t_0) + R_{ti}\ . \qquad (12.7)$$

This model means that the rates of increase have a random, individual-dependent component U_{1i}, in addition to the individual-dependent random deviations U_{0i} which affect all values Y_{ti} in the same way. The random effect of time can also be described as a random time-by-individual interaction.

The value t_0 is subtracted from t in order to let the intercept variance refer not to the (possibly meaningless) value $t = 0$ but to the reference point $t = t_0$, cf. p. 69. The variables (U_{0i}, U_{1i}) are assumed to have a joint bivariate normal distribution with expectations 0, variances τ_0^2 and τ_1^2, and covariance τ_{01}.

The variances and covariances of the measurements Y_{ti}, conditional on the explanatory variables, are given now by

$$\begin{aligned} \text{var}(Y_{ti}) &= \tau_0^2 + 2\tau_{01}\,(t - t_0) + \tau_1^2\,(t - t_0)^2 + \sigma^2 , \qquad (12.8) \\ \text{cov}(Y_{ti}, Y_{si}) &= \tau_0^2 + \tau_{01}\,\{(t - t_0) + (s - t_0)\} + \tau_1^2\,(t - t_0)(s - t_0) , \end{aligned}$$

where $t \neq s$. These formulae express the fact that the variances and covariances of the outcome variables are variable over time. This is called heteroscedasticity. The variance is minimal at $t = t_0 - \tau_{01}/\tau_1^2$ (if t would be allowed to assume any value). Further, the correlation between different measurements depends on their spacing (as well as on their position).

Extensions to more than one random slope are obvious; e.g., a second random slope could be given to the squared value $(t - t_0)^2$. In this way, one can perform a ***polynomial trend analysis*** to improve the fit of the random part. This means that one fits random slopes for a number of powers of $(t - t_0)$ to obtain a model that has a good fit to the data and where unexplained differences between individuals are represented as random individual-dependent regressions of Y on $(t - t_0)$, $(t - t_0)^2$, $(t - t_0)^3$, etc. Other functions than polynomials can also be used, e.g., splines (see Section 12.2.2). Polynomial trend analysis is discussed also in Maas and Snijders (1999).

Example 12.3 *Random slope of time in teachers' evaluations.*
Continuing the earlier example of teachers' evaluations, note that the observations are spaced a year apart, and a natural variable for the occasions is $s(t) = t$, the occasion number. This variable is a time dimension which counts the number of years of experience and will be referred to as 'experience'. A random slope of experience now is added (Model 3). The reference value for the time dimension, denoted t_0 in formula (12.7), is taken as $t_0 = 0$, corresponding to the first measurement occasion, where the teachers do not yet have any experience.

It is investigated now if the teacher's gender can explain part of the differences between teachers in average level and in the rate of change. Thus, gender is used as a between-subjects variable. A dummy Z is used with value 0 for men and 1 for women. The effect of gender on the rate of change, i.e., the interaction effect of gender with experience, is represented by the product of Z with t. The resulting model is an extension of a model of the type of (12.4) with a random slope:

$$Y_{ti} = \mu_t + \alpha\, z_i + \gamma\, z_i\, t + U_{0i} + U_{1i}\, t + R_{ti} \,. \qquad (12.9)$$

Parameter α is the main effect for gender, while γ is the interaction effect between gender and experience.

The results are in Table 12.2. Comparing the deviances of Models 2 and 3 shows that the random slope of experience is quite significant: $\chi^2 = 15.69$, $d.f. = 2$, $p < 0.001$. This implies that the compound symmetry model, Model 2 of Table 12.1, is not an adequate model for these data. The effect of gender and the gender × experience interaction, however, are not significant as can be concluded from the t-ratios $-0.133/0.169 = -0.79$ and $0.089/0.057 = 1.56$.

The fitted covariance for the complete data vector under Model 3 has elements given by (12.8). Filling in the estimated parameters τ_0^2, τ_1^2, and τ_{01}

Table 12.2 Estimates for random slope models.

		Model 3		Model 4	
Fixed Effect		Coefficient	S.E.	Coefficient	S.E.
μ_0	Effect time 0	0.574	0.076	0.610	0.090
μ_1	Effect time 1	0.721	0.066	0.733	0.075
μ_2	Effect time 2	0.665	0.057	0.655	0.065
μ_3	Effect time 3	0.633	0.058	0.601	0.065
α	Main effect gender			−0.133	0.169
γ	Interaction gender × experience			0.089	0.057
Random Effect		Parameter	S.E.	Parameter	S.E.
Level-two variation:					
τ_0^2	Intercept variance	0.233	0.056	0.239	0.057
τ_1^2	Slope variance	0.017	0.006	0.017	0.007
τ_{01}	Intercept-slope covariance	−0.052	0.017	−0.053	0.017
Level-one (i.e., occasion) variance:					
σ^2	Residual variance	0.048	0.009	0.047	0.009
Deviance		102.66		99.91	

from Table 12.2 into (12.8) yields the covariance matrix

$$\widehat{\Sigma}(Y^c) = \begin{pmatrix} 0.281 & 0.181 & 0.129 & 0.077 \\ 0.181 & 0.194 & 0.111 & 0.076 \\ 0.129 & 0.111 & 0.141 & 0.075 \\ 0.077 & 0.076 & 0.075 & 0.122 \end{pmatrix}, \qquad (12.10)$$

and the correlation matrix

$$\hat{R}(Y^c) = \begin{pmatrix} 1.000 & 0.775 & 0.648 & 0.416 \\ 0.775 & 1.000 & 0.671 & 0.494 \\ 0.648 & 0.671 & 1.000 & 0.572 \\ 0.416 & 0.494 & 0.572 & 1.000 \end{pmatrix}.$$

These matrices show that the variance becomes smaller as experience grows and correlations are smaller between wider separated time points. However, these values are conditional on the validity of the model with one random slope. We return to these data below, and then will investigate the adequacy of this model by testing it against the fully multivariate model.

12.1.3 The fully multivariate model

What is the use of restrictions, such as compound symmetry, on the covariance matrix of a vector of longitudinal measurements? There was a time (say, before 1980) when the compound symmetry model was used for repeated measures because, in practice, it was impossible to get results for other models. In that time, also a complete data matrix was required. These limitations were overcome gradually between 1970 and 1990.

A more compelling argument for restrictions on the covariance matrix is that when the amount of data is limited, the number of statistical parameters should be kept small to refrain from overfitting and to avoid convergence problems in the calculation of the estimates. Another, more appealing, argument is that sometimes the parameters of the models with restricted covariance matrices have nice interpretations. This is the case, e.g., for the random slope variance of model (12.7). But when there is enough data, one can also fit a model without restrictions on the covariance matrix, the ***fully multivariate model***. This also provides a benchmark to assess the goodness of fit of the models that do have restrictions on the covariance matrix.

Some insight in the extent to which a given model constrains the covariance matrix is obtained by looking at the number of parameters. The covariance matrix of a m-dimensional vector has $m(m+1)/2$ free parameters (m variances and $m(m-1)/2$ covariances). The compound symmetry model has only 2 parameters. The model with one random slope has 4 parameters; the model with q random slopes has $\{(q+1)(q+2)/2\}+1$ parameters, namely, $(q+1)(q+2)/2$ parameters for the random part at level two and 1 parameter for the variance of the random residual. This shows that using some random slopes will quickly increase the number of parameters and thereby lead to a better fitting covariance matrix. The maximum number of random slopes in a conventional random slope model for the fixed occasions design is $q = m-2$, because with $q = m-1$ there is one parameter too many.

This suggests how to achieve a perfect fit for the covariance matrix. The fully multivariate model is formulated as a model with a random intercept and $m-1$ random slopes at level two, and without a random part at level one. Alternatively, the random part at level two may consist of m random slopes and no random intercept. It is clear that, when all variances and covariances between m random slopes are free parameters, the number of parameters in the random part of the model is indeed $m(m+1)/2$.

The fully multivariate model is little more than a tautology,

$$Y_{ti} = \text{ fixed part } + U_{ti} \; . \tag{12.11}$$

This model is reformulated more recognizably as a hierarchical linear model by the use of dummy variables indicating the measurement occasions. These dummies were defined already in (12.2):

$$d_{hti} = \begin{cases} 1 & t = h \, , \\ 0 & t \neq h \, . \end{cases} \tag{12.2}$$

This leads to the formulation

$$Y_{ti} = \text{ fixed part } + \sum_{h=1}^{m} U_{hi} \, d_{hti} \; . \tag{12.12}$$

The variables U_{ti} for $t = 1, ..., m$ are random at level two, with expectations 0 and an unconstrained covariance matrix. (This means that all variances and all covariances must be freely estimated from the data.) This model does not have a random part at level one! It follows immediately from (12.11) that the covariance matrix of the complete data vector, conditional on the explanatory variables, is identical to the covariance matrix of $(U_{1i}, ..., U_{pi})$. This model for the random part is ***saturated*** in the sense that it yields a perfect fit for the covariance matrix.

The multivariate model of this section can be applied also to data that do not have a longitudinal nature. This means that the m variables can be completely different, measured on different scales, provided that they have (approximately) a multivariate normal distribution. Thus, this model and the corresponding multilevel software offers a way for the multivariate analysis of incomplete data (provided that missingness is at random as defined, e.g., in Little and Rubin, 1987).

Example 12.4 *Incomplete paired data.*
For the comparison of the means of two variables measured for the same individuals, the paired-samples t-test is a standard procedure. The fully multivariate model provides the possibility to carry out a similar test if the data are incomplete. This is an example of the most simple multilevel data structure: all level-two units (individuals) contain either 1 or 2 level-one units (measurements). Such a data structure may be called an incomplete pretest–post-test design if one measurement was taken before, and the other after, some kind of intervention.

In the example of the teachers' evaluations, consider the test of the null hypothesis that the average evaluation on the proximity scale is equal on the two time points $t = 0$ and $t = 1$. Of the 49 respondents for whom a measurement on at least one of these moments is available, there are 35 who have measurements for both moments, 11 who have a measurement only at $t = 0$ and 3 who have a measurement only at $t = 1$. With the multilevel approach to multivariate analysis, all these data can be used. The REML estimation method is used (cf. Section 4.6) because this will reproduce exactly the conventional t-test if the data are complete (i.e., all individuals have measurements for both variables).

The model fitted is

$$Y_{ti} = \gamma_0 + \gamma_1 \, d_{1ti} + U_{ti}$$

where the dummy variable d_1 equals 1 or 0, respectively, depending on whether or not $t = 1$. The null hypothesis that the means are identical for the two time points can be represented by '$\gamma_1 = 0$'.

Table 12.3 Estimates for incomplete paired data.

Fixed Effect		Coefficient	S.E.
γ_0	Constant term	0.577	0.081
γ_1	Effect time 1	0.151	0.064
Deviance		95.05	

The results are in Table 12.3. The estimated covariance matrix of the complete data vector is:

$$\widehat{\Sigma}(Y^c) = \begin{pmatrix} 0.309 & 0.166 \\ 0.166 & 0.184 \end{pmatrix} .$$

The test of the equality of the two means, which is the test of γ_1, is significant ($t = 0.151/0.064 = 2.36$, two-sided $p < 0.05$).

Example 12.5 *Fully multivariate model for teachers' evaluations.*
Continuing the example of teachers' evaluations, we now fit the fully multivariate model to the data for the four time points $t = 0, 1, 2, 3$. First we consider the model that may be called the *two-level multivariate empty model*, of which the fixed part contains only the dummies for the effects of the measurement occasions. This is an extension of Model 3 of Table 12.2. The estimates of the fixed effects are presented in Table 12.4.

Table 12.4 The empty multivariate model for the teachers' evaluations.

Fixed Effect		Coefficient	S.E.
μ_0	Mean at time 0	0.574	0.079
μ_1	Mean at time 1	0.717	0.066
μ_2	Mean at time 2	0.650	0.052
μ_3	Mean at time 3	0.657	0.057
Deviance		91.10	

The estimated covariance matrix of the complete data vector is:

$$\widehat{\Sigma}(Y^c) = \begin{pmatrix} 0.303 & 0.173 & 0.111 & 0.112 \\ 0.173 & 0.196 & 0.098 & 0.118 \\ 0.111 & 0.098 & 0.120 & 0.109 \\ 0.112 & 0.118 & 0.109 & 0.141 \end{pmatrix} . \qquad (12.13)$$

These estimates are the REML (residual maximum likelihood) estimates (cf. Section 4.6) of the parameters of a multivariate normal distribution with incomplete data. They differ slightly from the estimates that would be obtained in the usual way, where means and variances are calculated from available data with pairwise deletion of missing values. The REML estimates are more efficient in the sense of having smaller standard errors.

An eyeball comparison with the fitted covariance matrix for the model with one random slope, given in (12.10), suggests that the differences are minor. The deviance difference is $\chi^2 = 102.66 - 91.10 = 11.56$, $d.f. = 6$ (the covariance matrix of Model 3 has 4 free parameters, this covariance matrix has 10) with $0.05 < p < 0.10$. Thus, the fully multivariate model results in a fit that is almost, but not quite, significantly better than the fit of the model with one random slope. Covariance matrix 12.13 suggests that the main difference between the measurements is that the first measurement ($t = 0$) has a larger variance (0.303) than the later measurements. The second measurement also has a somewhat larger variance (0.196) than the last two measurements (0.120 and 0.141, respectively).

The fully multivariate model may have a slightly better fit than the random slope model, but an advantage of the random slope model is the clearer interpretation of the random part in terms of between-subject differences. The rate of change in the models of Table 12.2 has a random component U_{1i} with standard deviation $\sqrt{0.017} = 0.13$. This reflects substantial unexplained differences between the subjects, and suggests that it is worthwhile to look for within–between interaction effects. Such an interpretation is not directly obvious from the results of the fully multivariate model.

As a next step the effect of the teachers' self-perception on the same proximity scale is studied. This is a changing explanatory variable, measured at the same moments as the pupils' evaluations which constitute the dependent variable.

Table 12.5 The multivariate model with the teachers' self-evaluation.

Fixed Effect		Coefficient	S.E.
μ_0	Effect time 0	0.212	0.087
μ_1	Effect time 1	0.375	0.073
μ_2	Effect time 2	0.257	0.075
μ_3	Effect time 3	0.288	0.070
α	Self-evaluation	0.469	0.070
Deviance		56.42	

The estimates of the fixed effects are presented in Table 12.5. The estimated residual covariance matrix of the complete data vector is:

$$\widehat{\Sigma}(Y^c) = \begin{pmatrix} 0.228 & 0.115 & 0.070 & 0.044 \\ 0.115 & 0.120 & 0.049 & 0.057 \\ 0.070 & 0.049 & 0.092 & 0.061 \\ 0.044 & 0.057 & 0.061 & 0.084 \end{pmatrix} . \tag{12.14}$$

The effect of the self-evaluation is strongly significant ($t = 0.469/0.070 = 6.70, p < 0.0001$), which indicates that teachers and pupils have at least some agreement in their perceptions of the teacher's degree of cooperation. The time effects in this model are controlled for the effect of the self-evaluation. The residual effect of time, controlling for the teachers' own perceptions, is highest at time $t = 1$.

Concluding remarks about the fully multivariate model

There is nothing special about this formulation as a hierarchical linear model of a fully multivariate model for repeated measures with a fixed occasion design. It is just a mathematical formula. What is special, is that available algorithms and software for multilevel analysis accept this formulation, even without a random part at level one, and can calculate ML or REML parameter estimates. In this way, multilevel software can compute ML or REML estimates for multivariate normal distributions with incomplete data, and also for multivariate regression models with incomplete data and with sets of explanatory variables that are different for the different dependent variables. Methods for such models have been in existence for a longer period (see Little and Rubin, 1987), but software was not readily available before the development of multilevel software.

The use of the dummy variables (12.2) has the nice feature that the covariance matrix obtained for the random slopes is exactly the covariance matrix for the complete data vector. However, one could also give the random slopes to other variables. The only requirement is that there are random slopes for m linearly independent variables, depending only on the measurement occasion and not on the individual. For example, one could use powers of $(t - t_0)$ where t_0 may be any meaningful reference point, e.g., the average of all values for t. This means that one uses variables

$$d_{1ti} = 1$$
$$d_{hti} = (t - t_0)^{h-1} \ , \quad (h = 2, ..., m) \ ,$$

still with model specification (12.12). Since the first 'variable' is constant, this effectively means that one uses a random intercept and $m - 1$ random slopes.

Each model for the random part with a restricted covariance matrix is a submodel of the fully multivariate model. Therefore the fit of such a restricted model can be tested by comparing it to the fully multivariate model by means of a likelihood ratio (deviance) test (see Chapter 6).

For complete data, an alternative to the hierarchical linear model exists in the form of multivariate analysis of variance (MANOVA) and multivariate regression analysis. This is documented in many textbooks (e.g., Maxwell and Delaney, 1990; Stevens, 1996) and implemented in standard software such as SPPS and SAS. The advantage of these methods is the fact that the tests are exact whereas the tests in the hierarchical linear model formulation are approximate. For incomplete multivariate data, however, exact methods are not available. Maas and Snijders (1999) (also see Snijders and Maas, 1996) elaborate the correspondence between the MANOVA approach and the hierarchical linear model approach.

12.1.4 Multivariate regression analysis

Because of the unrestricted multivariate nature of the fully multivariate model, it is not required that the outcome variables Y_{ti} are repeated measurements of conceptually the same variable. This model is applicable to any set of multivariate measurements for which a multivariate normal distribution is an adequate model. Thus, the multilevel approach yields estimates and tests for normally distributed multivariate data with randomly missing observations (also see Maas and Snijders, 1999).

The fully multivariate model is also the basis for multivariate regression analysis, but then the focus usually is on the fixed part. There are supposed to be individual-dependent explanatory variables Z_1 to Z_q. If the regression coefficients of the m dependent variables on the Z_h are allowed all to be different, the regression coefficient of outcome variable Y_t on explanatory variable Z_h being denoted γ_{ht}, then the multivariate regression model with possibly incomplete data can be formulated by

$$Y_{ti} = \mu_t + \sum_{t=1}^{m} \sum_{h=1}^{q} \gamma_{ht} \, d_{hti} \, z_{hi} + \sum_{h=1}^{m} U_{hi} \, d_{hti} \ . \qquad (12.15)$$

This shows that the occasion dummies are fundamental, not only because they have the random slopes, but also because in the fixed part all variables are multiplied by these dummies. If some of the Z_k variables are not to be used for all dependent variables, then the corresponding cross-product terms $d_{hti} z_{hi}$ can be dropped from (12.15).

12.1.5 Explained variance

Explained variance for longitudinal data in fixed occasion designs can be defined analogous to explained variance for grouped data as defined in Section 7.1.

The proportion of explained variance at level one can be defined as the proportional reduction in prediction error for individual measurements, averaged over the m measurement occasions. A natural baseline model is provided by a model which can be represented by

$$Y_{it} = \mu_t + \text{random part},$$

i.e., the fixed part depends on the measurement occasion but not on other explanatory variables, and the random part is chosen so as to provide a good fit to the data.

The proportion of explained variance at level one, R_1^2, then is the proportional reduction in the average residual variance,

$$\frac{1}{m}\sum_{t=1}^{m} \text{var}(Y_{ti}) \ ,$$

when going from the baseline model to the model containing the explanatory variables.

The proportion of explained variance at level two, denoted R_2^2, can be defined as the proportional reduction in prediction error for the average over the m measurements. Thus, it is given by the proportional reduction in the residual variance of the average,

$$\text{var}(\overline{Y}_{.i}) \ .$$

If the random part of the compound symmetry model is adequate, the baseline model is (12.1). In this case the definitions of R_1^2 and R_2^2 are just like in Section 7.1, taking $n = m$ in the definition of R_2^2.

If the compound symmetry model is not adequate one could use, at the other end of the spectrum, the multivariate random part. This yields the baseline model

$$Y_{ti} = \mu_t + U_{ti} = \sum_{h=1}^{m} \mu_h \, d_{hti} + \sum_{h=1}^{m} U_{hi} \, d_{hti}$$

which is the fully multivariate model without covariates, cf. (12.11) and (12.12).

In all models for fixed occasion designs, the calculation of the proportions of explained variance can be related to the fitted complete data covariance matrix, $\widehat{\Sigma}(Y^c)$. The value of R_1^2 is the proportional reduction in the sum of diagonal values of this matrix when going from the baseline model to the model including the explanatory variables. For the explained variance at level two, note that the variance of the average of a vector of random variables is equal to the average of all elements of the covariance matrix. Therefore, R_2^2 is equal to the proportional reduction in the sum of *all* values of the fitted complete data covariance matrix.

Example 12.6 *Explained variance for teacher's evaluations.*
Recall that the dependent variable on the examples of this chapter is the evaluation of the teacher by the pupils on the proximity scale. We continue this example, now computing the proportion of variance explained by the teacher's own evaluation of him- or herself on this same scale, and the interaction of the self-evaluation with experience.

First suppose that the compound symmetry model is used. The baseline model then is Model 2 in Table 12.1. The variance per measurement is 0.121 + 0.072 = 0.193. The variance of the average over the $m = 4$ measurements is 0.121 + 0.072/4 = 0.139.

Table 12.6 Estimates for models with self-evaluation.

Fixed Effect		Coefficient	S.E.
μ_0	Effect time 0	0.188	0.082
μ_1	Effect time 1	0.345	0.082
μ_2	Effect time 2	0.232	0.090
μ_3	Effect time 3	0.233	0.088
α	Effect of self-evaluation	0.515	0.081
Random Effect		**Parameter**	**S.E.**
Level-two (i.e., individual) variance:			
$\tau_0^2 = \text{var}(U_{0i})$		0.069	0.019
Level-one (i.e., occasion) variance:			
$\sigma^2 = \text{var}(R_{ti})$		0.066	0.009
Deviance		84.26	

The results of the compound symmetry model with the effect of self-evaluation are in Table 12.6. In comparison with Model 1 of Table 12.1, inclusion of the fixed effect of self-evaluation has led especially to a smaller variance at the individual level. The residual variance per measurement is 0.069 + 0.066 = 0.135. The variance of the average over the $m = 4$ measurements is 0.069 + 0.066/4 = 0.085. Thus, the proportion of variance explained for level one is $R_1^2 = 1 - (0.135/0.193) = 0.30$, while the proportion of variance explained for level two is $R_2^2 = 1 - (0.085/0.139) = 0.39$.

Next suppose that the fully multivariate model is used. The estimated covariance matrix in the fully multivariate model is given in (12.13), the residual covariance matrix when controlling for the effect of self-evaluation is (12.14). The sum of diagonal values is 0.760 for the first covariance matrix and 0.524 for the second. Hence, calculated on the basis of the fully multivariate model, $R_1^2 = 1 - (0.524/0.760) = 0.31$. The sum of all values is 2.202 for the first and 1.316 for the second matrix. This leads to $R_2^2 = 1 - (1.316/2.202) = 0.40$. Comparing this to the values obtained above shows that for the calculations of R^2 it does not make much difference which random part is being used. The calculations using the fully multivariate model are more reliable, but in most cases the simpler calculations using the compound symmetry ('random intercept') model will lead to almost the same values for R^2.

12.2 Variable occasion designs

In data collection designs with variable measurement occasions, there is not such a thing as a complete data vector. The data are ordered according to some underlying dimension, e.g., time, and for each individual data are recorded at some set of time points which is not necessarily related to the time points at which the other individuals are observed. For example, body lengths are recorded at a number of moments during childhood and adolescence, the moments being determined by convenience rather than strict planning.

The notation can be the same as in the preceding section: for individual i, the dependent variable Y_{ti} is measured at occasions $s = 1, ..., m_i$. The time of measurement for Y_{ti} is t. This 'time' variable can refer to age, clock time, etc., but also to some other dimension such as one or more spatial dimensions, the concentration of a poison, etc.

The number of measurements per individual, m_i, can be anything. It is not a problem that some individuals contribute only one observation. This number must not in itself be informative about the studied process, however. Therefore it is not allowed that the observation times t are defined as the moments when some event occurs (e.g., change of job); such data collection designs should be investigated by means of event history models. Greater numbers m_i give more information about intra-individual differences, of course, and with larger average m_i's one will be able to fit models with a more complicated, and more precise, random part.

Because of the unbalanced nature of the data set, the random intercept and slope models are easily applicable, but the other models of the preceding section either have no direct analogue, or an analogue that is considerably more complicated. This section is restricted to random slope models, and follows the approach of Snijders (1996), where further elaboration and background material may be found. There exist more introductions to multilevel models for longitudinal data with variable occasion designs, e.g., Raudenbush (1995).

12.2.1 Populations of curves

An attractive way to view repeated measures in a variable occasion design, ordered according to an underlying dimension t (referred to as time), is as observations on a ***population of curves.*** (This approach can be extended to two- or more-dimensional ordering principles.) The variable Y for individual i follows a development represented by the function $F_i(t)$, and the population of interest is the population of curves F_i. The observations yield snapshots of this function on a finite set of time points, with superimposed residuals R_{ti} which represent incidental deviations and measurement error:

$$Y_{ti} = F_i(t) + R_{ti} \qquad (s = 1, ..., m_i). \tag{12.16}$$

Statistical modeling consists of determining an adequate class of functions F_i and investigating how these functions depend on explanatory variables.

12.2.2 Random functions

It makes sense here to consider models with functions of t as the only explanatory variables. This can be regarded as a model for random functions that has the role of a baseline model comparable to the role of the empty model in Chapter 4. Modeling can proceed by first determining an adequate random function model and subsequently incorporating the individual-based explanatory variables.

One could start by fitting the empty model, i.e., the random intercept model without any explanatory variables, but this is such a trivial model when one is modeling curves, that it may also be skipped. The next simplest model is a linear function,

$$F_i(t) = \beta_{0i} + \beta_{1i}\,(t - t_0)\ .$$

The value t_0 is subtracted for the same reasons as in model (12.7) for the fixed occasion design. It can be some reference value within the range of observed values for t.

If the β_{hi}'s ($h = 0, 1$) are split into their population average[1] γ_{h0} and the individual deviation $U_{hi} = \beta_{hi} - \gamma_{h0}$, similar to (5.2), this gives the following model for the observations:

$$Y_{ti} = \gamma_{00} + \gamma_{10}\,(t - t_0) + U_{0i} + U_{1i}\,(t - t_0) + R_{ti}\ . \tag{12.17}$$

The population of curves now is characterized by the bivariate distribution of the intercepts β_{0i} at time $t = t_0$, and the slopes β_{1i}. The average intercept at $t = t_0$ is γ_{00} and the average slope is γ_{10}. The intercepts have variance $\tau_0^2 = \text{var}(U_{0i})$, the slopes have variance $\tau_1^2 = \text{var}(U_{1i})$, and the intercept-slope covariance is $\tau_{01} = \text{cov}(U_{0i}, U_{1i})$. In addition, the measurements exhibit random deviations from the curve with variance $\sigma^2 = \text{var}(R_{ti})$. This level-one variance represents deviations from linearity together with measurement inaccuracy, and can be used to assess the fit of linear functions to the data.

Even when this model fits only moderately well, it can have a meaningful interpretation, because the estimated slope and slope variance still will give an impression of the average increase of Y per unit of time.

Example 12.7 *Retarded growth.*

In several examples in this section we present results for a data set of children with retarded growth. More information about this research can be found in Rekers-Mombarg et al. (1997). The data set concerns children who visited a paediatrician-endocrinologist because of growth retardation. Length measurements are available at irregular ages varying between 0 and 30 years. In the present example we consider the measurements for ages from 5.0 to 10.0 years. The linear growth model represented by (12.17) is fitted to the data.

For this age period there are data for 336 children available, establishing a total of 1,886 length measurements. Length is measured in centimeters (cm). Age is measured in years and the reference age is chosen as $t_0 = 5$ years, so the intercept refers to lengths at 5 years of age. Parameter estimates for model (12.17) are presented in Table 12.7.

[1] The notation with the parameters γ has nothing to do with the γ parameters used earlier in this chapter in (12.4), but is consistent with the notation in Chapter 5.

Table 12.7 Linear growth model for 5–10-year-old children with retarded growth.

Fixed Effect		Coefficient	S.E.
γ_{00}	Intercept	96.32	0.285
γ_{10}	Age	5.53	0.08
Random Effect		Parameter	S.E.
Level-two (i.e., individual) random effects:			
τ_0^2	Intercept variance	19.79	1.91
τ_1^2	Slope variance for age	1.65	0.16
τ_{01}	Intercept-slope covariance	−3.26	0.46
Level-one (i.e., occasion) variance:			
σ^2	Residual variance	0.82	0.03
Deviance		7099.87	

The level-one standard deviation is so low ($\hat{\sigma} = 0.9$ cm) that it can be concluded that the fit of the linear growth model for the period from 5 to 10 years is quite adequate. Deviations from linearity will be usually smaller than 2 cm (which is slightly more than two standard errors). This is notwithstanding the fact that, given the large sample size, it is possible that the fit can be improved significantly from a statistical point of view by including non-linear growth terms.

The intercept parameters show that at 5 years, these children have an average length of 96.3 cm with a variance of $\hat{\tau}_0^2 + \hat{\sigma}^2 = 19.79 + 0.82 = 20.61$ and an associated standard deviation of $\sqrt{20.61} = 4.5$ cm. The growth per year is $\gamma_{01} + U_{0i}$, which has an estimated mean of 5.5 cm and standard deviation of $\sqrt{1.65} = 1.3$ cm. This implies that 95 percent of these children have growth rates, averaged over this five-year period, between $5.5 - 2 \times 1.3 = 2.9$ and $5.5 + 2 \times 1.3 = 8.1$ cm per year. The slope-intercept covariance is negative: children who are relatively short at 5 years grow relatively fast between 5 and 10 years.

Polynomial functions

To obtain a good fit, however, one can try and fit more complicated random parts. One possibility is to use a ***polynomial random part.*** This means that one or more powers of $(t - t_0)$ are given a random slope. This corresponds to polynomials for the function F_i,

$$F_i(t) = \beta_{0i} + \beta_{1i}(t - t_0) + \beta_{2i}(t - t_0)^2 + \ldots + \beta_{ri}(t - t_0)^r, \quad (12.18)$$

where r is a suitable number, called the degree of the polynomial. For example, one may test the null hypothesis that individual curves are linear by testing the null hypothesis '$r = 1$' against the alternative hypothesis that the curves are quadratic, '$r = 2$'. Any function for which the value is determined at m points can be represented exactly by a polynomial of degree $m - 1$. Therefore, in the fixed occasion model with m measurement occasions, it makes no sense to consider polynomials of degree higher than $m - 1$.

Usually the data contain more information about inter-individual differences (corresponding to the fixed part) than about intra-individual differences (the random part). For some of the coefficients β_{hi} in (12.18), there will be empirical evidence that they are non-zero, but not that they are variable across individuals. Therefore, as in Section 5.2.2, one can have a fixed part that is more complicated than the random part and give random slopes only to the lower powers of $(t - t_0)$. This yields the model

$$Y_{ti} = \gamma_{00} + \sum_{h=1}^{r} \gamma_{h0}(t - t_0)^h + U_{0i} + \sum_{h=1}^{p} U_{hi}(t - t_0)^h + R_{ti}, \quad (12.19)$$

which has the same structure as (5.14). For $h = p+1, ..., r$, parameter γ_{h0} is the value of the coefficient β_{hi}, constant over all individuals i. For $h = 1, ..., p$, the coefficients β_{hi} are individual-dependent, with population average γ_{h0} and individual deviations $U_{hi} = \beta_{hi} - \gamma_{h0}$. All variances and covariances of the random effects U_{hi} are estimated from the data. The mean curve for the population is given by

$$\mathcal{E}(F_i(t)) = \gamma_{00} + \sum_{h=1}^{r} \gamma_{h0}(t - t_0)^h \ . \quad (12.20)$$

Numerical difficulties appear less often in the estimation of models of this kind when t_0 has a value in the middle of the range of t values in the data set than when t_0 is outside this range or at one of the extremes of this range. Therefore, when convergence problems occur, it is advisable to try and work with a value of t_0 close to the average or median value of t. Changing the value of t_0 only amounts to a new parametrization, i.e., a different t_0 leads to different parameters γ_{h0} for which, however, formula (12.20) constitutes the same function and the deviance of the model (given that the software will calculate it) also is the same.

The number of random slopes, p, is not larger than r and may be considerably smaller. To give a rough indication, $r + 1$ may not be larger than the total number of distinct time points, i.e., the number of different values of t in the entire data set for which observations exist; also it should not be larger than a small fraction, say, 10 percent, of the total number of observations $\sum_i m_i$. On the other hand, p rarely will be much larger than the maximum number of observations per individual, $\max_i m_i$.[2]

Example 12.8 *Polynomial growth model for children with retarded growth.*
We continue the preceding example of children with retarded growth for which length measurements are considered in the period between the ages of 5 and

[2] In the fixed occasion design, the number of random effects, including the random intercept, cannot be larger than the number of measurement occasions. In the variable occasion design this strict upper bound does not figure, because the variability of time points of observations can lead to richer information about intra-individual variations. But if one obtains a model with clearly more random slopes than the maximum of all m_i, this may mean that one has made an unfortunate choice of the functions of time (polynomial or other) that constitute the random part at level two, and it may be advisable to try and find another, and smaller, set of functions of t for the random part with an equally good fit to the data.

10 years. In fitting polynomial models, convergence problems occurred when the reference age t_0 was chosen as 5 years, but not when it was chosen as the midpoint of the range, 7.5 years.

A cubic model, i.e., a model with polynomials of the third degree, turned out to yield a much better statistical fit than the linear model of Table 12.7. Parameter estimates are shown in Table 12.8 for the cubic model with $t_0 = 7.5$ years. So the intercept parameters refer to children of this age.

Table 12.8 Cubic growth model for 5–10-year-old children with retarded growth.

Fixed Effect		Coefficient	S.E.
γ_{00}	Intercept	110.40	0.22
γ_{10}	$t - 7.5$	5.23	0.12
γ_{20}	$(t - 7.5)^2$	−0.007	0.038
γ_{30}	$(t - 7.5)^3$	0.009	0.020
Random Effect		**Variance**	**S.E.**
Level-two (i.e., individual) random effects:			
τ_0^2	Intercept variance	13.80	1.19
τ_1^2	Slope variance $t - t_0$	2.97	0.32
τ_2^2	Slope variance $(t - t_0)^2$	0.255	0.032
τ_3^2	Slope variance $(t - t_0)^3$	0.066	0.009
Level-one (i.e., occasion) variance:			
σ^2	Residual variance	0.37	0.02
Deviance		6603.75	

For the level-two random slopes $(U_{0i}, U_{1i}, U_{2i}, U_{3i})$ the estimated correlation matrix is:

$$\widehat{R}_U = \begin{pmatrix} 1.0 & 0.17 & -0.27 & 0.04 \\ 0.17 & 1.0 & 0.11 & -0.84 \\ -0.27 & 0.11 & 1.0 & -0.38 \\ 0.04 & -0.84 & -0.38 & 1.0 \end{pmatrix} .$$

The fit is much better than that of the linear model (deviance difference 496.12 for 9 degrees of freedom). The random effect of the cubic term is significant (the model with fixed effects up to the power $r = 3$ and with $p = 2$ random slopes, not shown in the table, has deviance 6824.63, so the deviance difference for the random slope of $(t - t_0)^3$ is 221.12 with 4 degrees of freedom).

The mean curve for the population (cf. (12.20)) is given here by

$$\mathcal{E}(F_i(t)) = 110.40 + 5.23\,(t - 7.5) - 0.007\,(t - 7.5)^2 + 0.009\,(t - 7.5)^3 .$$

This deviates hardly, and not significantly, from a straight line. However, the individual growth curves do differ from straight lines, but the pattern of variation implied by the level-two covariance matrix is quite complex. We return to this data set in the next example.

Other functions

There is nothing sacred about polynomial functions. They are convenient, reasonably flexible, and any reasonably smooth function can be approximated by a polynomial if only you are prepared to use a polynomial of a sufficiently high degree. One argument for using other classes of functions is that some function shapes are approximated more parsimoniously by other functions than polynomials. Another argument is that polynomials are wobbly: when the value of a polynomial function $F(t)$ is changed a bit at one value of t, this may require coefficient changes that make the function change a lot at other values of t. In other words, the fitted value at any given value of t can depend strongly on observations for quite distant values of t. This kind of sensitivity often is undesirable.

One can use other functions instead of polynomials. If the functions used are called $f_1(t), ..., f_p(t)$, then instead of (12.19) the random function is modeled as

$$F_i(t) = \beta_{0i} + \sum_{h=1}^{r} \beta_{hi} f_h(t) \tag{12.21}$$

and the observations as

$$Y_{ti} = \gamma_{00} + \sum_{h=1}^{r} \gamma_{h0} f_h(t) + U_{0i} + \sum_{h=1}^{p} U_{hi} f_h(t) + R_{ti} \; . \tag{12.22}$$

What is a suitable class of functions, depends on the phenomenon under study and the data at hand. Random functions that can be represented by (12.22) are particularly convenient because this representation is a ***linear*** function of the parameters γ and the random effects U. Therefore, (12.22) defines an instance of the hierarchical linear model. Below we treat various classes of functions; the choice among them can be based on the deviance but also on the level-one residual variance σ_R^2, which indicates the size of the deviations of individual data points with respect to the fitted population of functions.

Sometimes, however, theory or data point to function classes where the statistical parameters enter in a non-linear fashion. In this chapter we restrict attention to linear models, i.e., models that can be represented as (12.22). Readers who are interested in non-linear models are referred to more specialized literature such as Davidian and Giltinan (1995) and Hand and Crowder (1996, Chapter 8).

Piecewise linear functions

A class of functions which is flexible, easy to comprehend, and for which the fitted functions values have a very restricted sensitivity for observations made at other values of t, is the class of continuous ***piecewise linear functions***. These are continuous functions whose slopes may change discontinuously at a number of values of t called ***nodes***, but which are linear (and hence have constant slopes) between these nodes. A disadvantage is their angular appearance.

The basic piecewise linear function is linear on a given interval (t_1, t_2) and constant outside this interval, as defined by

$$f(t) = \begin{cases} a & (t \le t_1) \\ a + (b-a)\dfrac{t - t_1}{t_2 - t_1} & (t_1 < t < t_2) \\ b & (t \ge t_2) \,. \end{cases} \tag{12.23}$$

Often one of the constant values a and b is chosen to be 0. The nodes are t_1 and t_2. Boundary cases are the functions (choosing $a = t_1$ and $b = 0$ and letting the lower node t_1 tend to minus infinity)

$$f(t) = \begin{cases} t - t_2 & (t < t_2) \\ 0 & (t \ge t_2) \end{cases}$$

and (choosing $a = 0$ and $b = t_2$ and letting the the upper node t_2 tend to plus infinity)

$$f(t) = \begin{cases} 0 & (t \le t_1) \\ t - t_1 & (t > t_1) \,, \end{cases}$$

and also linear functions like $f(t) = t$ (where both nodes are infinite). Each piecewise linear function can be obtained as a linear combination of these basic functions. The choice of nodes sometimes will be suggested by the problem at hand, in other situations has to be determined by trial and error.

Example 12.9 *Piecewise linear models for retarded growth.*
Let us try to improve on the polynomial models for the retarded growth data of the preceding example by using piecewise linear functions. Recall that the length measurements were considered for ages from 5.0 to 10.0 years. For ease of interpretation, the nodes are chosen at the birthdays of the children, i.e., at 6.0, 7.0, 8.0, and 9.0 years. This means that growth is assumed to proceed linearly during each year, but that growth rates may be different between the years. For the comparability with the polynomial model, the intercept again corresponds to length at the age of 7.5 years. This is achieved by using piecewise linear functions that all are equal to 0 for $t = 7.5$. Accordingly, the model is based on the following five basic piecewise linear functions:

$$\begin{aligned}
f_1(t) &= \begin{cases} -1 & (t \le 5) \\ t - 6 & (5 < t < 6) \\ 0 & (t \ge 6) \end{cases} \\
f_2(t) &= \begin{cases} -1 & (t \le 6) \\ t - 7 & (6 < t < 7) \\ 0 & (t \ge 7) \end{cases} \\
f_3(t) &= \begin{cases} -0.5 & (t \le 7) \\ t - 7.5 & (7 < t < 8) \\ 0.5 & (t \ge 8) \end{cases} \\
f_4(t) &= \begin{cases} 0 & (t \le 8) \\ t - 8 & (8 < t < 9) \\ 1 & (t \ge 9) \end{cases} \\
f_5(t) &= \begin{cases} 0 & (t \le 9) \\ t - 9 & (9 < t < 10) \\ 1 & (t \ge 10) \end{cases}
\end{aligned}$$

The results for this model are presented in Table 12.9.

Table 12.9 Piecewise linear growth model for 5–10-year-old children with retarded growth.

Fixed Effect		Coefficient	S.E.
γ_{00}	Intercept	110.40	0.22
γ_{10}	f_1 (5–6 years)	5.79	0.24
γ_{20}	f_2 (6–7 years)	5.59	0.18
γ_{30}	f_3 (7–8 years)	5.25	0.16
γ_{40}	f_4 (8–9 years)	5.16	0.15
γ_{50}	f_5 (9–10 years)	5.50	0.16
Random Effect		**Variance**	**S.E.**
Level-two (i.e., individual) random effects:			
τ_0^2	Intercept variance	13.91	1.20
τ_1^2	Slope variance f_1	3.97	0.82
τ_2^2	Slope variance f_2	3.80	0.57
τ_3^2	Slope variance f_3	3.64	0.50
τ_4^2	Slope variance f_4	3.42	0.45
τ_5^2	Slope variance f_5	3.77	0.53
Level-one (i.e., occasion) variance:			
σ^2	Residual variance	0.302	0.015
Deviance		6481.87	

The level-two random effects $(U_{0i}, \ldots, U_{5i})$ have as estimated correlation matrix:

$$\widehat{R}_U = \begin{pmatrix} 1.0 & 0.22 & 0.31 & 0.14 & -0.05 & 0.09 \\ 0.22 & 1.0 & 0.23 & 0.01 & 0.18 & 0.33 \\ 0.31 & 0.01 & 1.0 & 0.12 & -0.16 & 0.48 \\ 0.14 & 0.01 & 0.12 & 1.0 & 0.47 & -0.23 \\ -0.05 & 0.18 & -0.16 & 0.47 & 1.0 & 0.03 \\ 0.09 & 0.33 & 0.48 & -0.23 & 0.03 & 1.0 \end{pmatrix} .$$

The average growth rate is between 5 and 6 cm/year over the whole age range from 5 to 10 years. There is large variability in individual growth rates. All slope variances are between 3 and 4, so the between-child standard deviations in yearly growth rate are almost 2. The correlations between individual growth rates in different years are not very high, ranging from −0.23 (between ages 7–8 and 9–10) to 0.48 (between ages 6–7 and 9–10). This indicates that the growth rate fluctuates rather erratically from year to year around the average of about 5.5 cm/year.

The deviance for this piecewise linear model is 121.88 less than the deviance of the polynomial model of Table 12.8, while it has 13 parameters more. This is a large deviance difference for only 13 degrees of freedom. The residual level-one variance also is smaller. Although the polynomial model is not a submodel of the piecewise linear model, so that we cannot test the former against the latter by the usual deviance test, yet we may conclude that the piecewise model not only is more clearly interpretable but also has a better fit than the polynomial model.

Spline functions

Another flexible class of functions which is suitable for modeling longitudinal measurements within the framework of the hierarchical linear model, is the class of *spline function*. Spline functions are, briefly, smooth piecewise polynomials; more precisely, a number of points called *nodes* are defined on the interval where the spline function is defined; between each pair of adjacent nodes the spline function is a polynomial of degree p, while these polynomials are glued together so smoothly that the function itself and also its derivatives, of order up to $p-1$, are continuous also in the nodes. For $p = 1$ this leads again to piecewise linear functions, but for $p > 1$ this yields functions which are smooth and do not have the kinky appearance of piecewise linear functions. An example of a quadratic spline (i.e., a spline with $p = 2$) was given in Chapter 8 by equation (8.3) and Figure 8.1. This equation and figure represent a function of IQ which is a quadratic for IQ < 0 and also for IQ > 0, but which has different coefficients for these two domains. The point 0 here is the node. The coefficients are such that the function and its derivative are continuous also in the node, as can be seen from the graph. Therefore it is a spline. Cubic splines ($p = 3$) also are often used. We present here only a sketchy introduction to the use of splines. For a more elaborate introduction to the use of spline functions in (single-level) regression models, see Seber and Wild (1989, Section 9.5).

Suppose that one is investigating the development of some characteristic over the age of 12 to 17 years. Within each of the intervals 12 to 15 years and 15 to 17 years, the development curves might be approximately quadratic (this could be checked by a polynomial trend analysis for the data for these intervals separately), while they are smooth but not quadratic over the entire range from 12 to 17. In such a case it would be worthwhile to try a quadratic spline ($p = 2$) with one node, at 15 years. Defining $t_1 = 15$, the basic functions can be taken as

$$\begin{aligned} f_1(t) &= t && \text{(linear function)} \\ f_2(t) &= \begin{cases} (t-t_1)^2 & (t \le t_1) \\ 0 & (t > t_1) \end{cases} && \text{(quadratic left of } t_1) \\ f_3(t) &= \begin{cases} 0 & (t \le t_1) \\ (t-t_1)^2 & (t > t_1) \end{cases} && \text{(quadratic right of } t_1). \end{aligned} \tag{12.24}$$

The functions f_2 and f_3 are quadratic functions left of t_1 and right of t_1, and they are continuous and have continuous derivatives. That the functions are continuous and have continuous derivatives even in the node $t = t_1$ can be verified by elementary calculus or by drawing a figure.

The individual development functions are modeled as

$$F_i(t) = \beta_{0i} + \beta_{1i}\, f_1(t) + \beta_{2i}\, f_2(t) + \beta_{3i}\, f_3(t) \,. \tag{12.25}$$

If $\beta_{2i} = \beta_{3i}$, the curve for individual i is exactly quadratic. The freedom to have these two coefficients differ from each other allows to represent functions that look very different from quadratic functions, e.g., if these coefficients have opposite signs then the function will be concave on one side of t_1 and convex on the other side. Equation (8.3) and Figure 8.1 provide an example of exactly such a function, where t is replaced by SES and the node is the point SES $= 0$.

The treatment of this model within the hierarchical linear model approach is completely analogous to the treatment of polynomial models. The functions f_1, f_2, and f_3 constitute the fixed part of the model as well as the random part of the model at level two. If there is no evidence for individual differences with respect to the coefficients β_{2i} and/or β_{3i}, then these could be deleted from the random part.

Formula (12.25) shows that a quadratic spline with one node has one parameter more than a quadratic function (4 instead of 3). Each node added further will increase the number of parameters of the function by one. There is considerable freedom of choice in defining the basic functions, subject to the restriction that they are quadratic on each interval between adjacent nodes, and are continuous with continuous derivatives in the nodes. For two nodes, t_1 and t_2, a possible choice is the following. This representation employs a reference value t_0 that is an arbitrary (convenient or meaningful) value. It is advisable to use a t_0 within the range of observation times in the data. The basic functions are

$$\begin{aligned} f_1(t) &= t - t_0 && \text{(linear function)} \\ f_2(t) &= (t - t_0)^2 && \text{(quadratic function)} \\ f_3(t) &= \begin{cases} (t - t_1)^2 & (t \le t_1) \\ 0 & (t > t_1) \end{cases} && \text{(quadratic left of } t_1) \\ f_4(t) &= \begin{cases} 0 & (t \le t_2) \\ (t - t_2)^2 & (t > t_2) \end{cases} && \text{(quadratic right of } t_2). \end{aligned} \tag{12.26}$$

When these four functions are used in the representation

$$F_i(t) = \beta_{0i} + \beta_{1i}\, f_1(t) + \beta_{2i}\, f_2(t) + \beta_{3i}\, f_3(t) + \beta_{4i}\, f_4(t),$$

coefficient β_{2i} is the quadratic coefficient in the interval between t_1 and t_2, while β_{3i} and β_{4i} are the changes in the quadratic coefficient that occur when time t passes the nodes t_1 or t_2, respectively. The quadratic coefficient for $t < t_1$ is $\beta_{2i} + \beta_{3i}$, and for $t > t_2$ it is $\beta_{2i} + \beta_{4i}$.

The simplest cubic spline ($p = 3$) has one node. If the reference point t_0 is equal to the node, then the basic functions are

$$\begin{aligned} f_1(t) &= t - t_0 && \text{(linear function)} \\ f_2(t) &= (t - t_0)^2 && \text{(quadratic function)} \\ f_3(t) &= \begin{cases} (t - t_0)^3 & (t \le t_0) \\ 0 & (t > t_0) \end{cases} && \text{(cubic left of } t_0) \\ f_4(t) &= \begin{cases} 0 & (t \le t_0) \\ (t - t_0)^3 & (t > t_0) \end{cases} && \text{(cubic right of } t_0). \end{aligned} \tag{12.27}$$

For more than two nodes, and an arbitrary order p of the polynomials, the basic spline functions may be chosen as follows, for nodes denoted by

t_1 to t_M:

$$f_k(t) = (t - t_0)^k \qquad (k = 1, ..., p)$$
$$f_{p+k}(t) = \begin{cases} 0 & (t \le t_k) \\ (t - t_k)^p & (t > t_k) \end{cases} \qquad (k = 1, ..., M) \tag{12.28}$$

The choice of nodes is important to obtain a good fitting approximation. Since the spline functions are non-linear functions of the nodes, formal optimization of the node placement is more complicated than fitting spline functions with given nodes, and the interested reader is referred to the literature on spline functions for the further treatment of node placement. If the individuals provide enough observations (i.e., m_i is large enough), plotting observations for individuals, together with some trial and error, can lead to a good placement of the nodes.

Example 12.10 *Cubic spline models for retarded growth, 12–17 years*
In this example we again consider the length measurements of children with retarded growth studied by Rekers-Mombarg et al. (1997), but now the focus is on the age range from 12 to 17 years. After deletion of cases with missing data, in this age range there are a total of 1941 measurements which were taken for a sample of 321 children.

Some preliminary model fits showed that quadratic splines provide a better fit than piecewise linear functions, and cubic splines fit even better. A reasonable model is obtained by having one node at the age of 15.0 years. The basic functions accordingly are defined by 12.27 with $t_0 = 15.0$. All these functions are 0 for $t = 15$ years, so the intercept refers to length at this age. The parameter estimates are in Table 12.10.

Table 12.10 Cubic spline growth model for 12–17-year-old children with retarded growth.

Fixed Effect		Coefficient	S.E.
γ_{00}	Intercept	150.00	0.42
γ_{10}	f_1 (linear)	6.43	0.19
γ_{20}	f_2 (quadratic)	0.25	0.13
γ_{30}	f_3 (cubic left of 15)	−0.038	0.030
γ_{40}	f_4 (cubic right of 15)	−0.529	0.096
Random Effect		**Variance**	**S.E.**
Level-two (i.e., individual) random effects:			
τ_0^2	Intercept variance	52.07	4.46
τ_1^2	Slope variance f_1	6.23	0.71
τ_2^2	Slope variance f_2	2.59	0.34
τ_3^2	Slope variance f_3	0.136	0.020
τ_4^2	Slope variance f_4	0.824	0.159
Level-one (i.e., occasion) variance:			
σ^2	Residual variance	0.288	0.014
Deviance		6999.06	

The correlation matrix of the level-two random effects $(U_{0i},...,U_{4i})$ is estimated as:

$$\widehat{R}_U = \begin{pmatrix} 1.0 & 0.26 & -0.31 & 0.32 & 0.01 \\ 0.26 & 1.0 & 0.45 & -0.08 & -0.82 \\ -0.31 & 0.45 & 1.0 & -0.89 & -0.71 \\ 0.32 & -0.08 & -0.89 & 1.0 & 0.40 \\ 0.01 & -0.82 & -0.71 & 0.40 & 1.0 \end{pmatrix} .$$

Testing the random effects (not reported here) shows that the functions f_1 to f_4 all have significant random effects. The level-one residual standard deviation is only $\hat{\sigma} = \sqrt{0.288} = 0.54$ cm, which demonstrates that this family of functions fits rather closely to the length measurements. Notation is made more transparent by defining

$$(t-a)_- = \begin{cases} -(t-a) & (t < a) \\ 0 & (t \geq a), \end{cases}$$

$$(t-a)_+ = \begin{cases} 0 & (t \leq a) \\ t-a & (t > a). \end{cases}$$

Thus, we denote $f_3(t) = (t-15)^3_-$, $f_4(t) = (t-15)^3_+$. The mean length curve can be obtained by filling in the estimated fixed coefficients, which yields

$$\mathcal{E}(F_i(t)) = 150.00 + 6.43\,(t-15) + 0.25\,(t-15)^2$$
$$-\,0.038\,(t-15)^2_- - 0.529\,(t-15)^3_+ .$$

This function can be differentiated to yield the mean growth rate. Note that $d\,f_3(t)/dt = -3\,(t-15)^2_-$ and $d\,f_4(t)/dt = 3\,(t-15)^2_+$. This implies that the mean growth rate is estimated as

$$\text{mean growth rate } = 6.43 + 0.25\,(t-15) + 0.114\,(t-15)^2_-$$
$$-\,1.587\,(t-15)^2_+ .$$

For example, for the minimum of the age range considered, $t = 12$ years, this is $6.43 - (0.25 \times 3) + (0.114 \times 9) + 0 = 6.71$ cm/year. This is a little bit larger than the mean growth rate found in the preceding examples for ages from 5 to 10 years. For the maximum age in this range, $t = 17$ years, on the other hand, the average growth rate is $6.43 + (0.25 \times 2) + 0 - (1.587 \times 4) = 0.58$ cm/year, indicating that growth has almost stopped at this age.

The results of this model are illustrated more clearly by a graph. Figure 12.1 presents the average growth curve and a sample of 15 random curves from the population defined by Table 12.10 and the given correlation matrix.

The average growth curve does not deviate noticeably from a linear curve for ages below 16 years. It levels off after 16 years. Some of the randomly drawn growth curves are decreasing in the upper part of the range. This is an obvious impossibility for the real growth curves, and indicates that the model is not completely satisfactory for the upper part of this age range. This may be related to the fact that the number of measurements is rather low at ages over 16.5 years.

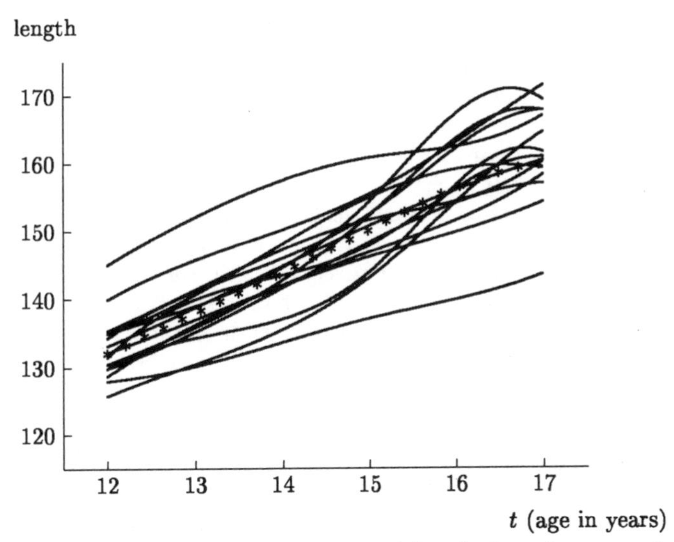

Figure 12.1 Average growth curve (∗) and 15 random growth curves for 12–17-year-olds for cubic spline model.

12.2.3 Explaining the functions

Whether one has been using polynomial, spline, or other functions, the random function model can be represented by (12.21), repeated here:

$$F_i(t) = \beta_{0i} + \sum_{h=1}^{r} \beta_{hi} f_h(t) . \tag{12.21}$$

The same can now be done as in Chapter 5 (see p. 75): the individual-dependent coefficients β_{hi} can be explained with individual-level (i.e., level-two) variables. Suppose there are q individual-level variables, and denote these variables by Z_1 to Z_q. The inter-individual model to explain the coefficients β_{0i} to β_{ri} then is

$$\beta_{hi} = \gamma_{h0} + \gamma_{h1} z_{1i} + \ldots + \gamma_{hq} z_{qi} + U_{hi} . \tag{12.29}$$

Substitution of (12.29) in (12.21) yields

$$Y_{ti} = \gamma_{00} + \sum_{h=1}^{p} \gamma_{h0} f_h(t) + \sum_{k=1}^{q} \gamma_{0k} z_{ki} + \sum_{k=1}^{q} \sum_{h=1}^{p} \gamma_{hk} z_{ki} f_h(t)$$

$$+ U_{0j} + \sum_{h=1}^{p} U_{hj} f_h(t) + R_{ij} . \tag{12.30}$$

We see that cross-level interactions here are interactions between individual-dependent variables and functions of time. The same approach can be followed as in Chapter 5. The reader may note that this model selection approach, where first a random function model is constructed and then the individual-based coefficients are approached as dependent variables in regression models at level two, is just what was described in Section 6.4.1 as 'working upward from level one'.

Example 12.11 *Explaining growth by gender and length of parents.*
We continue the analysis of the retarded growth data of the 12–17-year-olds, and now include the child's gender and the mean length of the child's parents. Leaving out children with missing value for mother's or father's length left 321 children with a total of 1,941 measurements. Gender is coded as +1 for girls and −1 for boys, so that the other parameters give values which are averages over the sexes. Parents' length is defined as the average length of father and mother minus 165 cm (this value is approximately the mean of the lengths of the parents). For parents' length, the main effect and the interaction effect with f_1 (age minus 15) was included. For gender, the main effect was included and also the interactions with f_1 and f_2. These choices were made on the basis of preliminary model fits. The resulting parameter estimates are in Table 12.11.

Table 12.11 Growth variability of 12–17-year-old children explained by gender and parents' length.

Fixed Effect		Coefficient	S.E.
γ_{00}	Intercept	150.20	0.47
γ_{10}	f_1 (linear)	5.85	0.18
γ_{20}	f_2 (quadratic)	0.053	0.124
γ_{30}	f_3 (cubic left of 15)	−0.029	0.030
γ_{40}	f_4 (cubic right of 15)	−0.553	0.094
γ_{01}	Gender	−0.385	0.426
γ_{11}	$f_1 \times$ gender	−1.266	0.116
γ_{21}	$f_2 \times$ gender	−0.362	0.037
γ_{02}	Parents' length	0.263	0.071
γ_{12}	$f_1 \times$ parents' length	0.0307	0.0152
Random Effect		**Variance**	**S.E.**
Level-two (i.e., individual) random effects:			
τ_0^2	Intercept variance	49.71	4.31
τ_1^2	Slope variance f_1	4.52	0.52
τ_2^2	Slope variance f_2	2.37	0.33
τ_3^2	Slope variance f_3	0.132	0.020
τ_4^2	Slope variance f_4	0.860	0.156
Level-one (i.e., occasion) variance:			
σ^2	Residual variance	0.288	0.013
Deviance		6885.18	

The correlation matrix of the level-two random effects $(U_{0i}, ..., U_{4i})$ is estimated as:

$$\widehat{R}_U = \begin{pmatrix} 1.0 & 0.22 & -0.38 & 0.35 & 0.07 \\ 0.22 & 1.0 & 0.38 & -0.05 & -0.81 \\ -0.38 & 0.38 & 1.0 & -0.91 & -0.75 \\ 0.35 & -0.05 & -0.91 & 1.0 & 0.48 \\ 0.07 & -0.81 & -0.75 & 0.48 & 1.0 \end{pmatrix}.$$

The fixed effect of gender (γ_{01}) is not significant, which means that at 15 years there is not a significant difference in length of boys and girls in this population with retarded growth. However, the interaction effects of gender with age, γ_{11} and γ_{21}, show that girls and boys do grow in different patterns during adolescence. The coding of gender implies that the average length difference between girls and boys is given by

$$\begin{aligned} &2\,(\gamma_{01} + \gamma_{11}\, f_1(t) + \gamma_{21}\, f_2(t)) \\ &\qquad = -0.77 - 2.532\,(t-15) - 0.724\,(t-15)^2 \,. \end{aligned}$$

The girl–boy difference in average growth rate, which is the derivative of this function, is equal to $-2.532 - 1.448\,(t-15)$. This shows that from about the age of 13 years (more precisely, for $t > 13.25$), girls grow on average slower than boys, while they grow faster before this age.

Parents length has a strong main effect: for each cm extra length of the parents, the children are on average 0.263 cm longer at 15 years of age. Moreover, for every cm extra of the parents, on average the children grow faster by 0.03 cm/year.

The intercept variance and the slope variances of f_1 and f_2 have decreased, compared to Table 12.10. The residual variance at level one remained the same, which is natural since the included effects explain differences between curves and do not yield better fitting curves. The deviance went down by 113.88 points, ($d.f. = 5$, $p < 0.0001$).

12.2.4 Changing covariates

Individual-level variables like the Z_k of the preceding section are referred to as *constant covariates* because they are constant over time. It is also possible that *changing covariates* are available, like changing social or economic circumstances, performance on tests, mood variables, etc. Fixed effects of such variables can be added to the model without problems, but the neat forward model selection approach is disturbed because the changing covariate normally will not be a linear combination of the functions f_h in (12.21). Depending on, e.g., the primacy of the changing covariates in the research question, one can employ the changing covariates in the fixed part right from the start (i.e., add this fixed effect to (12.19) or (12.22)), or incorporate them in the model at the point where also the constant covariates are considered (add the fixed effect to (12.30)).

Example 12.12 *Cortisol levels in infants.*
This example reanalyses data collected by De Weerth (1998, Chapter 3) in a study about stress in infants. In this example the focus is not on discovering the shape of the development curves, but on testing a hypothesized effect, while taking into account the longitudinal design of the study. Because of the complexity of the data analysis, we shall combine techniques described in various of the preceding chapters.

The purpose of the study was to investigate experimentally the effect of a stressful event on experienced stress as measured by the cortisol hormone, and whether this effect is stronger in certain hypothesized 'regression weeks'. The experimental subjects were 18 normal infants with gestational ages ranging from 17 to 37 weeks. Gestational age is age counted from the due day on which the baby should have been born after a complete pregnancy. Each infant was visited repeatedly, the number of visits per infant ranging from 8 to 13, with a total of 222 visits. The infants were divided randomly into an experimental group (14 subjects) and a control group (4 subjects). Cortisol was measured from saliva samples. The researcher collected a saliva sample at the start of the session; then a play session between the mother and the infant followed. For the experimental group, during this session the mother suddenly put the infant down and left the room. In the control group, the mother stayed with the child. After the session, a saliva sample was taken again. This provided a pretest and a post-test measure of the infant's cortisol level. In this example we consider only the effect of the experiment, not the effect of the 'regression weeks'. Further information about the study and about the underlying theories can be found in Chapter 3 of De Weerth (1998).

The post-test cortisol level was the dependent variable. Explanatory variables were the pretest cortisol level, the group (coded $Z = 1$ for infants in the experimental group and $Z = 0$ for the control group), and gestational age. Gestational age is the time variable t. In order to avoid very small coefficients, the unit is chosen as 10 weeks, so that t varies between 1.7 and 3.7. A preliminary inspection of the joint distribution of pretest and post-test showed that cortisol levels had a rather skewed distribution (as is usual for cortisol measurements). This skewness was reduced satisfactorily by using the square root of the cortisol level, both for the pretest and for the post-test. These variables are denoted X and Y, respectively. The pretest variable, X, is a changing covariate.

The law of initial value proposed by Wilder in 1956 implies that an inverse relation is expected between basal cortisol level and the subsequent response to a stressor (cf. De Weerth, 1998, p. 41). An infant with a low basal level of cortisol accordingly will react to a stressor by a cortisol increase, whereas an infant with a high basal cortisol level will react to a stressor by a cortisol decrease. The basal cortisol level is measured here by the pretest. The average of X (the square root of the pretest cortisol value) was 2.80. This implies that infants with a basal value x less than about 2.80 are expected to react to stress by a relatively high value for Y whereas infants with x more than about 2.80 are expected to react to stress by a relatively low value of Y. The play session itself may also lead to a change in cortisol value, so there may be a systematic difference between X and Y. Therefore, the stress reaction was investigated by testing whether the experimental group has a more strongly negative regression coefficient of Y on $X - 2.80$ than the control group; in other words, the research hypothesis is that there is a negative interaction between Z and $X - 2.80$ in their effect on Y.

It appeared that a reasonable first model for the square root of the post-test cortisol level, if the difference between the experimental and control group is not yet taken into account, is the model where the changing covariate, defined as $X - 2.80$, has a fixed effect, while time (i.e., gestational age) has a linear fixed as well as random effect:

$$Y_{ti} = \gamma_{00} + \gamma_{10}\, t + \gamma_{01}\, (x_{ti} - 2.80) + U_{0i} + U_{1i}\, t + R_{ti} \ .$$

The results are given as Model 1 in Table 12.12. This model can be regarded as the null hypothesis model, against which we shall test the hypothesis of the stressor effect.

The random effect of gestational age is significant. (The model without this effect, not shown in the table, has deviance 289.46, so that the deviance comparison yields $\chi^2 = 8.89$, $d.f. = 2$, $p < 0.01$; this test uses the halved p-value from the chi-squared distribution, see Section 6.2.1). The pretest has

Table 12.12 Estimates for two models for cortisol data.

		Model 1		Model 2	
Fixed Effect		Coefficient	S.E.	Coefficient	S.E.
γ_{00}	Intercept	3.21	0.33	3.16	0.34
γ_{10}	Gestational age	−0.201	0.117	−0.185	0.118
γ_{01}	$X - 2.80$	0.358	0.045	0.532	0.096
γ_{02}	Z			0.009	0.117
γ_{03}	$Z \times (X - 2.80)$			−0.224	0.108
Random Effect		Parameter	S.E.	Parameter	S.E.
Level-two (i.e., individual) random effects:					
τ_0^2	Intercept variance	1.25	0.66	1.32	0.68
τ_1^2	Slope variance	0.151	0.082	0.155	0.082
τ_{01}	Intercept-slope covariance	−0.43	0.23	−0.45	0.23
Level-one (i.e., occasion) variance:					
σ^2		0.175	0.018	0.172	0.018
Deviance		280.57		276.42	

a strongly significant effect ($t = 8.0$, $p < 0.0001$): children with a higher basal cortisol value tend to have also a higher post-test cortisol value. The fixed effect of gestational age is significant ($t = 1.67$, two-sided $p < 0.10$). The intercept variance is rather large because the time variable is not centered and $t = 0$ refers to the due date of birth, quite an extrapolation from the sample data.

To test the effect of the stressor, the fixed effect of the experimental group Z and the interaction effect of $Z \times (X - 2.80)$ were added to the model. The theory, i.e., Wilder's law of initial value, predicts the effect of the product variable $Z \times (X - 2.80)$ and therefore the test is focused on this effect. The main effect of Z is included only because including an interaction effect without the corresponding main effects can lead to errors in interpretation.

The result is presented as Model 2 in Table 12.12. The stressor effect (parameter γ_{03} in the table) is significant ($t = -2.07$ with many degrees of freedom because this is the effect of a level-one variable, one-sided $p < 0.025$). This confirms the hypothesized effect of the stressor in the experimental group.

The model fit is not good, however. The standardized level-two residuals defined by (9.9) were calculated for the 18 infants. For the second infant ($j = 2$) the value was $S_j^2 = 34.76$ ($d.f. = n_j = 13$, $p = 0.0009$). With the Bonferroni correction which takes into account that this is the most significant out of 18 values, the significance value still is $18 \times 0.0009 = 0.016$. Inspection of the data showed that this was a child which had been asleep shortly before many of the play sessions. Being just awake is known to have a potential effect on cortisol values. Therefore, for each session it had been recorded whether the child had been asleep in the half hour immediately preceding the session. Subsequent data analysis showed that having been asleep (represented by a dummy variable equal to 1 if the infant had been asleep in the half hour preceding the session and equal to 0 otherwise) had an important fixed effect, not a random effect at level two, and also was associated with heteroscedasticity at level one (see Chapter 8). Including these two effects led to the estimates presented in Table 12.13.

Table 12.13 Estimates for a model controlling for having been asleep.

		Model 3	
Fixed Effect		Coefficient	S.E.
γ_{00}	Intercept	3.04	0.31
γ_{10}	Gestational age	−0.142	0.099
γ_{01}	$X - 2.80$	0.548	0.084
γ_{02}	Z	−0.058	0.118
γ_{03}	$Z \times (X - 2.80)$	−0.136	0.096
γ_{04}	Sleeping	0.358	0.117
Random Effect		Parameter	S.E.
Level-two (i.e., individual) random effects:			
τ_0^2	Intercept variance	0.869	0.496
τ_1^2	Slope variance	0.092	0.057
τ_{01}	Intercept-slope covariance	−0.28	0.17
Level-one (i.e., occasion) variance parameters:			
σ_0^2	Basic residual variance	0.130	0.015
σ_{01}	Sleeping effect	0.116	0.045
Deviance		251.31	

The two effects of sleeping are jointly very significant (the comparison between the deviances of Models 2 and 3 yields $\chi^2 = 25.11$, $d.f. = 2$, $p < 0.0001$). The estimated level-one variance is 0.130 for children who had not slept and (using formula (8.1)) 0.362 for children who had slept. But the stressor ($Z \times (X - 2.80)$ interaction) effect now has lost its significance ($t = -1.42$, n.s.). The most significant standardized level-two residual defined by (9.9) for this model is obtained for the fourth child ($j = 4$), with the value $S_j^2 = 29.86$, $d.f. = n_j = 13$, $p = 0.0049$. Although this is a rather small p-value, the Bonferroni correction now leads to a significance probability of $18 \times 0.0049 = 0.09$, which is not alarmingly low. The fit of this model therefore seems satisfactory, and the estimates do not support the hypothesized stressor effect. However, these results cannot be interpreted as evidence *against* the stressor effect, because the parameter estimate does have the predicted negative sign and the number of experimental subjects is not very large, so that the power may have been low.

It can be concluded that it is very important to control for the infant having slept shortly before the play session, and in this case this control makes the difference between a significant and a non-significant effect. Having slept not only leads to a higher post-test cortisol value, controlling for the pretest value, but also triples the residual variance at the occasion level.

12.3 Autocorrelated residuals

Of the relevant extensions to the hierarchical linear model, we briefly want to mention the hierarchical linear model with ***autocorrelated residuals.*** In the fixed occasion design, the assumption that the level-one residuals R_{ti} are independent can be relaxed and replaced by the assumption of first-order autocorrelation,

$$R_{1i} = R_{1,i}^{(0)}, \ R_{t+1,i} = \rho R_{ti} + \sqrt{1-\rho^2}\, R_{t+1,i}^{(0)} \qquad (t \geq 1) \qquad (12.31)$$

where the variables $R_{t,i}^{(0)}$ are independent identically distributed variables. The parameter ρ is called the ***autocorrelation coefficient.*** This model represents a type of dependence between adjacent observations, which quickly dies out between observations which are further apart. The correlation between the residuals is

$$\rho(R_{ti}, R_{si}) = \rho^{|t-s|} \ . \qquad (12.32)$$

For variable occasion designs, level-one residuals with the correlation structure (12.32) also are called autocorrelated residuals, although they cannot be constructed by the relations (12.31). These models are discussed by Diggle, Liang, and Zeger (1994, Section 5.2), Goldstein, Healy, and Rasbash (1994), and Hedeker and Gibbons (1996b).

The extent to which over-time dependence can be modeled by random slopes, or rather by autocorrelated residuals, or a combination of both, depends on the phenomenon being modeled. This issue usually will have to be decided empirically; the multilevel computer programs MIXREG (see Hedeker and Gibbons, 1996b), MLwiN (see Yang et al., 1998), and HLM version 5, as well as the general statistical programs SAS (specifically, Proc Mixed) and BMDP (specifically, module 5V) can be used for this purpose. These programs are discussed in Chapter 15.

13 Multivariate Multilevel Models

This chapter is devoted to the multivariate version of the hierarchical linear model treated in Chapters 4 and 5. The term 'multivariate' refers here to the dependent variable: there are assumed to be two or more dependent variables. The model of this chapter is a three-level model, with variables as level-one units, individuals as level-two units, and groups as level-three units. It may also be regarded as an extension of the fully multivariate model, treated in Section 12.1.3, to the case where individuals are nested in groups. For the corresponding two-level model the reader is referred to Section 12.1.3. In the present chapter, the dependent variables are not necessarily supposed to be longitudinal measurements of the same variable (although they could be).

The nesting structure of Chapter 12, measurements nested within individuals, can be combined with the nesting structure considered earlier, individuals within groups. This leads to three levels: measurements within individuals within groups.

As an example of multivariate multilevel data, think of pupils (level-two units) i in classes (level-three units) j, with m variables, Y_1 to Y_m, being measured for each pupil if data are complete. The measurements are the level-one units and could refer to, e.g., test scores in various scholastic subjects but also variables measured in completely unrelated and incommensurable scales such as attitude measurements. The dependent variable is denoted

Y_{hij} is the measurement on the h'th variable
for individual i in group j.

It is not necessary that for each individual i in each group j, an observation of each of the m variables is available. (It must be assumed, however, that missingness is at random, i.e., the fact that a measurement is not available is not related to the value that this measurement would have taken, if it had been observed; if a model with explanatory variables is considered, then the availability of a measurement should be unrelated to its residual).

Just like in Chapter 12, the complete data vector can be defined as the vector of data for the individual, possibly hypothetical, who does have observations on all variables. This complete data vector is denoted by

$$Y^{c} = \begin{pmatrix} Y_{1ij} \\ \cdot \\ \cdot \\ \cdot \\ Y_{mij} \end{pmatrix}.$$

It is possible to analyse all m dependent variables separately. However, there are several reasons why it may be sensible to analyse the data jointly, i.e., as multivariate data.

1. Conclusions can be drawn about the correlations between the dependent variables, notably, the extent to which the correlations depend on the individual and on the group level. Such conclusions follow from the partitioning of the covariances between the dependent variables over the levels of analysis.

2. The tests of specific effects for single dependent variables are more powerful in the multivariate analysis. This will be visible in the form of smaller standard errors. The additional power is negligible if the dependent variables are only weakly correlated, but may be considerable if the dependent variables are strongly correlated while at the same time the data are very incomplete, i.e., the average number of measurements available per individual is considerably less than m.

3. Testing whether the effect of an explanatory variable on dependent variable Y_1 is larger than its effect on Y_2, when the data on Y_1 and Y_2 were observed (totally or partially) on the same individuals, is possible only by means of a multivariate analysis.

4. If one wishes to carry out a single test of the joint effect of an explanatory variable on several dependent variables, then also a multivariate analysis is required. Such a single test can be useful, e.g., to avoid the danger of chance capitalization which is inherent to carrying out a separate test for each dependent variable.

A multivariate analysis is more complicated than separate analyses for each dependent variable. Therefore, when one wishes to analyse several dependent variables, the greater complexity of the multivariate analysis will have to be balanced against the reasons listed above. Often it is advisable to start by analysing the data for each dependent variable separately.

13.1 The multivariate random intercept model

Suppose there are covariates X_1 to X_p, which may be individual-dependent or group-dependent. The random intercept model for dependent variable Y_h is expressed by the formula

$$Y_{hij} = \gamma_{0h} + \gamma_{1h}\, x_{1ij} + \gamma_{2h}\, x_{2ij} + \ldots + \gamma_{ph}\, x_{pij} + U_{hj} + R_{hij}\,. \quad (13.1)$$

In words: for the h'th dependent variable, the intercept is γ_{0h}, the regression coefficient on X_1 is γ_{1h}, the coefficient on X_2 is γ_{2h}, etc., the random part of the intercept in group j is U_{hj}, and the residual is R_{hij}. This is just a random intercept model like (4.5) and (4.8). Since the variables Y_1 to Y_m are measured on the same individuals, however, their dependence can be taken into account. In other words, the U's and R's are regarded as components of vectors,

$$\mathbf{R}_{ij} = \begin{pmatrix} R_{1ij} \\ \cdot \\ \cdot \\ \cdot \\ R_{mij} \end{pmatrix} , \quad \mathbf{U}_j = \begin{pmatrix} U_{1j} \\ \cdot \\ \cdot \\ \cdot \\ U_{mj} \end{pmatrix} .$$

Instead of residual variances at level 1 and 2, there are now residual covariance matrices,

$$\mathbf{\Sigma} = \text{cov}(\mathbf{R}_{ij}) \quad \text{and} \quad \mathbf{T} = \text{cov}(\mathbf{U}_j) .$$

The covariance matrix of the complete observations, conditional on the explanatory variables, is the sum of these,

$$\text{var}(Y^c) = \mathbf{\Sigma} + \mathbf{T} . \tag{13.2}$$

To represent the multivariate data in the multilevel approach, three nesting levels are used. The first level is that of the dependent variables indexed by $h = 1, ..., m$, the second level is that of the individuals $i = 1, ..., n_j$, and the third level is that of the groups, $j = 1, ..., N$. So each measurement of a dependent variable on some individual is represented by a separate line in the data matrix, containing the values i, j, h, Y_{hij}, x_{1ij}, and those of the other explanatory variables.

The multivariate model is formulated as hierarchical linear model using the same trick as in Section 12.1.3. Dummy variables d_1 to d_m are used to indicate the dependent variables, just like in formula (12.2). Dummy variable d_h is 1 or 0, depending on whether the data line refers to dependent variable Y_h or to one of the other dependent variables. Formally, this is expressed by

$$d_{shij} = \begin{cases} 1 & h = s , \\ 0 & h \neq s . \end{cases} \tag{13.3}$$

With these dummies, the random intercept models (13.1) for the m dependent variables can be integrated into one three-level hierarchical linear model by the expression

$$Y_{hij} = \sum_{s=1}^{m} \gamma_{0s}\, d_{shij} + \sum_{k=1}^{p}\sum_{s=1}^{m} \gamma_{ks}\, d_{shij}\, x_{kij} + \sum_{s=1}^{m} U_{sj}\, d_{shij} + \sum_{s=1}^{m} R_{sij}\, d_{shij} . \tag{13.4}$$

All variables (including the constant!) are multiplied by the dummy variables. Note that the definition of the dummy variables implies that in the sums over $s = 1, ..., m$ only the term for $s = h$ gives a contribution and all other terms disappear. So this formula is just a complicated way of rewriting formula (13.1).

The purpose of this formula is that it can be used to obtain a multivariate hierarchical linear model. The variable-dependent random residuals R_{hij} in this formula are random slopes *at level two* of the dummy variables, R_{sij} being the random slope of d_s, and the random intercepts U_{hj} become the random slopes *at level three* of the dummy variables. *There is no random part at level one.*

This model can be further specified, e.g., by omitting some of the variables X_k from the explanation of some of the dependent variables Y_h. This amounts to dropping some of the terms $\gamma_{ks}\, d_{shij}\, x_{kij}$ from (13.4). Another possibility is to include variable-specific covariates, in analogy to the changing covariates of Section 12.2.4. An example of this is a study of school pupils performance on several subjects (these are the multiple dependent variables), using the pupils' motivation for each of the subjects separately as explanatory variables.

Multivariate empty model

The multivariate empty model is the multivariate model without explanatory variables. For $m = 2$ variables, this is just the model of section 3.6.1. Writing out formulae (13.1) and (13.4) without explanatory variables (i.e., with $p = 0$) yields the following formula, which is the specification of the three-level multivariate empty model:

$$\begin{aligned} Y_{hij} &= \gamma_{0h} + U_{hj} + R_{hij} \\ &= \sum_{s=1}^{m} \gamma_{0s}\, d_{shij} + \sum_{s=1}^{m} U_{sj}\, d_{shij} + \sum_{s=1}^{m} R_{sij}\, d_{shij}\ . \end{aligned} \qquad (13.5)$$

This empty model can be used to decompose the raw variances *and covariances* into parts at the two levels. When referring to the multivariate empty model, the covariance matrix

$$\mathbf{\Sigma} = \text{cov}(\mathbf{R}_{ij})$$

may be called the *within-group covariance matrix* while

$$\mathbf{T} = \text{cov}(\mathbf{U}_j)$$

may be called the *between-group covariance matrix.* In the terminology of Chapter 3, these are the *population* (and not *observed*) covariance matrices. In Chapter 3 we saw that, if the group sizes n_j all are equal to n, then the covariance matrix between the group means is

$$\mathbf{T} + \frac{1}{n}\mathbf{\Sigma} \qquad (13.6)$$

(cf. Equation (3.9)).

Section 3.6.1 presented a relatively simple way to estimate the within- and between-group correlations from the intraclass coefficients and the observed total and within-group correlations. Another way is to fit the multivariate empty model (13.5) using multilevel software by the ML or the REML method. For large sample sizes these methods will provide virtually the same results, but for small sample sizes the ML and REML methods will provide more precise estimators. The gain in precision will be especially large if the correlations are high while there are relatively many missing data (i.e., there are many individuals who provide less than m dependent variables, but missingness still is at random).

Example 13.1 *Language and arithmetic scores in elementary schools.* The example used throughout Chapters 4 and 5 is elaborated by analysing not only the scores on the language test but also on the arithmetic test. So there are $m = 2$ dependent variables.

Table 13.1 Parameter estimates for multivariate empty model.

	Language $h = 1$		Arithmetic $h = 2$		(Covariance)	
Fixed Effect	Par.	S.E.	Par.	S.E.		
γ_{0h} = Intercept	40.34	0.42	18.93	0.34		
Random Effect	Par.	S.E.	Par.	S.E.	Par.	S.E.
Between-schools covariance matrix:						
$\tau_h^2 = \text{var}(U_{hj})$	19.00	2.86	12.76	1.84		
$\tau_{12} = \text{cov}(U_{1j}, U_{2j})$					14.54	2.17
Within-schools covariance matrix:						
$\sigma_h^2 = \text{var}(R_{hij})$	64.64	1.97	32.25	0.98		
$\sigma_{12} = \text{cov}(R_{1ij}, R_{2ij})$					28.62	1.16
Deviance			29674.1			

First the multivariate empty model, represented by (13.5), is fitted. The results are in Table 13.1.

The results for the language test may be compared with the univariate empty model for the language test, presented in Table 4.1 of Chapter 4. The parameter estimates are slightly different, which is a consequence of the fact that the estimation procedure now is multivariate. The extra result of the multivariate approach is the estimated covariance at level three, $\text{cov}(U_{1j}, U_{2j}) = 14.54$ and at level two, $\text{cov}(R_{1ij}, R_{2ij}) = 28.62$. The corresponding population correlation coefficients at the school level and the pupil level are, respectively,

$$\rho(U_{1j}, U_{2j}) = \frac{14.54}{\sqrt{12.76 \times 19.00}} = 0.93 ,$$

$$\rho(R_{1ij}, R_{2ij}) = \frac{28.62}{\sqrt{32.25 \times 64.64}} = 0.63 .$$

This shows that especially the random school effects for language and arithmetic are very strongly correlated.

For the correlations between observed variables, these estimates yield a correlation between individuals of (cf. (13.2))

$$\hat{\rho}(Y_{1ij}, Y_{2ij}) = \frac{14.54 + 28.62}{\sqrt{(19.00 + 64.64)(12.76 + 32.25)}} = 0.70$$

and, for groups of a hypothetical size $n = 30$, a correlation between group means of (cf. (13.6))

$$\hat{\rho}(\bar{Y}_{1.j}, \bar{Y}_{2.j}) = \frac{14.54 + 28.62/30}{\sqrt{(19.00 + 64.64/30)(12.76 + 32.25/30)}} = 0.90.$$

Explanatory variables included are IQ, the group mean of IQ, and group size. As in the examples in Chapters 4 and 5, the IQ measurement is the verbal IQ from the ISI test and the average value of 23.1 is subtracted from the group size. The correspondence with formulae (13.1) and (13.4) is that X_1 is IQ, X_2 is the group mean of IQ, X_3 is group size and $X_4 = X_1 X_3$ represents the interaction between IQ and group size. The results are in Table 13.2.

Table 13.2 Parameter estimates for multivariate model for language and arithmetic tests.

	Language $h = 1$		Arithmetic $h = 2$		(Covariance)	
Fixed Effect	Par.	S.E.	Par.	S.E.		
γ_{0h} Intercept	40.84	0.29	19.33	0.25		
γ_{1h} IQ	2.422	0.072	1.464	0.054		
γ_{2h} $\overline{\text{IQ}}$ (Group mean)	1.33	0.32	1.21	0.27		
γ_{3h} Group size	0.046	0.036	0.044	0.031		
γ_{4h} IQ × Group size	−0.021	0.010	−0.016	0.007		
Random Effect	Par.	S.E.	Par.	S.E.	Par.	S.E.
Residual between-schools covariance matrix:						
$\tau_h^2 = \text{var}(U_{hj})$	7.36	1.24	6.06	0.94		
$\tau_{12} = \text{cov}(U_{1j}, U_{2j})$					5.70	0.97
Residual within-schools covariance matrix:						
$\sigma_h^2 = \text{var}(R_{hij})$	42.12	1.28	24.07	0.73		
$\sigma_{12} = \text{cov}(R_{1ij}, R_{2ij})$					15.04	0.76
Deviance	28540.7					

Calculating t-statistics for the fixed effects shows that, except for the main effect of group size on either dependent variable, all effects are significant at the 0.05 significance level. The residual correlations are $\rho(U_{1j}, U_{2j}) = 0.85$ at the school level and $\rho(R_{1ij}, R_{2ij}) = 0.47$ at the pupil level. This shows that taking the explanatory variables into account has led to somewhat smaller, but still substantial residual correlations. Especially the school-level residual correlation is large. This suggests that, also when controlling for IQ, mean IQ, and group size, the factors at school level that determine language and arithmetic proficiency are the same. Such factors could be associated with school policy but also with aggregated pupil characteristics not taken into account here, such as average socio-economic status.

When the interaction effect of group size with IQ is to be tested for both dependent variables simultaneously, this can be carried out by fitting the model from which these interaction effects are excluded. In formula (13.4) this corresponds to the effects γ_{41} and γ_{42} of $d_1 X_4$ and $d_2 X_4$. The model from which these effects are excluded has a deviance of 28549.1, which is 8.4 lower than the model of Table 13.2. In a chi-squared distribution with $d.f. = 2$, this is a significant result ($p < 0.05$).

13.2 Multivariate random slope models

The notation is quite complex already, and therefore only the case of one random slope is treated here. More random slopes are, in principle, a straightforward extension.

Suppose that variable X_1 has a random slope for the various dependent variables. Denote for the h'th dependent variable the random intercept by U_{0hj} and the random slope of X_1 by U_{1hj}. The model for the h'th dependent variable then is

$$\begin{aligned} Y_{hij} &= \gamma_{0h} + \gamma_{1h} x_{1ij} + \ldots + \gamma_{ph} x_{pij} \\ &\quad + U_{0hj} + U_{1hj} x_{1hj} + R_{hij} . \end{aligned} \tag{13.7}$$

The three-level formulation of this model is

$$\begin{aligned} Y_{hij} = & \sum_{s=1}^{m} \gamma_{0s} d_{shij} + \sum_{k=1}^{p} \sum_{s=1}^{m} \gamma_{ks} d_{shij} x_{kij} + \\ & \sum_{s=1}^{m} U_{0sj} d_{shij} + \sum_{s=1}^{m} U_{1sj} d_{shij} x_{1ij} + \sum_{s=1}^{m} R_{sij} d_{shij} . \end{aligned} \tag{13.8}$$

This means that again there is no random part at level one, there are m random slopes at level two (of variables d_1 to d_m) and $2m$ random slopes at level three (of variables d_1 to d_m and of the product variables $d_1 X_1$ to $d_m X_1$). With this kind of model, an obvious further step is to try and model the random intercepts and slopes by group-dependent variables like in Section 5.2.

14 Discrete Dependent Variables

Up to now, it was assumed in this book that the dependent variable has a continuous distribution and that the residuals at all levels (U_{0j}, R_{ij}, etc.) have normal distributions. This provides a satisfactory approximation for many data sets. However, there also are many situations where the dependent variable is 'so discrete' that the assumption of a continuous distribution could lead to erroneous conclusions. This chapter treats the most frequently occurring kinds of discrete variable: dichotomous variables (with only two values), ordered categorical variables (with a small number of ordered categories, e.g., 1, 2, 3, 4) and counts (with natural numbers as values: 0, 1, 2, 3, etc.).

14.1 Hierarchical generalized linear models

Important instances of discrete dependent variables are dichotomous variables (e.g., success vs. failure of whatever kind) and counts (e.g., in the study of some kind of event, the number of events happening in a predetermined time period). It is (usually) unwise to apply linear regression methods to such variables, for two reasons.

The first reason is that the range of such a dependent variable is restricted, and the usual linear regression model might take its fitted value outside this allowed range. For example, dichotomous variables can be represented as having the values 0 and 1. A fitted value of, say, 0.7, still can be interpreted as a probability of 0.7 for the outcome 1 and a probability of 0.3 for the outcome 0. But what about a fitted value of −0.23 or 1.08? A meaningful model for outcomes that have only the values 0 or 1 should not allow fitted values that are negative or greater than 1. Similarly, a meaningful model for count data should not lead to negative fitted values.

The second reason is of a more technical nature, and is the fact that for discrete variables there is often some natural relation between the mean and the variance of the distribution. For example, for a dichotomous variable Y that has probability p for outcome 1 and probability $1 - p$ for outcome 0, the mean is

$$\mathcal{E}Y = p$$

and the variance is

$$\text{var}(Y) = p\,(1 - p). \tag{14.1}$$

Thus, the variance is not a free parameter but is determined already by the mean. In terms of multilevel modeling, this leads to a relation between the parameters in the fixed part and the parameters of the random part.

This has led to the development of regression-like models that are more complicated than the usual multiple linear regression model and that take account of the non-normal distribution of the dependent variable, its restricted range, and the relation between mean and variance. The best-known method of this kind is ***logistic regression***, a regression-like model for dichotomous data. *Poisson regression* is a similar model for count data. In the statistical literature, such models are known as ***generalized linear models***, cf. McCullagh and Nelder (1989) or Long (1997).

The present chapter gives an introduction into multilevel versions of some generalized linear models; these multilevel versions are aptly called hierarchical generalized linear models. Much research and also software development is still being carried out to develop and implement statistical procedures for these models, so the introduction and overview provided in this chapter are necessarily limited.

14.2 Introduction to multilevel logistic regression

Logistic regression (treated, e.g., in Hosmer and Lemeshow, 1989; Long, 1997; McCullagh and Nelder, 1989; Ryan, 1997) is a kind of regression analysis for dichotomous, or binary, outcome variables, i.e., outcome variables with two possibly values such as 'pass / fail', 'yes / no', or 'like / dislike'. The reader is advised to study an introductory text on logistic regression if he or she is not already acquainted with this technique.

14.2.1 Heterogeneous proportions

The basic data structure of two-level logistic regression is a collection of N groups ('units at level two'), with in group j ($j = 1, ..., N$) a random sample of n_j level-one units ('individuals'). The outcome variable is dichotomous and denoted by Y_{ij} for level-one unit i in group j. The two outcomes are supposed to be coded 0 and 1: 0 for 'failure', 1 for 'success'. The total sample size is denoted by $M = \sum_j n_j$. If one does not (yet) take explanatory variables into account, the probability of success is constant in each group. The success probability in group j is denoted P_j. In a random coefficient model (cf. Section 4.2.1), the groups are considered as being taken from a population of groups and the success probabilities in the groups, P_j, are regarded as random variables defined in this population. The dichotomous outcome can be represented as the sum of this probability and a residual,

$$Y_{ij} = P_j + R_{ij} \; . \qquad (14.2)$$

In words, the outcome for individual i in group j, which is either 0 or 1, is expressed as the sum of the probability (average proportion of successes) in this group plus some individual-dependent residual. This residual has (like all residuals) mean zero but for these dichotomous variables it has the peculiar property that it can assume only the values $-P_j$ and $1-P_j$, since (14.2) must be 0 or 1. A further peculiar property is the fact that, given the value of the probability P_j, the variance of the residual is

$$\text{var}(R_{ij}) = P_j\,(1-P_j)\ , \tag{14.3}$$

in accordance with formula (14.1).

Equation (14.2) is the dichotomous analogue of the empty (or unconditional) model defined for continuous outcomes in equations (3.1) and (4.6). Section 3.3.1 remains valid for dichotomous outcome variables, with P_j taking the place of $\mu+U_j$, except for one subtle distinction. In the empty model for continuous outcome variables it was assumed that the level-one residual variance was constant. This is not adequate here because, in view of formula (14.3), the groups have different within-group variances. Therefore the parameter σ^2 must be interpreted here as the *average residual variance*, i.e., the average of (14.3) in the population of all groups. With this modification in the interpretation, the formulae of section 3.3.1 still are valid. For example, the intraclass correlation coefficient still is defined by (3.2) and can be estimated by (3.12). Another definition of the intraclass correlation is also possible, however, as is mentioned below in Section 14.3.2.

Since the outcome variable is coded 0 and 1, the group average

$$\bar{Y}_{.j} = \frac{1}{n_j}\sum_{i=1}^{n_j} Y_{ij} \tag{14.4}$$

now is the proportion of successes in group j. This is an estimate for the group-dependent probability P_j. Similarly, the overall average

$$\hat{P}_{.} = \bar{Y}_{..} = \frac{1}{M}\sum_{j=1}^{N}\sum_{i=1}^{n_j} Y_{ij} \tag{14.5}$$

here is the overall proportion of successes.

Testing heterogeneity of proportions

To test whether there are indeed systematic differences between the groups, the well-known chi-squared test can be used. The test statistic of the chi-squared test for a contingency table is often given in the familiar form $\sum(O-E)^2/E$ where O is the observed and E the expected count in a cell of the contingency table. In this case it can be written also as

$$X^2 = \sum_{j=1}^{N} n_j\,\frac{(\bar{Y}_{.j}-\hat{P}_{.})^2}{\hat{P}_{.}(1-\hat{P}_{.})}\ . \tag{14.6}$$

It can be tested in the chi-squared distribution with $N-1$ degrees of freedom. This chi-squared distribution is an approximation valid if the expected numbers of successes and of failures in each group, $n_j\bar{Y}_{.j}$ and $n_j(1-\bar{Y}_{.j})$, respectively, all are at least 1 while 80 percent of them are at least 5 (cf. Agresti, 1990, p. 246). This condition will not always be satisfied, and the chi-squared test then may be seriously in error. For a large number of groups the null distribution of X^2 then can be approximated by a normal distribution with the correct mean and variance, cf. Haldane (1940) and

McCullagh and Nelder (1989), p. 244; or an exact permutation test may be used.

Another test of heterogeneity of proportions was proposed by Commenges and Jacqmin (1994). The test statistic is

$$T = \frac{\sum_{j=1}^{N} \left\{ n_j^2 \left(Y_{.j} - \hat{P}_{.} \right)^2 \right\} - M\hat{P}_{.}(1 - \hat{P}_{.})}{\hat{P}_{.}(1 - \hat{P}_{.})\sqrt{2 \sum_{j=1}^{N} n_j(n_j - 1)}} . \tag{14.7}$$

Large values of this statistic are an indication of heterogeneous proportions. This statistic can be tested in a standard normal distribution.

The fact that the numerator contains a weight of n_j^2 whereas the chi-squared test uses the weight n_j shows that these two tests combine the groups in different ways. When the group sizes n_j are different, it is possible that the two tests lead to different outcomes.

The theoretical advantage of test (14.7) over the chi-squared test is that it has somewhat higher power to test randomness of the observations against the empty model treated below, i.e., against the alternative hypothesis represented by (14.10) with $\tau_0^2 > 0$. The practical advantage is that it can be applied whenever there are many groups, even with small group sizes, provided that no single group dominates. A rule of thumb for the application of this test is that there should be at least $N = 10$ groups, the biggest group should not have a relative share larger than $n_j/M = 0.10$, and the ratio of the largest group size to the 10th largest group size should not be more than 10.

Estimation of between- and within-groups variance

The true variance between the group-dependent probabilities, i.e., the population value of $\text{var}(P_j)$, can be estimated by formula (3.11),

$$\hat{\tau}^2 = S^2_{\text{between}} - \frac{S^2_{\text{within}}}{\tilde{n}} , \tag{3.11}$$

where $\tilde{n}$ is defined as in (3.7) by

$$\tilde{n} = \frac{1}{N-1} \left\{ M - \frac{\sum_j n_j^2}{M} \right\} = \bar{n} - \frac{s^2(n_j)}{N\bar{n}} .$$

For dichotomous outcome variables, the observed between-groups variance is closely related to the chi-squared test statistic (14.6). They are connected by the formula

$$S^2_{\text{between}} = \frac{\hat{P}_{.}(1 - \hat{P}_{.})}{\tilde{n}(N-1)} X^2 .$$

The within-groups variance in the dichotomous case is a function of the group averages, viz.,

$$S^2_{\text{within}} = \frac{1}{M-N} \sum_{j=1}^{N} n_j \bar{Y}_{.j} (1 - \bar{Y}_{.j}) .$$

Example 14.1 *Experience with cohabitation in Norway.*
This example is about the influence of age and geographical region on whether inhabitants of Norway have experience with living together with a partner without being married, using data collected in the ISSP-1994 survey (International Social Survey Programme, 1994) on Family and Changing Gender Roles. The dependent variable is whether the respondent ever lived together with a partner without being married ('experience with cohabitation'). After deleting cases with missing or inconsistent answers, there were 2079 respondents classified into 19 geographical regions in Norway. The number of respondents per region ranged from 35 (in Finnmark) to 235 (in Oslo). It was expected that cultural differences will lead to differences between regions and between age groups in the proportion of individuals having experience with cohabitation. In the whole data set, 43.4 percent of the respondents reported that they had ever cohabitated with a partner.

A two-level structure is used with the region as the second-level unit. This is based on the idea that there may be differences between regions that are not captured by the explanatory variables and hence may be regarded as unexplained variability within the set of all regions. Figure 14.1 represents for each of the 19 regions the proportion of people in the sample who ever lived together without being married.

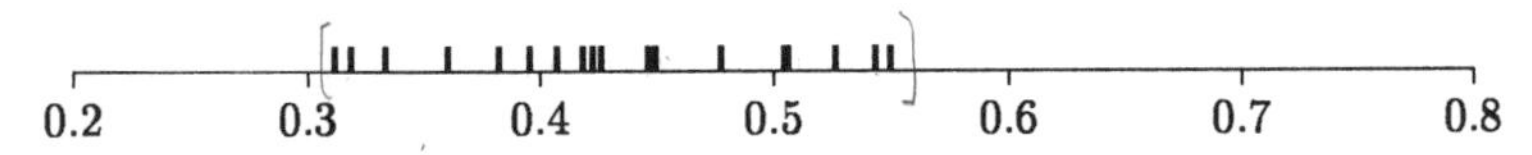

Figure 14.1 Proportion of experience with cohabitation.

Are these proportions, ranging from 0.31 to 0.55, very different? The chi-squared test (14.6) for equality of these 19 proportions yields $X^2 = 35.40$, *d.f.* $= 18$, $p < 0.01$. Thus, there is evidence for heterogeneity between the regions. The estimated true variance between the region-dependent proportions calculated from (3.11) is $\hat{\tau}^2 = 0.0022$, thus the estimated true between-region standard deviation is $\hat{\tau} = \sqrt{0.0022} = 0.047$. With an average probability of 0.434, this standard deviation is not very large but not quite negligible either.

14.2.2 The logit function: Log-odds

It can be relevant to include explanatory variables in models for dichotomous outcome variables. For example, in the example above, the individual's age will undoubtedly have a major influence on experience with cohabitation. When explanatory variables are included to model probabilities, a problem is that probabilities are restricted to the domain between 0 to 1, whereas a linear effect for an explanatory variable could take the fitted value outside this interval. This was mentioned already in the introduction.

Instead of the probability of some event, one may consider the ***odds***: the ratio of the probability of success to the probability of failure. When the probability is p, the odds are $p/(1-p)$. In contrast to probabilities, odds can assume any value from 0 to infinity, and odds can be considered to constitute a ratio scale.

The *logarithm* transforms a multiplicative to an additive scale and transforms the set of positive real numbers to the whole real line. Indeed, one of the most widely used transformations of probabilities is the *log odds*, defined by

$$\text{logit}(p) = \ln\left(\frac{p}{1-p}\right), \tag{14.8}$$

where $\ln(x)$ denotes the natural logarithm of the number x. The logit function, of which a graph is shown in Figure 14.2, is an increasing function defined for numbers between 0 and 1, and its range is from minus infinity to plus infinity. Figure 14.3 shows in a different way how probability values are transformed to logit values.

For example, $p = 0.269$ is transformed to $\text{logit}(p) = -1$ and $p = 0.982$ to $\text{logit}(p) = 4$. The logit of $p = 0.5$ is exactly 0.

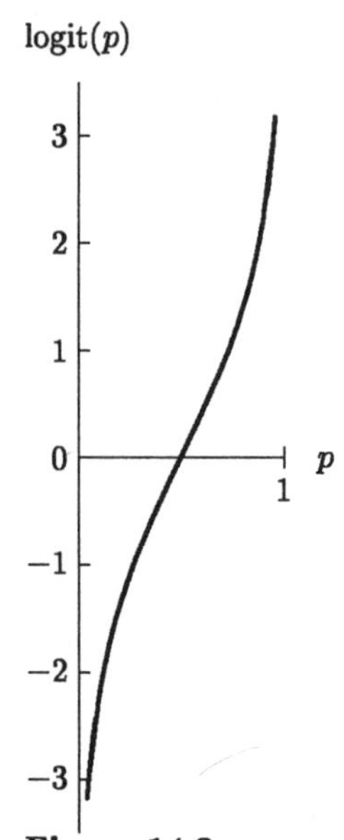

Figure 14.2
The logit function.

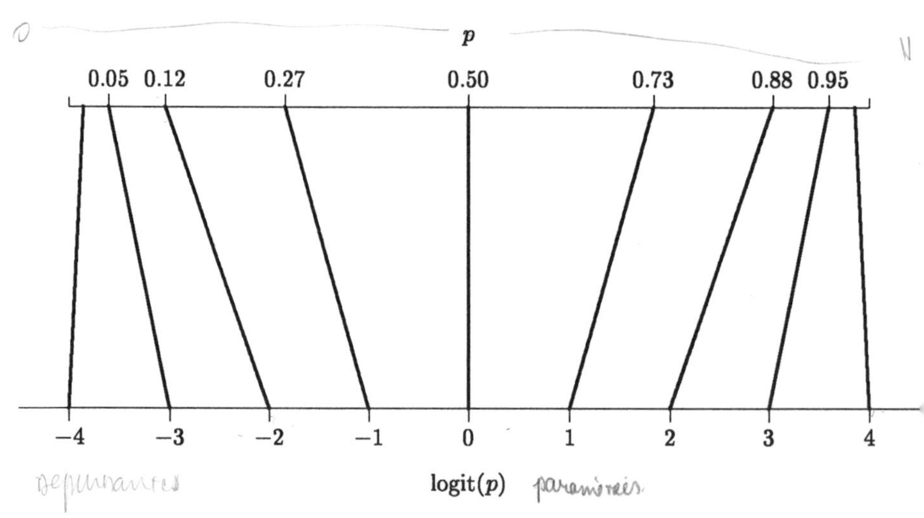

Figure 14.3 Correspondence between p and $\text{logit}(p)$.

The logistic regression model is a model where $\text{logit}(p)$ is a linear function of the explanatory variables. In spite of the attractive properties of the logit function, it is by no means the only suitable function for transforming probabilities to arbitrary real values. The general term for such a transformation function is the *link function*, as it links the probabilities (or more generally, the expected values of the dependent variable) to the explanatory variables. The probit function (which is the inverse cumulative distribution function of the standard normal distribution) also is often used as a link function for dichotomous variables (also see Section 14.3.2). A generalized linear model for a dichotomous outcome with the probit link function is called a probit regression model. For still other link functions see, e.g., Long (1997) or McCullagh and Nelder (1989).

The choice of the link function has to be guided by the empirical fit of the model, ease of interpretation, and convenience – e.g., availability of computer software. In this chapter the link function for a probability will be denoted by $f(p)$, and we shall concentrate on the logit link function.

14.2.3 The empty model

The empty two-level model for a dichotomous outcome variable refers to population of groups (level-two units) and specifies the probability distribution for the group-dependent probabilities P_j in (14.2), without taking further explanatory variables into account. Several such specifications have been proposed.

We focus on the model that specifies the transformed probabilities $f(P_j)$ to have a normal distribution. This is expressed, for a general link function $f(p)$, by the formula

$$f(P_j) = \gamma_0 + U_{0j} \ , \tag{14.9}$$

where γ_0 is the population average of the transformed probabilities and U_{0j} the random deviation from this average for group j. If $f(p)$ is the logit function, then $f(P_j)$ is just the log-odds for group j. Thus, for the logit link function, the log-odds have a normal distribution in the population of groups, which is expressed by

$$\text{logit}(P_j) = \gamma_0 + U_{0j} \ . \tag{14.10}$$

For the deviations U_{0j} it is assumed that they are independent random variables with a normal distribution with mean 0 and variance τ_0^2.

This model does not include a separate parameter for the level-one variance. This is because the level-one residual variance of the dichotomous outcome variable follows directly from the success probability, as indicated by equation (14.3).

Denote by π_0 the probability corresponding to the average value γ_0, as defined by

$$f(\pi_0) = \gamma_0 \ .$$

For the logit function, this means that π_0 is the so-called logistic transform of γ_0, defined by

$$\pi_0 = \text{logistic}(\gamma_0) = \frac{\exp(\gamma_0)}{1 + \exp(\gamma_0)} \ . \tag{14.11}$$

Here $\exp(\gamma_0) = e^{\gamma_0}$ denotes the exponential function, where e is the basis of the natural logarithm. The logistic and logit functions are each others' inverses, just like the exponential and the logarithmic functions. Figure 14.4 shows the shape of the logistic function.

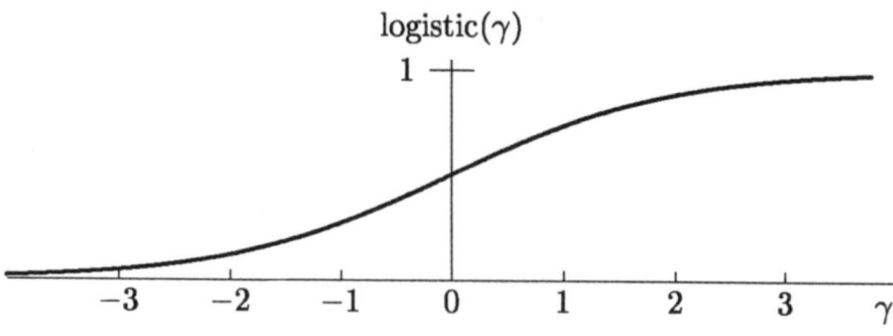

Figure 14.4 The logistic function.

This π_0 is close (but not quite equal) to the average value of the probabilities P_j in the population of groups. Because of the non-linear nature of the link function, there is not a simple relation between the variance of these probabilities and the variance of the deviations U_{0j}. There is an approximate formula, however, valid when the variances are small. The approximate relation (valid for small τ_0^2) between the population variances is

$$\mathrm{var}(P_j) \approx \frac{\tau_0^2}{(f'(\pi_0))^2} \ . \tag{14.12}$$

For the logit function, this yields

$$\mathrm{var}(P_j) \approx (\pi_0(1-\pi_0))^2 \ \tau_0^2 \ . \tag{14.13}$$

When τ_0^2 is not so small, the variance of the probabilities will be less than the right hand side of (14.13). (Note that these are population variances and not variances of the observed proportions in the groups; see Section 3.3 for this distinction.)

Example 14.2 *Empty model for the Norwegian cohabitation data.*
Fitting the empty model with normally distributed log-odds to the data of experience with cohabitation in Norway yields the results presented in Table 14.1.

Table 14.1 Estimates for empty logistic model.

Fixed Effect	Coefficient	S.E.
γ_0 = Intercept	−0.276	0.062
Random Effect	**Variance Component**	**S.E.**
Level-two variance:		
$\tau_0^2 = \mathrm{var}(U_{0j})$	0.032	0.023

The estimated average log-odds is −0.276, which corresponds (according to equation (14.11)) to a probability of $\pi_0 = 0.43$. The variance approximation formula (14.13) yields $\mathrm{var}(P_j) \approx 0.0019$, not too far from the non-parametric estimate calculated from (3.11), which is 0.0022. The difference between these values is caused by the facts that formula (14.13) is only an approximation and that the estimation methods are different. The deviance is not given, because the deviance given by most multilevel programs for this kind of model is an approximation that cannot be used reliably in tests as treated in Section 6.2 (cf. Section 14.2.7).

Figure 14.5 presents the observed log-odds of the 19 regions with the fitted normal distribution. These observed log-odds are the logit transformations of the proportions depicted in Figure 14.1. The fitted normal distribution has a smaller dispersion than the observed log-odds values, which can be seen from the fact that rather many of the observed log-odds (the small vertical lines) occur in the region where the probability density is very low. This is because the estimation takes into account that the dispersion of the observed log-odds is the consequence not only of true between-region differences but also of within-region random sampling fluctuation.

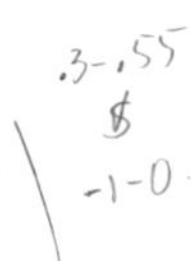

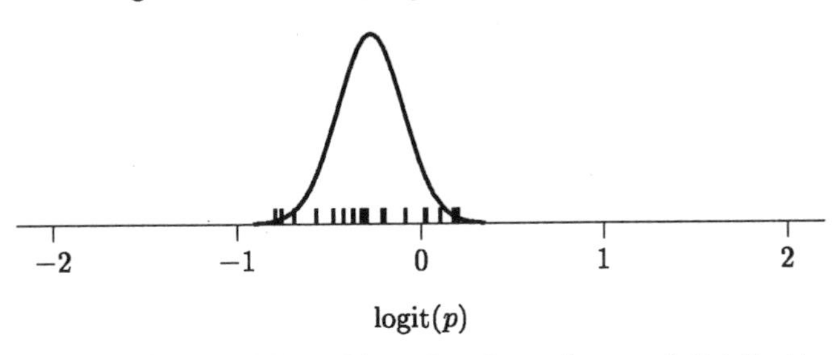

Figure 14.5 Observed log-odds and estimated normal distribution of population log-odds of cohabitation experience.

14.2.4 The random intercept model

In logistic regression analysis, linear models are constructed for the log-odds. The multilevel analogue, random coefficient logistic regression, is based on linear models for the log-odds that include random effects for the groups or other higher-level units. As mentioned above, Y_{ij} denotes the dichotomous outcome variable for level-one unit i in level-two unit j. There are n_j level-one units in the j'th level-two unit. The outcome Y_{ij} is coded as 0 or 1, also referred to as 'failure' and 'success'. We shall use the terms 'individual' and 'group' to refer to the level-one units and level-two units.

We now assume that there are variables which are potential explanations for the observed success or failure. These variables are denoted by X_1 to X_r. The values of X_h $(h = 1, ..., r)$ are indicated in the usual way by x_{hij}. Since some (or all) of these variables could be level-one variables, the success probability is not necessarily the same for all individuals in a given group. Therefore the success probability now depends on the individual as well as on the group, and is denoted by P_{ij}. Accordingly, equation (14.2) expressing how the outcome is split into an expected value and a residual now is replaced by

$$Y_{ij} = P_{ij} + R_{ij} \; . \tag{14.14}$$

The logistic random intercept model expresses the log-odds, i.e., the logit of P_{ij}, as a sum of a linear function of the explanatory variables[1] and a

[1]Rather than the double-subscript notation γ_{hk} used earlier in Chapters 4 and 5, we now use – to obtain a relatively simple notation – a single-subscript notation γ_h analogous to (5.14).

random group-dependent deviation U_{0j}:

$$\text{logit}(P_{ij}) = \gamma_0 + \sum_{h=1}^{r} \gamma_h \, x_{hij} + U_{0j} \, . \tag{14.15}$$

Thus, a unit difference between the X_h values of two individuals in the same group is associated with a difference of γ_h in their log-odds, or equivalently, a ratio of $\exp(\gamma_h)$ in their odds. The deviations U_{0j} are assumed to have zero mean (given the values of all explanatory variables) and a variance of τ_0^2. Formula (14.15) does not include a level-one residual because it is an equation for the probability P_{ij} rather than for the outcome Y_{ij}. The level-one residual is included already in Formula (14.14).

Example 14.3 *Random intercept model for the Norwegian cohabitation data.*

Age is expected to have a major influence on experience with cohabitation. One reason is that older persons have had a longer period in which they had the occasion to practice cohabitation. Another reason is that normative attitudes with respect to cohabitation have changed over the years, so that older people, who grew up in a period where cohabitation was less accepted than in more recent years, might have a different attitude than younger people.

Ages of respondents range from 16 to 79 years. The effect of age, however, cannot be expected to be linear over this whole range. Therefore the age effect is modeled with a more complicated function. In this case we choose for a quadratic spline, because this is a flexible and easily used family of functions (cf. Section 12.2.2). Quadratic splines are functions which are quadratic on a number of intervals, defined in such a way that the function is smooth also at the transitions between the intervals. It turned out that for this data set it was suitable to use intervals with endpoints (the so-called nodes) at 30 and 40 years. Indicating age by t, the age effect was modeled by using the following four transformed functions of age as explanatory variables:

$$\begin{aligned}
X_1(t) &= t - 20 && \text{(linear function)} \\
X_2(t) &= (t - 20)^2 && \text{(quadratic function)} \\
X_3(t) &= \begin{cases} (t-30)^2 & (t > 30) \\ 0 & (t \le 30) \end{cases} && \text{(quadratic function for } t > 30) \\
X_4(t) &= \begin{cases} (t-40)^2 & (t > 40) \\ 0 & (t \le 40) \end{cases} && \text{(quadratic function for } t > 40).
\end{aligned}$$

The age $t = 20$ is used as the reference value, so that the intercept refers to the age of 20 years. Table 14.2 contains as Model 1 the estimation results for this model.

The coefficients of all these functions of age cannot be interpreted in isolation from each other. Together they represent the following function of age, which here is the model's fixed part:

$$\begin{aligned}
&-1.200 + 0.539\,(t-20) \; - \; 0.029\,(t-20)^2 \\
&\qquad + 0.0241((t-30)_+)^2 \; + \; 0.0068((t-40)_+)^2 \, ,
\end{aligned}$$

where $(t-a)_+$ is defined by $(t-a)$ for $t > a$ and by 0 otherwise. This function is represented graphically in Figure 14.6.

Table 14.2 Estimates for two models for cohabitation data.

	Model 1		Model 2	
Fixed Effect	**Coefficient**	**S.E.**	**Coefficient**	**S.E.**
γ_0 Intercept	−1.200	0.154	−1.107	0.153
γ_1 Age - 20	0.539	0.051	0.544	0.052
γ_2 (Age - 20)^2	−0.0290	0.0037	−0.0289	0.0038
γ_3 (Age - 30)^2 for Age > 30	0.0241	0.0054	0.0235	0.0055
γ_4 (Age - 40)^2 for Age > 40	0.0068	0.0025	0.0075	0.0025
γ_5 Religion			−1.85	0.24
Random Effect	**Var. Comp.**	**S.E.**	**Var. Comp.**	**S.E.**
Random intercept:				
$\tau_0^2 = \text{var}(U_{0j})$	0.061	0.037	0.049	0.034

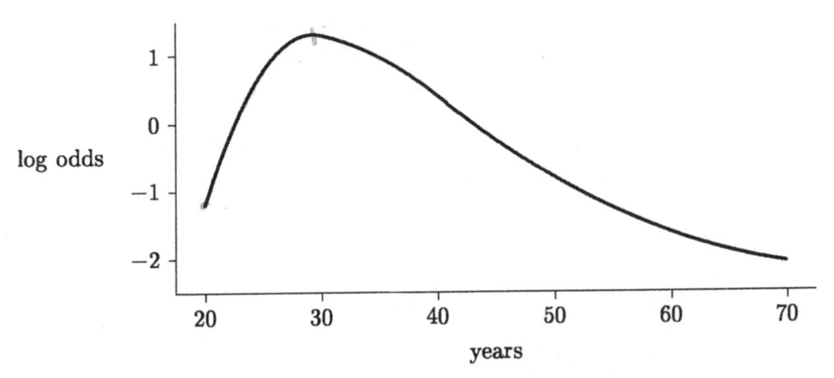

Figure 14.6 Age effect on cohabitation experience (fixed part).

The figure shows that the probability of experience with cohabitation swiftly increases to a maximum attained at about 30 years, after which it decreases again. The fitted logits at 20, 30 and 70 years are −1.2, 1.29, and −2.07, respectively, corresponding (cf. Figure 14.3) to probabilities of 0.23, 0.78, and 0.11.

When age is being controlled for in this way, there remains a random region effect. The estimated variance, 0.061, corresponding to a standard deviation of the random intercepts of 0.25, is even larger than it was in the empty model. In Section 7.1 we remarked that this is possible for the hierarchical linear model with normal distributions, and it may be a sign of misspecification of the model. In the hierarchical linear model for dichotomous data this is not necessarily the case. In Section 14.3.5 it will be explained that adding level-one variables with strong effects will tend to increase estimated level-two variances, and to make the regression coefficients of already included variables, if these are uncorrelated with the newly included variables, larger in absolute size.

In addition to age, the religious conviction of the individual can also have an effect on cohabitation. Many Christian religions are not in favor of cohabitation without being married. This is represented by including in the model a dummy variable called 'Religion', with value 1 if the respondent reported to attend religious services at least once a month, and value 0 otherwise. The resulting estimates are given in Table 14.2 as Model 2. It appears that religious activity has a strong effect on cohabitation, with an effect of -1.85 on the log-odds and a t-ratio of $-1.85/0.24 = -7.71$ (highly significant). Including this effect hardly changes the age effect but does lead to a smaller random intercept variance. This is because the regions differ in the proportion of religious inhabitants. Including the regional average of this dummy variable, which is just the regional proportion of religious inhabitants, showed that the within-region regression coefficient does not differ significantly from the between-region regression coefficient (parameter estimates not shown here).

14.2.5 Estimation

Parameter estimation in hierarchical generalized linear models is more complicated than in hierarchical linear models. Inevitably some kind of approximation is involved, and various kinds of approximation have been proposed. Reviews were given by Rodríguez and Goldman (1995) and Davidian and Giltinan (1995). We mention some references and some of the terms used without explaining them. The reader who wishes to study these algorithms is referred to the mentioned literature. More information about the computer programs mentioned in this section is given in Chapter 15.

The most frequently used methods are based on a first- or second-order Taylor expansion of the link function. When the approximation is around the estimated fixed part, this is called marginal quasi-likelihood (MQL), when it is around an estimate for the fixed plus the random part it is called penalized or predictive quasi-likelihood (PQL) (Breslow and Clayton, 1993; Goldstein, 1991; Goldstein and Rasbash, 1996). This procedure is implemented in the programs MLn/MLwiN, HLM, and VARCL. A Laplace approximation, which is supposed to be more precise, was proposed by Raudenbush, Yang, and Yosef (1999). This is implemented in HLM, version 5. Other approximation methods were proposed by Wong and Mason (1985) and by Engel and Keen (1994).

Numerical integration is used in procedures proposed by Stiratelli, Laird and Ware (1984), Anderson and Aitkin (1985), Gibbons and Bock (1987), and Longford (1994). An example is given in Gibbons and Hedeker (1994). This was extended to three-level models by Gibbons and Hedeker (1997). This numerical integration method is implemented in the program MIXOR and its relatives. An important practical advantage of this method is the production of a deviance statistic (see Section 6.2) that can be used for hypothesis testing. The deviance statistic produced by MQL and PQL methods cannot be used in this way. It may be noted that MIXOR estimates for the random part parameters the standard deviations τ_0 etc. rather than the variances τ_0^2, and that formula (5.15) describes how the standard errors of the standard deviations can be transformed to those of the variances.

Computer-intensive methods related to the bootstrap and the Gibbs sampler were proposed by Zeger and Karim (1991), Kuk (1995), Meijer et al.

(1995), and McCulloch (1997). Some of such procedures are implemented in MLwiN and MLA. A method based on the principle of indirect inference was proposed by Mealli and Rampichini (1999). Another computer-intensive method is the method of simulated moments. This method is applied to the models of this chapter by Gouriéroux and Montfort (1996, Section 3.1.4 and Chapter 5) and an overview of some recent work is given by Baltagi (1995, Section 10.4.3).

The estimates produced by these methods differ primarily with respect to the parameters of the random part. The estimates for the fixed parameters often are not strongly different. But especially if the variance components are rather large, these methods may produce quite different estimates for the random part parameters, which in turn will have an effect on the estimated standard errors for the fixed part parameters.

The first-order MQL and PQL estimates of the variance parameters of the random part have an appreciable downward bias (Rodríguez and Goldman, 1995). The second-order MQL and PQL methods produce parameter estimates with less bias but, it seems, a higher mean squared error.

The numerical integration approach and the Laplace approximation seem to produce statistically more satisfactory estimates than the MQL and PQL approaches. They have the additional advantage that the deviance produced is reliable, in contrast to the deviance produced by the MQL and PQL methods. But they are implemented less widely in computer software.

The estimation procedures for these models still are in a state of active development. The choice between these procedures should be based on stability of the algorithm (will the algorithm converge to a valid estimate?), statistical efficiency, availability of software, and the possibility to obtain parameter tests.

The currently available algorithms are not perfectly stable; whether they will converge depends on the data set, the complexity of the fitted model, and the starting values. Small group sizes can contribute to instability of the algorithm. Of the first and second order MQL and PQL methods, the first-order MQL method is the most stable.

14.2.6 Aggregation

If the explanatory variables assume only few values, then it is advisable to aggregate the individual 0–1 data to success counts, depending on the explanatory variables, within the level-two units. This will improve the speed and stability of the algorithm as well as, for some estimation methods, the statistical efficiency. This is carried out as follows.

For a random intercept model with a small number of discrete explanatory variables X_1 to X_r, let L be the total number of combinations of values $(x_1, ..., x_r)$. All individuals with the same combination of values $(x_1, ..., x_r)$ are treated as one subgroup in the data. They all have a common success probability, given by (14.15). Thus, each level-two unit includes L subgroups, or less if some of the combinations do not occur in this level-two unit. Aggregation is advantageous if L is considerably less than the average group size n_j.

Denote by

$$n_j^+(x_1, ..., x_r)$$

the number of individuals in group j with the values $x_1, ..., x_r$ on the respective explanatory variables, and denote by

$$Y_j^+(x_1, ..., x_r)$$

the number of individuals among these who yielded a success, i.e., for whom $Y_{ij} = 1$. Then $Y_j^+(x_1, ..., x_r)$ has the binomial distribution with binomial denominator ('number of trials') $n_j^+(x_1, ..., x_r)$ and success probability given by (14.15), which is the same for all individuals i in this subgroup. The multilevel analysis is now applied with these subgroups as the level-one units. Subgroups with $n_j^+(x_1, ..., x_r) = 0$ can be omitted from the data set.

14.2.7 Testing the random intercept

The parameters in the fixed part of a multilevel logistic model can be tested by t-tests or Wald tests in the way indicated in Section 6.1. Testing parameters of the random part, however, is more difficult than in a hierarchical linear model with normal distributions. If one is using a program that implements numerical integration (such as MIXOR, cf. Section 14.2.5) or the Laplace approximation (such as HLM version 5), the deviance can be used to produce chi-squared tests just like in Section 6.2. But the deviance values produced by the MQL and PQL methods are so crude approximations that they cannot be used in reliable deviance tests.

Another test for the random intercept in the case that there are some explanatory variables was proposed by Commenges et al. (1994). When there are no explanatory variables (i.e., the empty model is being considered), the test statistic reduces to (14.7). This is a test for the null hypothesis that there is no random intercept, i.e., $\tau_0^2 = 0$, while controlling for the fixed effects of the explanatory variables. This means that the null hypothesis is the usual logistic regression model with explanatory variables X_1 to X_r, while the alternative hypothesis adds to this a random intercept as in model (14.15). The test is based on the so-called score principle, which means that it requires only the estimation of the parameters under the null model. (Therefore it can be calculated from the results of fitting a logistic regression model, without random intercept, either by multilevel or by conventional (non-multilevel) software. So one might wait with turning to multilevel software until this test turns out to be significant.) For the details of this test we refer to the paper cited.

14.3 Further topics about multilevel logistic regression

14.3.1 Random slope model

The random intercept logistic regression model of Section 14.2.4 can be extended to a random slope model just like in Chapter 5. We only give the formula for one random slope; the extension to several random slopes is straightforward. The remarks made in Chapter 5 remain valid, given the appropriate changes, and will not be repeated here.

Like in the random intercept model, assume that there are r explanatory variables X_1 to X_r. Assume that the effect of the first one, X_1, is variable across groups, and accordingly has a random slope. Expression (14.15) for the logit of the success probability is extended with the random effect $U_{1j}\, x_{1ij}$, which leads to

$$\text{logit}(P_{ij}) = \gamma_0 + \sum_{h=1}^{r} \gamma_h x_{hij} + U_{0j} + U_{1j} x_{1ij} \; . \tag{14.16}$$

There now are two random group effects, the random intercept U_{0j} and the random slope U_{1j}. It is assumed that both have a zero mean. Their variances are denoted, respectively, by τ_0^2 and τ_1^2 and their covariance is denoted by τ_{01}.

Testing a random slope in multilevel logistic regression models can be based on the deviance, provided that the deviance statistic is based on numerical integration rather than on an approximation method (see Section 14.2.5). Such a deviance statistic is produced by MIXOR and HLM version 5 (see Chapter 15). The deviance statistic currently provided by MLn/MLwiN cannot be used for this purpose, because it is based on a different type of approximation. Testing procedures for random slopes in hierarchical generalized linear models are still a matter of active research, see, e.g., Lin (1999).

Example 14.4 *Friends and foes in the former GDR.*
Personal networks are data about the relations of individuals: each respondent is asked to mention other persons to whom she or he is related according to some specified criterion, and to give further information about the relation with this person. In this context, respondents are referred to as 'egos' and the mentioned relations as 'nominees' or 'alters'. In the resulting data set, alters are nested within egos. The application of multilevel analysis to personal networks is discussed in Snijders, Spreen, and Zwaagstra (1994) and in Van Duijn, Van Busschbach, and Snijders (1999).

Völker (1995) collected personal network data in the former German Democratic Republic (GDR) and investigated changes in personal relations associated with the downfall of communism and the German reunification. The data was collected retrospectively in 1992 and 1993. Based on these data, Völker and Flap (1997) studied neighborhood relations in the former GDR on the basis of data about relationships existing in 1989 (before the breaking of the Berlin Wall). This example reanalyses data presented in the latter paper, focusing on distrust in relations. The question is, what influences whether an individual distrusts another person to whom he or she is related.

The level-two units are the respondents (also called 'egos'). For each respondent the personal network is delineated in a standardized way, which leads to a list of related persons ('alters', 'network members'). This list is the group of level-one units. The dichotomous dependent variable indicates whether ego distrusts alter. Since family relations and friends were almost never distrusted, these relations were omitted from the data set, which left the following role relations: acquaintance, neighbor, colleague, superior, and subordinate. This resulted in a set of 426 egos with a total of 1,683 alters. The analyses were carried out using MIXOR.

There are two political variables. Membership of the communist party (SED) for ego is denoted by $X_1 = 1$ (yes) or 0 (no). It was unknown whether alter is a party member himself, but the variable X_2 indicates whether his function requires party membership or is under some kind of political supervision with the values $X_2 = 1$ (yes) and 0 (no).

Model 1 presented in Table 14.3 includes the fixed effects of these variables and a random slope for X_2. The fixed effects of the political variables are non-significant. The deviance of the model without the random slope for X_2 is 1587.77. Hence the test for the random slope has $\chi^2 = 5.48, d.f. = 2$, which yields (one-sided, see Section 6.2.1) $p < 0.05$. The estimated slope variance is quite large (5.76) but its standard error is even larger. It can be concluded that although the random slope is significant, there is a large uncertainty about the slope variance.

Table 14.3 Parameter estimates for two models for distrust relations.

	Model 1		Model 2	
Fixed Effect	Par.	S.E.	Par.	S.E.
γ_0 Intercept	−1.92	0.13	−2.96	0.24
Political variables:				
γ_1 Party member ego	0.30	0.24	0.26	0.25
γ_2 Political function alter	−0.40	0.93	−0.15	0.78
γ_3 Party member × pol. function			−1.09	1.07
Role relations:				
γ_4 Colleague			1.19	0.21
γ_5 Superior			1.34	0.24
γ_6 Subordinate			−0.22	0.90
γ_7 Neighbor			2.31	0.31
Random Effect	Par.	S.E.	Par.	S.E.
Random intercept:				
$\tau_0^2 = \text{var}(U_{0j})$ intercept variance	1.55	0.37	1.68	0.44
Random slope:				
$\tau_2^2 = \text{var}(U_{2j})$ slope variance of political function	5.76	7.43	5.54	6.27
$\tau_{04} = \text{cov}(U_{0j}, U_{2j})$ intercept-slope covariance	−0.06	1.02	0.22	1.04
Deviance	1528.29		1519.95	

Model 2 in this table also includes the interaction between party membership of ego and political function of alter, and the fixed effect of the role relation between respondent and network member. The interaction variable is defined as $X_3 = X_1 \times X_2$. The role relation is captured by four dummy variables derived from a categorical variable with five values: acquaintance, colleague, subordinate, superior, neighbor. The 'acquaintance' relation is used as the reference category. The four resulting dummy variables, X_4 to X_7, are 1 if the relation is in the indicated category and 0 if it is not.

The results for Model 2 show that the interaction between the political variables for ego and for alter is not significant. With respect to the role relations, neighbors are distrusted most, the superiors and colleagues occupy a middle position in this respect, while acquaintances (which as the reference category have an effect 0.0) and subordinates are distrusted least. The random slope for alter's political function is hardly affected by the inclusion of the interaction effect and the role relations.

14.3.2 Representation as a threshold model

The multilevel logistic regression can also be formulated as a so-called threshold model. The dichotomous outcome Y, 'success' or 'failure', then is conceived as the result of an underlying non-observed continuous variable. When Y denotes passing or failing some test or exam, the underlying continuous variable could be the scholastic aptitude of the subject; when Y denotes whether the subject behaves in a certain way, the underlying variable could be a variable representing benefits minus costs of this behavior; etc. Denote the underlying variable by $\breve{Y}$. Then the threshold model states that Y is 1 if $\breve{Y}$ is larger than some threshold, and 0 if it is less than the threshold. Since the model is about unobserved entities, it is not a restriction to assume that the threshold is 0. This leads to the representation

$$Y = \begin{cases} 1 & \text{if } \breve{Y} > 0 \\ 0 & \text{if } \breve{Y} \leq 0 \, . \end{cases} \tag{14.17}$$

For the unobserved variable $\breve{Y}$, a usual random intercept model is assumed:

$$\breve{Y}_{ij} = \gamma_0 + \sum_{h=1}^{r} \gamma_h \, x_{hij} + U_{0j} + R_{ij} \, . \tag{14.18}$$

To represent a logistic regression model, the level-one residual of the underlying variable $\breve{Y}$ must have a ***logistic distribution***. This means that, when the level-one residual is denoted by R_{ij}, the cumulative distribution function of R_{ij} must be the logistic function,

$$\mathrm{P}(R_{ij} < x) = \text{logistic}(x) \qquad \text{for all } x \, , \tag{14.19}$$

defined in (14.11). This is a symmetric probability distribution, so that also

$$\mathrm{P}(-R_{ij} < x) = \text{logistic}(x) \qquad \text{for all } x \, .$$

Its mean is 0 and its variance is $\pi^2/3 = 3.29$. When it is assumed that R_{ij} has this distribution, the logistic random intercept model 14.15 is equivalent to the threshold model defined by (14.17) and (14.18).

To represent the random slope model (14.16) as a threshold model, we define

$$\breve{Y}_{ij} = \gamma_0 + \sum_{h=1}^{r} \gamma_h x_{hij} + U_{0j} + U_{1j} \, x_{1ij} + R_{ij} \, , \tag{14.20}$$

where R_{ij} has a logistic distribution. It then follows that

$$\mathrm{P}(Y_{ij} = 1) = \mathrm{P}\left(\breve{Y}_{ij} > 0\right) =$$

$$= \mathrm{P}\left(-R_{ij} < \gamma_0 + \sum_{h=1}^{r} \gamma_h \, x_{hij} + U_{0j} + U_{1j} \, x_{1ij}\right)$$

$$= \text{logistic}\left(\gamma_0 + \sum_{h=1}^{r} \gamma_h\, x_{hij} + U_{0j} + U_{1j}\, x_{1ij}\right) .$$

Since the logit and the logistic functions are each others inverses, the last equation is equivalent with (14.16).

If the residual R_{ij} has a standard normal distribution with unit variance, then the probit link function is obtained. Thus, the threshold model which specifies that the underlying variable $\breve{Y}$ has a distribution according to the hierarchical linear model of Chapters 4 and 5, with a normally distributed level-one residual, corresponds exactly to the multilevel probit regression model. Since the standard deviation of R_{ij} is $\sqrt{\pi^2/3} = 1.81$ for the logistic and 1 for the probit model, the fixed estimates for the logistic model will tend to be about 1.81 times as big as for the probit model and the variance parameters of the random part about $\pi^2/3 = 3.29$ times as big (but see Long (1997, p. 48), who notes that in practice the proportionality constant for the fixed estimates is closer to 1.7).

14.3.3 Residual intraclass correlation coefficient

The intraclass correlation coefficient for the multilevel logistic model can be defined in at least two ways. The first definition is by applying the definition in Section 3.3 straightforwardly to the binary outcome variable Y_{ij}. This approach was also mentioned in Section 14.2.1. It was followed, e.g., by Commenges and Jacqmin (1994).

The second definition is by applying the definition in Section 3.3 to the unobserved underlying variable $\breve{Y}_{ij}$. Since the logistic distribution for the level-one residual implies a variance of $\pi^2/3 = 3.29$, this implies that for a two-level logistic random intercept model with an intercept variance of τ_0^2, the intraclass correlation is

$$\rho_{\text{I}} = \frac{\tau_0^2}{\tau_0^2 + \pi^2/3} .$$

These two definitions are different and will lead to somewhat different outcomes. For example, for the empty model for the Norwegian cohabitation data presented in Table 14.1, the first definition yields $0.00222/(0.00222+0.244) = 0.009$, whereas the second definition leads to the value $0.032/(0.032+3.29) = 0.010$. In this case, the difference is immaterial and both values of the intraclass correlation are very small.

An advantage of the second definition is that it can be directly extended to define the residual intraclass correlation coefficient, i.e., the intraclass correlation which controls for the effect of explanatory variables. The example can be continued by moving to the two models in Table 14.2. The residual intraclass correlation controlling for age (Model 1) is $0.061/(0.061+3.29) = 0.018$, controlling for age as well as religion (Model 2) it is $0.049/(0.049+3.29) = 0.015$. Thus, introducing a level-one variable increases the residual intraclass correlation, while controlling for religion brings the intraclass correlation down again. This can be understood from the fact that religion is a level-one variable with, however, an appreciable variation at level two.

For the multilevel probit model, the second definition for the intraclass correlation (and its residual version) leads to

$$\rho_{\mathrm{I}} = \frac{\tau_0^2}{\tau_0^2 + 1},$$

since this model fixes the level-one residual variance of the unobservable variable $\breve{Y}_{ij}$ to 1.

14.3.4 Explained variance

There are several definitions of the explained proportion of variance (R^2) in single-level logistic and probit regression models. Reviews are given by Hagle and Mitchell (1992), Veall and Zimmermann (1992), and Windmeijer (1995). Long (1997, Section 4.3) presents an extensive overview. One of these definitions, the R^2 measure of McKelvey and Zavoina (1975), which is based on the threshold representation treated in Section 14.3.2, is considered very attractive in each of these reviews. In this section we propose a measure for the explained proportion of variance which extends McKelvey and Zavoina's measure to the logistic and probit random intercept model.

It is assumed that it makes sense to conceive of the dichotomous outcomes Y as being generated through a threshold model with underlying variable $\breve{Y}$. In addition, it is assumed that the explanatory variables X_h are random variables. (This assumption is always made for defining the explained proportion of variance, see the introduction of Section 7.1.) Therefore the explanatory variables are, like in Chapter 7, indicated by capital letters. For the underlying variable $\breve{Y}$, equation (14.18) gives the expression

$$\breve{Y}_{ij} = \gamma_0 + \sum_{h=1}^{r} \gamma_h X_{hij} + U_{0j} + R_{ij}.$$

Denote the fixed part by

$$\hat{Y}_{ij} = \gamma_0 + \sum_{h=1}^{r} \gamma_h X_{hij}.$$

This variable is also called the ***linear predictor*** for Y. Its variance is denoted by σ_F^2. The intercept variance is $\mathrm{var}(U_{0j}) = \tau_0^2$ and the level-one residual variance is denoted $\mathrm{var}(R_{ij}) = \sigma_R^2$. Recall that σ_R^2 is fixed to $\pi^2/3 = 3.29$ for the logistic and to 1 for the probit model.

For a randomly drawn level-one unit i in a randomly drawn level-two unit j, the X-values are randomly drawn from the corresponding population and hence the total variance of $\breve{Y}_{ij}$ is equal to

$$\mathrm{var}(\breve{Y}_{ij}) = \sigma_F^2 + \tau_0^2 + \sigma_R^2.$$

The explained part of this variance is σ_F^2 and the unexplained part is $\tau_0^2 + \sigma_R^2$. Of this unexplained variation, τ_0^2 resides at level two and σ_R^2 at level one. Hence the proportion of explained variation can be defined by

$$R^2_{\mathrm{dicho}} = \frac{\sigma_F^2}{\sigma_F^2 + \tau_0^2 + \sigma_R^2}. \tag{14.21}$$

The corresponding definition of the residual intraclass correlation,

$$\rho_{\mathrm{I}} = \frac{\tau_0^2}{\tau_0^2 + \sigma_R^2}, \tag{14.22}$$

was also given in Section 14.3.3.

To estimate (14.21), one can first compute the linear predictor $\hat{Y}_{ij}$ using the estimated coefficients $\hat{\gamma}_0$, $\hat{\gamma}_1$, ..., $\hat{\gamma}_r$ and then estimate the variance σ_F^2 by the observed variance of this computed variable. It is then easy to plug this value into (14.21) together with the estimated intercept variance $\hat{\tau}_0^2$ and the fixed value of σ_R^2.

In the interpretation of this R^2 value it should be kept in mind that such values are known for single-level logistic regression to be usually considerably lower than the OLS R^2 values obtained for predicting continuous outcomes.

Example 14.5 *Taking a science subject in high school.*
This example continues the analysis of the data set of Example 8.3 about the cohort of pupils entering secondary school in 1989, studied by Dekkers, Bosker, and Driessen (1998). The focus is now on whether the pupils chose at least one science subject for their final examination. The sample is restricted to pupils in general education (excluding junior vocational education), and to only those who progressed to their final examination (excluding drop-outs and pupils who repeated grades once or twice). This left 3,432 pupils distributed over 240 secondary schools. There were 736 pupils who took no science subjects, 2,696 who took one or more.

The multilevel logistic regression model was estimated by the MIXOR program, with gender (0 for boys, 1 for girls), and minority status (0 for children of parents born in industrialized countries, 1 for other countries) as explanatory variables. The results are presented in Table 14.4.

Table 14.4 Estimates for probability to take a science subject.

	Model 1	
Fixed Effect	Coefficient	S.E.
γ_0 Intercept	2.487	0.110
γ_1 Gender	−1.515	0.102
γ_2 Minority status	−0.727	0.195
Random Effect	Var. Comp.	S.E.
Level-two variance:		
$\tau_0^2 = \mathrm{var}(U_{0j})$	0.481	0.082
Deviance	3238.27	

The linear predictor for this model is

$$\hat{Y}_{ij} = 2.487 - 1.515\,\mathrm{Gender}_{ij} - 0.727\,\mathrm{Minority}_{ij}$$

and the variance of this variable in the sample is $\hat{\sigma}_F^2 = 0.582$. Therefore the explained proportion of variation is

$$R^2_{\mathrm{dicho}} = \frac{0.582}{0.582 + 0.481 + 3.29} = 0.13\ .$$

In other words, gender and minority status explain about 13 percent of the variation in whether the pupil takes at least one science subject for the high school exam.

The unexplained proportion of variation, $1 - 0.13 = 0.87$, can be written as

$$\frac{0.481}{0.582 + 0.481 + 3.29} + \frac{3.29}{0.582 + 0.481 + 3.29} = 0.11 + 0.76 = 0.87,$$

which represents the fact that 11 percent of the variation is unexplained variation at the school level and 76 percent is unexplained variation at the pupil level. The residual intraclass correlation is $\rho_{\mathrm{I}} = 0.481/(0.481 + 3.29) = 0.13$.

14.3.5 Consequences of adding effects to the model

When a random intercept is added to a given logistic or probit regression model, and also when variables with fixed effects are added to such a model, the effects of earlier included variables may change. The nature of this change, however, can be different from such changes in OLS or multilevel linear regression models for continuous variables.

This phenomenon can be illustrated by continuing the example of the preceding section. Table 14.5 presents three other models for the same data, in all of which some elements were omitted from Model 1 as presented in Table 14.4.

Table 14.5 Three models for taking a science subject.

	Model 2		Model 3		Model 4	
Fixed Effect	Par.	S.E.	Par.	S.E.	Par.	S.E.
γ_0 Intercept	2.246	0.090	2.440	0.109	1.448	0.058
γ_1 Gender	-1.397	0.102	−1.507	0.102		
γ_2 Minority status					−0.644	0.174
Random Effect	Par.	S.E.	Par.	S.E.	Par.	S.E.
Level-two variance:						
$\tau_0^2 = \mathrm{var}(U_{0j})$			0.514	0.084	0.293	0.043
Deviance	3345.15		3251.86		3476.06	

Models 2 and 3 include only the fixed effect of gender; Model 2 does not contain a random intercept and therefore is a single-level logistic regression model, Model 3 does include the random intercept. The deviance difference ($\chi^2 = 3345.15 - 3251.86 = 93.29, d.f. = 1$) indicates that the random intercept is very significant. But the sizes of the fixed effects increase in absolute value when adding the random intercept to the model, both by about 8 percent. Gender is evenly distributed across the 240 schools, and one may wonder why the absolute size of the effect of gender increases when the random school effect is added to the model.

Model 4 differs from Model 1 in that the effect of gender is excluded. The fixed effect of minority status in Model 4 is −0.644, whereas in Model 1 it is −0.727. The intercept variance in Model 4 is 0.293 and in Model 1 it is 0.481. Again, gender is evenly distributed across schools and across the majority and minority pupils, and the question is how to interpret the fact that the intercept variance, i.e., the unexplained between-school variation, rises, and that also effect of minority status becomes larger in absolute value, when the effect of gender is added to the model.

The explanation can be given on the basis of the threshold representation. When all fixed effects γ_h and also the random intercept U_{0j} and the level-one residual R_{ij} would be multiplied by the same positive constant c, then the unobserved variable $\breve{Y}_{ij}$ would also be multiplied by c. This corresponds also to multiplying the variances τ_0^2 and σ_R^2 by c^2. However, it follows from (14.17) that the outcome Y_{ij} would not be affected because when $\breve{Y}_{ij}$ is positive then so is $c\breve{Y}_{ij}$. This shows that the regression parameters and the random part parameters of the multilevel logistic and probit models are meaningful only because the level-one residual variance σ_R^2 has been fixed to some value, and that this value is more or less arbitrary.

The meaningful parameters in these models are the *ratios* between the regression parameters γ_h, the random effect standard deviations τ_0 (and possibly τ_1, etc.), and the level-one residual standard deviation σ_R. Armed with this knowledge, we can understand the consequences of adding a random intercept or a fixed effect to a logistic or probit regression model.

When a single-level logistic or probit regression model has been estimated, the random variation of the unobserved variable $\breve{Y}$ in the threshold model is σ_R^2. When subsequently a random intercept is added, this random variation becomes $\sigma_R^2 + \tau_0^2$. For explanatory variables that are evenly distributed between the level-two units, the ratio of the regression coefficients to the standard deviation of the (unexplained) random variation will remain approximately constant. This means that the regression coefficients will be multiplied by about the factor

$$\sqrt{1 + \frac{\tau_0^2}{\sigma_R^2}} .$$

In the comparison between Models 2 and 3 above, this factor is $\sqrt{1 + (0.514/3.29)} = 1.08$. This is indeed approximately the number by which the regression coefficients were multiplied, when going from Model 2 to Model 3.

It can be concluded that, compared to single level logistic or probit regression analysis, including random intercepts tends to increase (in absolute value) the regression coefficients. In the biostatistical literature, this is known as the phenomenon that population-averaged effects (i.e., effects in models without random effects) are closer to zero than cluster-specific effects (which are the effects in models with random effects). Further discussions can be found in Neuhaus et al. (1991), Neuhaus (1992), and Diggle et al. (1994, Section 7.4).

Now suppose that a multilevel logistic or probit regression model has been estimated, and the fixed effect of some level-one variable X_{r+1} is added to the model. One would think that this would lead to a decrease in the level-one residual variance σ_R^2. However, this is impossible as this residual variance is fixed so that instead, the estimates of the other regression coefficients will tend to become larger in absolute value and the intercept variance (and slope variances, if any) will also tend to become larger. If the level-one variable X_{r+1} is uncorrelated with the other included fixed effects and also is evenly distributed across the level-two units (i.e., the intraclass correlation of X_{r+1} is about nil), then the regression coefficients γ_h and the standard deviations τ_0 (etc.) of the random effects will all increase by about the same factor. Correlations between X_{r+1} and other variables or positive intraclass correlation of X_{r+1} may distort this pattern to a greater or smaller extent.

This explains that the effect of minority status and the intercept variance increase when going from Model 4 to Model 1. The standard deviation of the random intercept increases by a larger factor than the regression coefficient of minority status, however. This might be related to an interaction between the effects of gender and minority status and to a very even distribution of the sexes across the schools (cf. Section 7.1).

The same can be observed in the example of the Norwegian cohabitation data when going from the empty model in Table 14.1 to Model 1 in Table 14.2. The level-one variables added (age and non-linear transformations of age) led to a substantial increase in the intercept variance (from 0.032 to 0.061). This is explained by the argument above. That adding religion, when going from Model 1 to Model 2 in Table 14.2, does not lead to a further increase in the intercept variance must be due to the fact that the composition of the regions (level-two units) with respect to religion is uneven.

14.3.6 Bibliographic remarks

Random effects models for binary data have been studied by many authors. Anderson and Aitkin (1985), Stiratelli, Laird, and Ware (1984), Wong and Mason (1985), and Longford (1994, based on his earlier papers) were among the first to propose estimation methods for the logistic-normal model treated in Section 14.2. Gibbons and Bock (1987), McCulloch (1994), and Ochi and Prentice (1984) proposed probit-normal models. A concise review of work up to 1992 is given by Searle et al. (1992, Chapter 10). Further references can be found in these papers and in Section 14.2.5.

A review of fixed and random coefficient models for discrete data is given by Hamerle and Ronning (1995). General multilevel models for non-normally distributed variables were considered, e.g., by Goldstein (1991). A textbook that treats various approaches to estimation and contains many further references is Davidian and Giltinan (1995).

14.4 Ordered categorical variables

Variables that have as outcomes a small number of ordered categories are quite common in the social and biomedical sciences. Examples of such variables are outcomes of questionnaire items (with outcomes, e.g., 'completely disagree', 'disagree', 'agree', 'completely agree'), a test scored by a teacher as 'fail', 'satisfactory', or 'good', etc. This section is about multilevel models where the dependent variable is such an ordinal variable.

When the number of categories is two, the dependent variable is dichotomous and Section 14.2 applies. When the number of categories is rather large (5 or more), it may be possible to approximate the distribution by a normal distribution and apply the hierarchical linear model for continuous outcomes. The main issue in such a case is the homoscedasticity assumption: is it reasonable to assume that the variances of the random terms in the hierarchical linear model are constant? (The random terms in a random intercept model are the level-one residuals and the random intercept, R_{ij} and U_{0j} in (4.7).) To check this, it is useful to investigate the skewness of the distribution. If in some groups, or for some values of the explanatory variables, the dependent variable assumes outcomes that are very skewed toward the lower or upper end of the scale, then the homoscedasticity assumption is likely to be violated.

If the number of categories is small (3 or 4), or if it is between 5 and, say, 10, and the distribution cannot well be approximated by a normal distribution, then statistical methods for ordered categorical outcomes can be useful. For single-level data such methods are treated, e.g., in McCullagh and Nelder (1989) and Long (1997).

It is usual to assign numerical values to the ordered categories, taking into account that the values are arbitrary. To have a notation that is compatible with the dichotomous case of Section 14.2, the values for the ordered categories are defined as $0, 1, ..., c-1$, where c is the number of categories. Thus, in the four point scale mentioned above, 'completely disagree' would get the value 0, 'disagree' would be represented by 1, 'agree' by 2, and 'completely agree' by the value 3. The dependent variable for level-one unit i in level-two unit j is again denoted Y_{ij}, so that Y_{ij} now assumes values in the set $\{0, 1, 2, ..., c-1\}$.

A very useful model for this type of data is the ***multilevel ordered logistic regression model***, also called the ***multilevel ordered logit model*** or the ***multilevel proportional odds model***; and the closely related ***multilevel ordered probit model***. These models are discussed, e.g., by Agresti and Lang (1993), Gibbons and Hedeker (1994), Hedeker and Gibbons (1994), and Goldstein (1995, Section 7.7). A three-level model was discussed by Gibbons and Hedeker (1997). They can be formulated as threshold models like in Section 14.3.2, now with $c-1$ thresholds rather than one. The real line is divided by the thresholds into c intervals (of which the first and the last have infinite length), corresponding to the c ordered categories. The first threshold is $\theta_0 = 0$, the higher thresholds are denoted θ_1, θ_2, ..., θ_{c-2}. Threshold θ_k defines the boundary between the intervals corresponding to observed outcomes k and $k+1$ (for $k = 0, ..., c-2$). The assumed unobserved underlying continuous variable is again denoted by $\check{Y}$ and the observed categorical

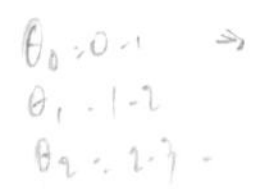

variable Y is related to $\check{Y}$ by the 'measurement model' defined as

$$Y = \begin{cases} 0 & \text{if } \check{Y} \leq \theta_0 \\ 1 & \text{if } \theta_0 \quad < \check{Y} \leq \theta_1 \, , \\ k & \text{if } \theta_{k-1} < \check{Y} \leq \theta_k \quad (k = 2, ..., c-2), \\ c-1 & \text{if } \theta_{c-2} < \check{Y} \, . \end{cases} \tag{14.23}$$

For $c = 2$ categories this reduces to just the two-category threshold representation in (14.17).

The random intercept ordered category model with explanatory variables X_1 to X_r is based on the 'structural model' for the unobserved underlying variable,

$$\check{Y}_{ij} = \gamma_0 + \sum_{h=1}^{r} \gamma_h \, x_{hij} + U_{0j} + R_{ij} \, . \tag{14.24}$$

The structural model and the measurement model (14.23) together determine the distribution of Y_{ij}. If R_{ij} has the logistic distribution (14.19) this results in the ***multilevel ordered logistic regression model***, also called the multilevel ordered logit model or multilevel proportional odds model. If R_{ij} has the standard normal distribution, this leads to the ***multilevel ordered probit model***. The differences between these two models are minor and the choice between them is a matter of fit and convenience.

The parameters of the structural model can be interpreted in principle just as in the hierarchical linear model. The intraclass correlation coefficient can be defined as in Section 14.3.3. This definition referred only to the model for the underlying variable $\check{Y}$, and therefore can be applied immediately to the multi-category case. Similarly, the proportion of explained variance can be defined, as in Section 14.3.4, by the formula

$$R^2_{\text{poly}} = \frac{\sigma_F^2}{\sigma_F^2 + \tau_0^2 + \sigma_R^2} \, , \tag{14.25}$$

where σ_F^2 is the variance of the fixed part (or the linear predictor) while σ_R^2 is $\pi^2/3 = 3.29$ for the logistic model and 1 for the probit model.

The threshold parameters are usually of secondary importance and reflect the marginal probabilities of the outcome categories: if category k has a low probability then θ_{k-1} will be not much less than θ_k. For more discussion about the interpretation of the fixed parameters we refer to the literature on the single-level version of this model, such as Long (1997).

The model can be extended with a random slope in a straightforward manner. However, estimation algorithms for these models are less stable than for the standard hierarchical linear model, and it is not uncommon that it is impossible to obtain converging parameter estimates for models with even only one random slope.

What was said in Section 14.3.5 about the effect of adding level-one variables to a multilevel logistic regression model is valid also for the multilevel ordered logit and probit models. When some model has been fitted and an important level-one variable is added to this model, this will tend to increase the level-two variance parameters (especially if the newly added variable explains mainly within-group variation), the threshold parameters, and the absolute sizes of the regression coefficients (especially for variables that are uncorrelated with the newly added variable).

These models can be estimated by the programs MLn/MLwiN (Yang, Rasbash, and Goldstein, 1998), HLM version 5, and MIXOR (Hedeker and Gibbons, 1996a), see Chapter 15. The MIXOR program utilizes a procedure that provides a credible deviance statistic, which is not provided by MLn/MLwiN. MLn/MLwiN does not put the first threshold equal to 0, but the intercept. Thus, $\gamma_0 = 0$ and this parameter is not estimated, but instead θ_0 is not equal to 0 and is estimated. This is a reparametrization of the model and yields parameters which can simply be translated into one another, since it follows from (14.23) and (14.24) that subtracting the same number of all thresholds as well as of γ_0 yields the same distribution of the observed variables Y_{ij}.

Dichotomization of ordered categories

Models for ordered categorical outcomes are more complicated to fit and to interpret than models for dichotomous outcomes. Therefore it can make sense also to analyse the data after dichotomizing the outcome variable. For example, if there are 3 outcomes, one could analyse the dichotomization 1 versus $\{2, 3\}$ and also $\{1, 2\}$ versus 3. Each of these analyses separately is based of course on less information, but may be easier to carry out and to interpret.

Suppose that indeed the multilevel c-category logistic or probit model, defined by (14.23) and (14.24), is valid. Then for each of the dichotomizations, the population parameters of the structural model (14.24), except for the fixed intercept γ_0, are the population parameters also of the multilevel logistic regression model. (The fixed intercept in the analysis of the dichotomized outcomes depends on the fixed intercept for the multicategory outcomes together with the threshold θ_k for this dichotomization.) This implies that the analyses of the dichotomized outcomes also provide insight into the fit of the model for the c categories. If the estimated regression parameters γ_1 to γ_r depend strongly on the dichotomization point, then it is likely that the multilevel multicategory logistic or probit model does not fit well.

Example 14.6 *The number of science subjects taken in high school.*
This example continues the analysis of Example 14.5, but now analyses the number of science subjects instead of only whether this number was larger than 0.

The number of science subjects ranges from 0 to 3 (maths, chemistry, physics). There were 736 pupils who took no science subjects, 1,120 who took one, 873 who took two, and 703 who took all three.

The multilevel 4-category logistic regression model was estimated by the MLwiN and MIXOR programs, with gender (0 for boys, 1 for girls), socio-economic status (an ordered scale with values 1–6, from which the mean, 3.68, was subtracted), and minority status (0 for children of parents born in industrialized countries, 1 for other countries) as explanatory variables. Both programs produced very similar results. The MIXOR results are presented to have the deviance available.

Table 14.6 contains the result for the empty model (which only has the thresholds and the intercept variance as parameters) and for the model with these three explanatory variables. The deviance is minus twice the log-likelihood reported by MIXOR. Further, MIXOR reports standard deviations of random effects rather than variances. To obtain consistency with the other tables in this book, we report the intercept variance instead. The standard error of the variance estimate is obtained from the formula $\text{S.E.}(\hat{\sigma}^2) = 2\,\hat{\sigma}\,\text{S.E.}(\hat{\sigma})$, see equation (5.15).

Table 14.6 Multilevel 4-category logistic regression model for the number of science subjects.

		Model 1		Model 2	
Threshold parameters		Threshold	S.E.	Threshold	S.E.
θ_1	Threshold 1 - 2	1.541	0.041	1.763	0.045
θ_2	Threshold 2 - 3	2.784	0.046	3.211	0.054
Fixed Effect		Coefficient	S.E.	Coefficient	S.E.
γ_0	Intercept	1.370	0.057	2.591	0.079
γ_1	Gender girls			−1.680	0.066
γ_2	SES			0.117	0.037
γ_3	Minority status			−0.514	0.156
Level-two Random Effect		Parameter	S.E.	Parameter	S.E.
τ_0^2	Intercept variance	0.243	0.034	0.293	0.040
Deviance		9308.8		8658.2	

In the empty model the intercept variance is 0.243. In the model with explanatory variables this parameter has increased to 0.293. This increase is in accordance with what was said on p. 229 about the effect of including level-one variables in the model. The same effect is responsible for the fact that the threshold parameters have increased. All three fixed effects in Model 2 are significant: girls tend to take considerably less science subjects than boys, the number of science subjects is in increasing functions of socio-economic status, and minority pupils tend to take less science subjects than non-minority pupils. The proportion of explained variance can be calculated from (14.25). The linear predictor is given here by

$$-1.680\,X_1 + 0.117X_2 - 0.514X_3\,,$$

and has variance 0.745. Hence the explained variance is

$$\frac{0.745}{0.745 + 0.293 + 3.29} = 0.17\,.$$

The thresholds are approximately equidistant (recall that the first threshold is $\theta_0 = 0$), with in Model 2 a difference between them of about 1.6. The gender effect has about the same size in absolute value, −1.68, so girls tend to take about one science subject less than boys. The effect of minority status, −0.514, amounts to an average difference of about 0.3 subject. The random intercept standard deviation is $\sqrt{0.293} = 0.54$, so the difference between schools is quite large (as the span from schools with the few percent lowest U_{0j}

to schools with the few percent highest U_{0j} is four standard deviations). The fact that the intercepts (fixed coefficients) are so different between Models 1 and 2 is because of the explanatory variables, only SES is centered: for Model 1 the intercept corresponds to the average $\check{Y}$ value of all pupils, whereas for Model 2 it corresponds to the average $\check{Y}$ value of the non-minority boys.

14.5 Multilevel Poisson regression

Another important type of discrete data is count data. For example, for a population of road crossings one might count the number of accidents in one year; or for a population of doctors, one could count how often in one year they are confronted with a certain medical problem. The set of possible outcomes of count data is the set of natural numbers: 0, 1, 2, 3, 4, ... The standard distribution for counts is the Poisson distribution (see textbooks on probability theory, or, e.g., Long, 1997, Section 8.1). Box et al. (1978, Section 5.6) gives a nice introduction to the analysis of count data on the basis of the Poisson distribution in cases where there are no explanatory variables. An extensive treatment of Poisson and other models for count data is given by Cameron and Trivedi (1998).

The Poisson distribution is an approximation of the binomial distribution for the situation that the number of trials is large and the success probability is low. For the road accident example, one could consider a division of the year into 8,760 one-hour periods, and assume that the chance is negligible that more than one accident will happen in any single hour, that the occurrence of accidents is independent for different hours, and that the probability of an accident in any given hour, say, p, is very low. The total number of accidents would then be binomially distributed with parameters 8,760 and p, and this distribution is extremely close to the Poisson distribution with mean $8{,}760 \times p$. Even when the 'success' probabilities are variable over trials (the accident probability is indeed variable from one hour to the next), the number of 'successes' will still have approximately a Poisson distribution.

Just like we described in Section 14.1 for the binomial distribution, also for the Poisson distribution there is a natural relation between the mean and the variance: its mean and the variance are equal. If Y has a Poisson distribution with mean λ, then

$$\mathcal{E}(Y) = \text{var}(Y) = \lambda \,. \tag{14.26}$$

If the counts tend to be large, their distribution can be approximated by a continuous distribution. If all counts are large enough, say, more than 8, then it is advisable to use the square root of the counts, $\sqrt{Y}$, as the dependent variable and apply the hierarchical linear model (with the usual assumption checks). The reason that this is a good approach resides in the fact that the square root transformation succeeds very well in transforming the Poisson distribution to an approximately homoscedastic normal distribution (the square root is the so-called variance-stabilizing transformation for the Poisson distribution; see Box et al. (1978), p. 144). An even better transformation to a homoscedastic distribution is the Freeman-Tukey transformation defined by $\sqrt{Y} + \sqrt{Y+1}$ (see Bishop, Fienberg, and Holland, 1975, Section 14.6.2).

If all or some of the counts are small, a normal distribution will not be satisfactory and a hierarchical generalized linear model can be considered. This model is analogous to the multilevel logistic regression model. However, instead of the probability of success of Section 14.2 we now model the *expected value of the count.* Denote by Y_{ij} the count for level-one unit i in group j, and by L_{ij} the expected (i.e., population average) count, given that unit i is in group j and given the values of the explanatory variables (if any). Then L_{ij} is necessarily a non-negative number, which could lead to difficulties if we considered linear models for this value (cf. Section 14.2.2). Just as the logit function was used as the link function for probabilities in Section 14.2.2, so the natural logarithm is mostly used as the link function for expected counts. For single level data this leads to the Poisson regression model (Long 1997, Chapter 8) which is a linear model for the natural logarithm of the counts, $\ln(L_{ij})$. For multilevel data, hierarchical linear models are considered for the logarithm of L_{ij}. The random intercept Poisson regression model thus is formulated, analogous to (14.15), as a regression model plus a random intercept for the logarithm of the expected count:

$$\ln(L_{ij}) = \gamma_0 + \sum_{h=1}^{r} \gamma_h \, x_{hij} + U_{0j} \ . \tag{14.27}$$

The variance of the random intercept is denoted again by τ_0^2. This model is treated in Diggle, Liang, and Zeger (1994, Section 9.4) and briefly in Goldstein (1995, Section 7.6).

The multilevel Poisson regression model can be estimated by various multilevel software packages (see Chapter 15): HLM, MIXPREG, MLwiN, and VARCL. The MIXPREG program employs a better estimation method, based on numerical integration rather than the Taylor series approximation used by the other programs. The numerical integration method also provides, unlike the Taylor series approximation, a deviance statistic which can be used for hypothesis testing. The MIXPREG documentation includes an extensive introduction to the multilevel Poisson regression model.

To transform the linear model back to the expected counts, the inverse transformation of the natural logarithm must be used, which is the exponential function $\exp(x) = e^x$. This function has the property that it transforms sums into products:

$$e^{a+b} = e^a \times e^b \ .$$

Therefore the explanatory variables and the level-two random effects in the (additive) multilevel Poisson regression model have multiplicative effects on the expected counts. For example, if there is only $r = 1$ explanatory variable, equation (14.27) is equivalent to

$$
\begin{aligned}
L_{ij} &= \exp\left(\gamma_0 + \gamma_1\, x_{1ij} + U_{0j}\right) \\
&= \exp\left(\gamma_0\right) \times \exp\left(\gamma_1\, x_{1ij}\right) \times \exp\left(U_{0j}\right) \,. \qquad (14.28)
\end{aligned}
$$

Therefore, each additional unit of X_1 will have the effect of ***multiplying*** the expected count by e^{γ_1}. Similarly, in a group with a high intercept, e.g., two standard deviations so that $U_{0j} = 2\,\tau_0$, the expected count will be $e^{2\tau_0}$ times as high as in a group with an average value, $U_{0j} = 0$, of the intercept.

In models for counts it is quite usual that there is a variable D that is known to be proportional to the expected counts. For example, if the count Y_{ij} is the number of events in some time interval of non-constant length d_{ij}, it often is natural to assume that the expected count is proportional to this length of the time period. If in the example of counts of medical problems there are several doctors each with his or her own population of patients, then D could be the size of the patient population of the doctor. In view of equation (14.28), in order to let the expected count be proportional to D, there should be a term $\ln(d_{ij})$ in the linear model for $\ln(L_{ij})$, with a regression coefficient fixed to 1. Such a term is called an ***offset*** in the linear model (see McCullagh and Nelder, 1989, and Goldstein, 1995, Section 7.6).

Example 14.7 *Number of doctor calls outside office hours.*
The distribution of the number of doctor calls outside office hours is of importance for the organization of the emergency duty service. In this example the number of calls outside office hours (house visits by the doctor and surgery visits by the patient) is considered for five general medical practices in the town of Groningen. The counts were collected per day during the years 1996 and 1997. The patient population can be considered to be fixed in size, with a total of 15,145 patients. The data were registered by the 'Registration Network Groningen', see Van der Werf et al. (1998). Special holidays (like Christmas, etc.) were deleted from the data, and there remained 704 days. Differences between the practices were not important, so that the data for the five practices were combined.

The number of calls ranges from 0 to 11. The average is 1.89 and the standard deviation 2.05. Thus, the variance is considerably greater than the mean, which indicates that a Poisson distribution with constant expected value over the year is not an adequate assumption. The main difference is obviously between weekends (where every call is outside office hours) and working days (with usually 9 office hours and 15 hours of emergency service). The average number of calls ranges from 0.99 on Wednesdays to 4.29 on Saturdays. There are also some differences between the months. In July (the holiday season) there are on average 1.50 calls per day, the average for December is 2.33. The other months are in between.

This data defines a time series of daily counts. In addition to the purely random fluctuations from day to day (represented by the Poisson distribution) and the systematic variation between weekdays and between months (to be represented by fixed effects), there are random fluctuations over time in the amount of health problems. Such fluctuations can be due to the weather, infection epidemias, etc., and will lead to a higher or lower level of health problems lasting for a few days to a couple of weeks. These fluctuations result in some pattern of autocorrelation of this time series (e.g., correlation between the number of calls for consecutive days). The fluctuations could be modeled in various ways (e.g., autocorrelation or moving average models for the residuals in a Poisson regression model). However, time series models for Poisson distributions are not very well-known. A practical way for multilevel modelers is to use the multilevel Poisson regression model with nesting of the days in longer periods such as weeks, fortnights, or months. When the length of this period is chosen in accordance with the length of the fluctuations in amount of health problems, the main part of the autocorrelation in the number of calls outside office hours will be represented by the random part at level two. Some trials with periods of different lengths revealed that the highest intercept variance is obtained for one-week periods. This implies that the random epidemiological fluctuations in calls outside office hours have a length in the order of magnitude of a week: some weeks are busier than others.

Therefore we present results of an analysis where the level-two units are the weeks, defined to run from Monday to Sunday. So Y_{ij} is defined as the number of calls outside office hours for days i in week j. There are data for a total of 704 days in 105 weeks. The data were analysed by means of MIXPREG. The results are in Table 14.7.

Table 14.7 Two Poisson models for number of calls outside office hours.

		Model 1		Model 2	
Fixed Effect		Coefficient	S.E.	Coefficient	S.E.
γ_0	Intercept	0.062	0.041	0.049	0.048
γ_1	Weekend	1.297	0.056		
γ_2	Saturdays			1.381	0.065
γ_3	Sundays			1.199	0.068
γ_4	June			−0.197	0.104
γ_5	July			−0.170	0.104
γ_6	August			0.147	0.087
γ_7	September			0.174	0.089
γ_8	December			0.215	0.088
Level-two Random Effect		Variance	S.E.	Variance	S.E.
τ_0^2	Intercept variance	0.028	0.007	0.012	0.004
Deviance		2293.87		2272.64	

The first model only distinguishes between weekend days and other days, and already captures most of the variation. Equation (14.28), and its extension to more than one explanatory variable, is used for the interpretation of the parameters. For working days during average weeks, the expected count is $e^{0.062} = 1.06$ calls per day. During weekends this is $e^{0.062+1.297} = 3.89$. The variation between weeks has a standard deviation of $\tau_0 = \sqrt{0.028} = 0.17$ on the logarithmic scale. Since the year has 52 weeks, it may be expected that about once per year there is a week that is at least two standard deviations below average, and also a week that is at least two standard deviations above average. Two standard deviations is $2\,\tau_0 = 0.34$, which corresponds to a multiplicative effect of $e^{0.34} = 1.40$. In other words, about once every year one may expect a week that is at least 1.40 times more busy than average, as well as a week that is at least 1.40 times less busy than average. The deviance of the Poisson regression model without a random intercept is 2299.03, so the deviance test for the random intercept has $\chi^2 = 2299.03 - 2293.87 = 5.16, d.f. = 1$, one-sided (see Section 6.2.1) $p < 0.02$. Thus, there is significant unexplained between-week variability.

The second model distinguishes more precisely between weekdays and also between months. The days and months appearing in the table were selected after fitting a model with separate effects for all days and all months, and deleting very small effects. It turned out that differences between working days were minor, and that also between the months of January to May, October, and November, the differences were quite small. So the intercept refers to working days in the mentioned months. On such days the average number of calls is $e^{0.049} = 1.05$. The busiest day is Saturday ($e^{1.381} = 3.98$ times busier than working days), the busiest month is December ($e^{0.215} = 1.24$ times busier than January - May, October, and November). Finally, one the basis of the estimated intercept variance, a 'randomly very busy' week, as occurring once every year, is expected to be $e^{2\sqrt{0.012}} = 1.24$ times more busy than would be expected on the basis of the month in which it falls.

However, the deviance obtained when omitting the random intercept from Model 2 is 2273.65, so the random intercept is not significant ($\chi^2 = 1.01, d.f. = 1$, one-sided $p > 0.10$). Thus, the weekdays and months as used in Model 2 are sufficient to explain the variability between the daily number of calls outside office hours.

15 Software

Almost all procedures treated in this book can be carried out by standard software for multilevel statistical models. This of course is intentional, since this book covers those parts of the theory of the multilevel model that can be readily applied in everyday research practice. This chapter briefly reviews the available software in three sections.

The first one is devoted to special purpose programs, such as HLM and MLwiN, specifically designed for multilevel modeling. The second section treats modules in general purpose software packages, such as SAS and SPSS, that allow for (some) multilevel modeling. SAS has full-fledged possibilities for multilevel modeling. SPSS and BMDP have very limited possibilities for multilevel analysis, but can help absolute beginners in this field to learn to apply the basics of multilevel modeling, mainly the techniques treated in Chapter 4: random intercept models, but no random slopes. BMDP includes, in addition, special modules for longitudinal models of the kind treated in Chapter 12. Stata has some possibilities, especially for tests of fixed effects that take into account the two-level structure. The third section, finally, mentions some specialized software programs, built for specific research purposes.

Details on the specialized multilevel software packages can be found via the links provided on the Multilevel Models Project homepage, with internet address http://www.ioe.ac.uk/multilevel/ and via its mirror sites (with identical contents) with addresses http://www.medent.umontreal.ca/multilevel/ and http://www.edfac.unimelb.edu.au/multilevel/. A review of many multilevel programs is at http://www.stat.ucla.edu./~deleeuw/software.pdf.

To keep things simple, we only use two-level examples in this chapter. Generalizations to more levels are straightforward, and the interested reader can find enough support in the manuals which accompany the packages. Commands to reproduce some of the examples in this book using HLM, MLn/MLwiN, MIXOR, and MIXPREG can be found at the web site, http://stat.gamma.rug.nl/snijders/multilevel.htm.

15.1 Special software for multilevel modeling

Multilevel software packages aim at researchers who specifically seek to apply multilevel modeling techniques. Other statistical techniques are not available in these specific multilevel programs (leaving exceptions aside).

Of these programs, only MLn/MLwiN currently contains facilities for data manipulation. Each of the programs described in this section was designed by pioneers in the field of multilevel modeling.

15.1.1 HLM

HLM was written by Bryk, Raudenbush and Congdon (1996), and the theoretical background behind most applications can be found in Bryk and Raudenbush (1992). The main exception is the case of discrete dependent variables which is covered in the program but not in the textbook. The main features of HLM are its interactive operation (although one can run the program in batch mode as well) and the fact that it is rather easy to learn. Therefore it is well suited for undergraduate courses and for postgraduate courses for beginners. The many options available make it a good tool for professional researchers as well. Information is obtainable from the web site, http://www.ssicentral.com/hlm/mainhlm.htm, which also features a demo version.

Input consists of separate files for each level in the design, linked by common identifiers. In a simple two-level case, e.g., with data about students in schools, one file contains all the school data with a school identification code, while another file contains all the student data with the school identification code for each student. The input can come from system files of SPSS, SAS, SYSTAT or STATA, or may be in the form of ASCII text files. Once data have been read and stored into a sufficient statistics file (a kind of system file), there are three ways to work with the program. One way is to run the program interactively (answer to questions posed by the program). Another is to run the program in batch mode. Batch and interaction modes also can be combined. And finally, since there is a Windows version, one can make full use of the graphical interface. In each case the two-step logic of Section 6.4.1 is followed. We present the main commands for analysing a simple two-level example with two explanatory variables, X at level one and Z at level two. The model is

$$\begin{aligned} Y_{ij} &= \beta_{0j} + \beta_{ij}\, x_{ij} + R_{ij} \\ \beta_{0j} &= \gamma_{00} + \gamma_{01}\, z_j + U_{0j} \\ \beta_{1j} &= \gamma_{10} + \gamma_{11}\, z_j + U_{1j} \ , \end{aligned}$$

which of course can also be written in one equation as

$$\begin{aligned} Y_{ij} = \gamma_{00} &+ \gamma_{10}\, x_{ij} + \gamma_{01}\, z_j + \gamma_{11}\, z_j\, x_{ij} \\ &+ x_{ij}\, U_{1j} + U_{0j} + R_{ij} \ . \end{aligned} \tag{15.1}$$

This example will be used throughout this chapter, sometimes with some other explanatory variables added to the data set. In the example of the HLM commands, we present the commands at the left and our comments at the right.

Of course the first thing is to invoke the program.

`HLM2L data.ssm commandfile` — `data.ssm` contains the sufficient statistics and `commandfile` – a name which should be replaced by your own file name – contains the commands.

The basic part of the command file then could be as follows.

`Level1:y=intrcpt1+x+random` — the level-one model: Y is dependent, regresses on an intercept and on X, and there is a random residual;

`Level2:intrcpt1=intrcpt2+Z+random` — one part of the level-two model: the intercept of the first equation, β_{0j}, is regressed on a general intercept and on a level-two variable Z, while it also has a random component; so this model contains a random intercept;

`Level2:x=intercpt2+Z+random` — the second part of the level-two model: the regression coefficient β_{1j} of X on Y is itself regressed on an intercept (which is the fixed part of the effect of X on Y), a level-two variable Z, and it has a random component, so this model also contains a random slope.

In interactive mode, however, one does not need to know the syntax and one would simply answer to questions like (for didactical purposes we only take the most relevant questions):

```
Please specify a level-1 outcome variable:

   The choices are:
   For  Y enter 1    For  X enter 2 For  GENDER enter 3

What is the outcome variable:

Which level-1 predictors do you wish to use?

   The choices are:
   For  X enter  2   For  GENDER enter  3

Level-1 predictor? (Enter 0 to end)
```

```
Which level-2 predictors do you wish to use?

   The choices are:
   For  SIZE   enter 1    For  SECTOR enter 2
   For  BUDGET enter 3    For  Z      enter 4

Which level-2 predictors to model INTERCPT  ?
Level-2 predictor? (Enter 0 to end)

Which level-2 predictors to model   X  ?
Level-2 predictor? (Enter 0 to end)

Do you want to constrain the variances in any of the
level-2 random effects to zero?
```

The program assumes that you are potentially interested in any cross-level interaction effects: it asks for level-two variables that may be predictors of the X-effect on the dependent variable. Therefore you do not need to create product terms before running the analysis. If you are interested, however, in an interaction effect of variables defined at the same level, then you need to calculate the product beforehand, and put this new variable in the dataset. Moreover, unless you answer 'No' to the last question ('do you want to constrain the variances?'), all level-one predictor variables are assumed to have random slopes. Once one has estimated the model, the HLM session is closed.

The Windows version of HLM gives the program a graphical interface. It only helps to construct the command file, but it does this in a very convenient way indeed. In the Windows interface, one does not leave the HLM session after having fitted a model.

HLM does not allow for data manipulation, but both the input and output can come from, and can be fed into, SPSS, SAS, SYSTAT, or STATA. HLM does not go beyond three levels. It can be used for practically all the analyses presented in this book. Almost all examples in this book can be reproduced using HLM version 5. Some interesting features of the program are the possibility to test model assumptions directly, e.g., by test (9.4) for level-one heteroscedasticity, and the help provided to construct contrast tests. Furthermore the program routinely asks for centering of predictor variables, but the flipside of the coin is – in case one opts for group mean centering – that group means themselves must have been calculated outside HLM, if one whishes to use these as level-two predictor variables.

A special feature of the program is that it allows for statistical meta-analysis of research studies that are summarized by only an effect size estimate and its associated standard error (called in Bryk and Raudenbush, 1992, a 'V-known problem'). Other special features added to version 5 are the analysis of data where explanatory variables are measured with error (explained in Raudenbush and Sampson, 1999) and the analysis of multiply imputed data (see, e.g., Rubin, 1996).

15.1.2 MLn / MLwiN

MLn and MLwiN are the most extensive multilevel packages, written by researchers working at the multilevel models project at the London Institute of Education (Rasbash and Woodhouse, 1995; Goldstein et al., 1998). Current information can be obtained form the web site, http://www.ioe.ac.uk/mlwin/. MLwiN is based on the older DOS program MLn. It has a Windows-95 interface and current methodological developments are added to MLwiN, not to MLn. Almost all examples in this book can be reproduced using MLn/MLwiN. For heteroscedastic models, the term used in the MLn/MLwiN documentation is 'complex variation'. For example, level-one heteroscedasticity is complex level-one variation.

A nice feature of both MLn and MLwiN is that the programs make use of a statistical environment (NANOSTAT) which allows for data manipulation, graphing, simple statistical computations, file manipulation (like sorting), etc. Data manipulation procedures include some handy procedures relating to the multilevel data structure.

Input for MLn is one ASCII text file that contains all the data, including the level-one and level-two identifiers. This dataset is read and put in a 'worksheet', a kind of system file. This worksheet, which can include model specifications, variable labels, and results, can be saved and used in later sessions. There are again three ways to work with the program: in an interactive command mode, in batch mode, or interactively using the graphical interface of Windows 95. The interactive and batch modes are operated from the 'command interface' within the Windows interface.

To start with the first mode: the program does not ask any question, the user should know the commands he or she wants to feed into the program. What is interactive, however, is the answer to the command which prompts up immediately on the screen. To show how this works we construct the same model as we did with HLM, represented in (15.1). We start in this case from the point that the data have been read, and labels have been assigned to variables.

`Iden 2 'school'`	the identification for the level-two units is found in the variable labeled *school*;
`Iden 1 'pupil'`	identifier for the level-one units;
`Resp 'y'`	declare Y as the dependent variable;
`Expl 'cons'`	declare the variable *cons* as an explanatory variable; this variable *cons* is equal to 1 for every level-one unit, and its use is a trick to declare the intercept;
`Expl 'x'`	declare the variable X as an explanatory variable;
`Expl 'z'`	declare the variable Z as an explanatory variable; whether Z is a variable defined at level one or level two does not matter for the commands or computations;

`Calc c11='x'*'z'`	calculate a new variable, the product of X and Z (a cross-level interaction);
`Name c11 'x*z'`	give this cross product the name $'x * z'$;
`Expl 'x*z'`	declare this cross product as an explanatory variable;
`Setv 1 'cons'`	the intercept is random at level one, which is another way to express that the model contains a random residual at level one;
`Setv 2 'cons'`	the intercept is random at level two as well;
`Setv 2 'x'`	the effect of X on Y is random at level two, so this a random slope model;
`Sett`	show the model settings;
`Summ`	show the nesting structure;
`Star`	start the estimation procedure;
`Fixe`	report the estimates for the fixed part;
`Rand`	report the estimates for the random part;
`Like`	report the deviance.

A drawback of running the program in this command mode is that the user should know beforehand what he or she is going to do and which of the (more than 200) commands will do the tricks. The manuals that go with the software, however, are written in such a form (leading users through worked examples) that one easily learns to handle the most important commands. Moreover one can run parts of the program (like setting up the model) in batch form to save oneself the tedium of typing the same command over and over again. The batch file then has to contain the same sequence of commands as shown above. Since one does not leave the session after having fitted a model, one can freely change between working in command mode, through the interactive Windows interface, and in batch mode.

The graphical interface in the Windows 95 version MlwiN makes it possible to build models interactively, without knowing the commands. In this case the commands are available on the screen, as are the variables to be included in the model. It is still possible to give commands or batch commands in a dialog box. For some purposes, like transformation of variables (including the calculation of group means, within-group standard deviations, etc.), it is necessary to use the command mode. MlwiN makes full use of the possibility that Windows offers, like immediate graphical display, setting of coefficients by double-clicking them, drag-and-drop, etc.

MLn and MLwiN are very flexible software programs, also because the user can create his or her own macros (batch files), and a series of macros (batch files) is delivered with the software, e.g., for fitting models with ordered categorical outcomes and models with cross-classification. Since data manipulation can be carried out within the program itself one can quite easily create data sets for multivariate multilevel models. The worked examples, moreover provide the user with clear guidance how to do specific things like creating residuals, setting up contrast tests, etc. Of the available multilevel packages, MLwiN is the most flexible one, but it may take some time to get acquainted with its features. It is an excellent tool for professional researchers and statisticians.

15.1.3 VARCL

VARCL contains two programs, VARL3 and VARL9, written by Longford (1993b). VARL9 allows to model hierarchically structured data with up to nine levels, but is restricted to models in which all slopes are fixed, whereas VARL3 goes up to three levels and does allow for random slopes. The software is very easy to handle, being completely interactive, and one can fit multilevel models quite easily by just answering a limited series of questions.

The input for VARCL contains several sets of data: one dataset for each level (well sorted) and a 'basic information file'. The latter contains file format information, variable labels, an indication for the measurement level of variables, and an awkward part (a count of case numbers) that defines which series of cases from the level-one data file belong to one unit in the level-two data file.

The logic of VARCL is that one starts with defining a 'maximal model', which means that one defines a dependent variable, and in addition one takes from the dataset the potentially interesting predictors. If one wishes to do so, one can save this file (as a sort of system file.) After this, the model fitting can start. At this stage one should correctly know the numbering of the variables (the intercept is defined by default as the first variable by the program itself), and either by picking variables, including them all, or excluding some, the model is specified. The crucial part of the program is shown below.

```
THE  M O D E L  TO  BE  FITTED
--------------------------------


  VARIABLE                                          NESTING
 No. & NAME    CATEGORY    IN/OUT/CONSTR             LEVEL

    1   G. MEAN             IN                          1
    5   X                   IN                          1
   14   GENDER              IN                          1
   17   X*Z                 IN                          1
   18   Z                   IN                          2
```

```
        R A N D O M   P A R T   -   L E V E L    2

  VARIABLE
 No. & NAME

   1   G. MEAN       1

   5   X             0      0
```

The last lines give the representation of the specification of the random part at level two in the form of a lower triagonal matrix, where the entry '1' indicates that the corresponding element of the covariance matrix is included in the model, while the entry '0' indicates that it is not included.

```
 ENTER:      I . . .  INCLUDE A VARIABLE
             E . . .  EXCLUDE A VARIABLE
             X . . .  EXIT AND LIST THE VARIABLES
```

In this stage one defines the fixed part of the model. The random part can be specified following the question:

```
        ALTERATIONS  ON  THE  LEVEL  2  SCHOOL
 ENTER:    I . . .  INCLUDE A VARIABLE
           E . . .  EXCLUDE A VARIABLE
           X . . .  NO (MORE) ALTERATIONS ON THIS LEVEL
```

Now one can define variables that should have random slopes. Although VARCL is very easy to learn, it has drawbacks in the awkwardness of the basic information file and the impossibility of data manipulation. The dataset should be well constructed (e.g., product terms reflecting interaction effects) before one starts. Moreover, the only output is a text file which contains the regression equation, the standard errors, the variance components and their standard errors, and the deviance. Included in this file may also be the posterior means (Empirical Bayes level-two residuals). Facilities for model checks are very restricted: level-one residuals are not available, and to use the level-two residuals one must edit the textfile that contains them.

Many of the examples presented in this book could have been analysed in VARL3/VARL9, including models for binary outcomes and count data (except for the deviances for the discrete outcome models).

The program is designed in a very friendly way, so that it is ideal for didactical purposes. During the last ten years, however, the program has not gone through any evolution, which implies that especially the estimation of models for discrete outcomes does not meet the latest developments.

15.1.4 MIXREG, MIXOR, MIXNO, MIXPREG

D. Hedeker has constructed several modules for multilevel modeling, focusing at special models that go beyond the basic hierarchical linear model. Windows versions of these programs are freely available with manuals and examples at http://www.uic.edu/~hedeker/mix.html. PowerMac and Sun/Solaris versions are available at http://www.stat.ucla.edu/~deleeuw/mixfoo.

These programs have no facilities for data manipulation and use ASCII text files for data input. The algorithms are based on numerical integration. This contrasts with the Taylor expansions which are the basis for the algorithms for discrete outcome variables used in other multilevel computer programs. The difference is discussed in Section 14.2.5. Currently, the only publicly available programs for multilevel analysis of discrete outcomes that provide a credible log-likelihood statistic for deviance tests are these programs, joined for dichotomous variables only by HLM version 5.

In the interpretation of the output of these programs, it should be noted that the parametrization used for the random part is not the covariance matrix but its Cholesky decomposition. This is a lower triangular matrix C with the property that $C\,C' = \Sigma$, where Σ is the covariance matrix. The output does also give the estimated variance and covariance parameters, but (in the present versions) not their standard errors. If you want to know these standard errors, some additional calculations are necessary.

If the random part only contains the random intercept, the parameter in the Cholesky decomposition is the standard deviation of the intercept. The standard error of the intercept variance can be calculated with formula (5.15). For other parameters, the relation between the standard errors is more complicated. We treat only the case of one random slope. The level-two covariance matrix then is

$$T = \begin{pmatrix} \tau_0^2 & \tau_{01} \\ \tau_{01} & \tau_1^2 \end{pmatrix}$$

with the Cholesky decomposition denoted by

$$C = \begin{pmatrix} c_{00} & 0 \\ c_{01} & c_{11} \end{pmatrix} .$$

The correspondence between these matrices is

$$\begin{aligned} \tau_0^2 &= c_{00}^2 \\ \tau_{01} &= c_{00}\,c_{01} \\ \tau_1^2 &= c_{01}^2 + c_{11}^2 \, . \end{aligned}$$

Denote the standard errors of the estimated elements of C by s_{00}, s_{01}, and s_{11}, respectively, and the correlations between these estimates by $r_{00,01}$, $r_{00,11}$, and $r_{01,11}$. These standard errors and correlations are given in the output of the programs. Approximate standard errors for the elements of the covariance matrix of the level-two random part then are given by

$$\begin{aligned} \text{S.E.}(\hat\tau_0^2) &\approx 2\,c_{00}\,s_{00} \\ \text{S.E.}(\hat\tau_{01}) &\approx \sqrt{c_{00}^2\,s_{01}^2 + c_{01}^2\,s_{00}^2 + 2\,c_{00}\,c_{01}\,s_{00}\,s_{01}\,r_{00,01}} \\ \text{S.E.}(\hat\tau_1^2) &\approx 2\sqrt{c_{01}^2\,s_{01}^2 + c_{11}^2\,s_{11}^2 + 2\,c_{01}\,c_{11}\,s_{01}\,s_{11}\,r_{01,11}} \, . \end{aligned}$$

For the second and further random slopes, the formulae for the standard errors are even more complicated. These formulae can be derived with the multivariate delta method, explained, e.g., in Bishop, Fienberg, and Holland (1975, Section 14.6.3).

MIXREG (Hedeker and Gibbons, 1996b) is a computer program for mixed effects regression analysis with autocorrelated errors. This is suitable especially for longitudinal models (see Chapter 12). Various correlation patterns for the level-one residuals R_{ti} are allowed, such as autocorrelation or moving average dependence.

MIXOR (Hedeker and Gibbons, 1996a) provides estimates for multilevel models for dichotomous and ordinal outcome variables (cf. Chapter 14) allowing probit, logistic, and complementary log-log link functions. These models can include multiple random slopes. The present version (number 2) also permits right-censoring of the ordinal outcome (useful for analysis of multilevel grouped-time survival data), non-proportional odds (or hazards) for selected covariates, and the possibility for the random effect variance terms to vary by groups of level-one or level-two units. This allows estimation of many types of Item Response Theory models (where the variance parameters vary by the level-one items) as well as models where the random effects vary by groups of subjects (e.g., males versus females).

MIXNO implements a multilevel multinomial logistic regression model. As such, it can be used to analyse two-level categorical outcomes without an ordering. As in MIXOR (version 2), the random effect variance terms can also vary by groups of level-one or level-two units. This program has an extensive manual with various examples.

Finally, MIXPREG is a program to estimate the parameters of the multilevel Poisson regression model (treated in Section 14.5). This program also has an extensive manual with examples.

15.2 Modules in general purpose software packages

Several of the main general purpose statistical packages have incorporated modules for multilevel modeling. These usually are presented as modules for mixed models, random effects, random coefficients, or variance components. As it is, most researchers may be used to one of these packages and may feel reluctant to learn to handle the specialized software discussed above. Especially for these people the threshold to multilevel modeling may be lowered when they can stick to their own software. One should bear in mind, however, that many (maybe even most) of the things treated in this book cannot be done in these general purpose packages, with the exception of SAS. Moreover, the algorithms used in these general purpose programs are not necessarily very efficient for the specific case of multilevel models.

15.2.1 SAS, procedure MIXED

The SAS procedure MIXED has been up and running since 1996 (Littell et al., 1996). For experienced SAS users a quick introduction to multilevel modeling using SAS can be found in Singer (1998a, 1998b). Verbeke and Molenberghs (1997) present a SAS-oriented practical introduction to mixed linear models, i.e., the hierarchical linear model.

Unlike the other general purpose packages SAS allows for fitting very complex multilevel models and calculating all corresponding statistics (like deviance tests). PROC MIXED is a procedure oriented toward general mixed linear models, and allows to analyse practically all hierarchical linear model examples for continuous outcome variables presented in this book. The general mixed model orientation has the advantage that crossed random coefficients can be easily included, but the disadvantage that this procedure does not provide the specific efficiency for the nested random coefficients of the hierarchical linear model that is provided by dedicated multilevel programs.

Available are also the macros GLMMIX and NLINMIX, that allow for fitting the models for discrete outcome variables described in Chapter 14.

To get a basic idea of how the procedure works we take a bit of a syntax example from Singer (1998a, p. 8) (slightly altered for didactical purposes), used to fit model (15.1).

`Proc mixed`	invoke the procedure;
`Class school;`	the random factor or level-two identifier is *school*;
`Model y = x Z x*Z/solution`	fit the model with y as dependent and x, Z and $x \times Z$ as predictors; the intercept is implied by default;
`random intercept x/`	both the intercept and the predictor x have random effects, so this is a model with random intercept and random slope;
`sub=school type=un;`	level-one units are nested within ('**sub**') level-two units identified by *school*; estimate an unrestricted ('**un**') level-two covariance matrix.

15.2.2 SPSS, command VARCOMP

The most simple introduction to multilevel modeling for SPSS users is the module (in SPSS-language: 'command') VARCOMP, which is available in the latest versions (e.g., SPSS 7 for Windows). One can get no further, however, than the random intercept model, as described in Chapter 4 of this book, with the possibility also to include crossed random effects (Chapter 11).

In analysis of variance terms, the module allows to select from the list of variables the dependent variable, the factors (i.e., categorical explanatory variables), and to specify for each of the factors whether it has fixed or random effects. To specify a random intercept model the random factor will be the identification code for the level-two units, e.g., schools or neighborhoods. Continuous covariates can be added to the model, but only with fixed effects.

A possible use for researchers might be to find out whether an employed two-stage sampling strategy has caused design effects (i.e., whether there is substantial variation between the level-two units so that ignoring this leads to standard errors in one-level models being highly underestimated), even when controlling for relevant covariates. Bear in mind that VARCOMP, as is implied by the name of the command, only estimates variance components and not random slopes.

15.2.3 BMDP-V modules

BMDP provides a series of modules (BMDP-3V and following) that allow to do a bit of multilevel modeling. Since BMDP also has a Windows version with a graphical interface, the accessibility of the multilevel modeling procedures is quite high. Documentation is given by Dixon (1992) and Dixon and Merdian (1992). The relevant modules of BMDP allow to do the following kinds of analysis.

BMDP-3V allows, like the SPSS command VARCOMP, to fit models with nested and crossed random effects and with fixed effects of covariates. This implies that the basic models of Chapters 4 and 11 can be estimated, but random slopes cannot be included.

BMDP-4V and BMDP-5V are intended for longitudinal models, as described in Chapter 12. BMDP-5V includes the possibility to fit random slope models and more complicated residual covariance matrices, like those with autocorrelated level-one residuals. Although BMDP-5V is intended for longitudinal data, its facility for random slope models can also be used more generally to estimate models with one random slope: use the variable with the random slope as the time variable.

15.2.4 Stata

Stata (StataCorp, 1997) contains some modules that permit the estimation of certain multilevel models.

Module **loneway** ('long oneway') gives estimates for the empty model.

The **xt** series of modules are designed for the analysis of longitudinal data (cf. Chapter 12), but can be used (like BMDP-5V) for analysing any two-level random intercept model. Command **xtreg** estimates the random intercept model while **xtpred** calculates posterior means. Commands **xtpois** and **xtprobit**, respectively, provide estimates of the multilevel Poisson regression and multilevel probit regression models. These estimates are based on the so-called generalized estimating equations method.

A special feature of Stata is the so-called sandwich variance estimator, also called robust or Huber estimator. This estimator can be applied in many Stata modules that are not specifically intended for multilevel analysis. For statistics calculated in a single-level framework (e.g., estimated OLS regression coefficients), the sandwich estimator when using the keyword 'cluster' computes standard errors that are asymptotically correct under two-stage sampling. In terms of our Chapter 2, this solves many instances of 'dependence of a nuisance', although it does not help to get a grip on 'interesting dependence'.

15.3 Other multilevel software

There are other programs available for special purposes, and which can be useful to supplement the software mentioned above.

15.3.1 PinT

PinT is a specialized program for calculations of *P*ower *in* *T*wo-level designs, implementing the results of Snijders and Bosker (1993). This program can be used for *a priori* estimation of standard errors of fixed coefficients. This is useful in the design phase of a multilevel study, as discussed in Chapter 10 of this book. Being shareware, it can be downloaded with the manual from http://stat.gamma.rug.nl/snijders/multilevel.htm.

15.3.2 Mplus

A program with very general facilities for covariance structure analysis is M*plus* (Muthén and Muthén, 1998). Information about this program is available at http://www.StatModel.com. This program allows the analysis of univariate and multivariate two-level data not only with the hierarchical linear model but also with path analysis, factor analysis, and other structural equation models. Introductions to this type of model are given by Muthén (1994) and Kaplan and Elliott (1997).

15.3.3 MLA

The MLA program (see Meijer et al., 1995) can calculate estimates for two-level models by resampling methods. Various bootstrap implementations are included (based on random drawing from the data set or on random draws from an estimated population distribution) and also implementations of the jackknife (based on deletion of cases). It can be downloaded from http://www.fsw.leidenuniv.nl/www/w3_ment/medewerkers/busing/mla.htm.

15.3.4 BUGS

A special program which uses the Gibbs sampler is BUGS (Gilks et al., 1996). Gibbs sampling is a simulation-based procedure for calculating Bayesian estimates. This program can be used to estimate a large variety of models, including hierarchical linear models, possibly in combination with models for structural equations and measurement error. However, it requires balanced data (i.e., groups of equal sizes). It is mainly used as a research tool for statisticians, and available with manuals from http://www.mrc-bsu.cam.ac.uk/bugs/.

References

Agresti, A. (1990) *Categorical Data Analysis.* New York: Wiley.

Agresti, A. and Lang, J. (1993) 'A proportional odds model with subject-specific effects for repeated ordered categorical responses'. *Biometrika*, 80, 527–534.

Aitkin, M., Anderson, D., and Hinde, J. (1981) 'Statistical modeling of data on teaching styles'. *Journal of the Royal Statistical Society, Ser. A*, 144, 419–461.

Aitkin, M. and Longford, N. (1986) 'Statistical modelling issues in school effectiveness studies' (with discussion). *Journal of the Royal Statistical Society, Ser. A*, 149, 1–43.

Alker, H.R. (1969) 'A typology of ecological fallacies'. In: Dogan and Rokkan (eds.), pp. 69–86.

Anderson, D. and Aitken, M. (1985) 'Variance components models with binary response: Interviewer variability'. *Journal of the Royal Statistical Society, Ser. B*, 47, 203–210.

Arminger, G., Clogg, C.C., and Sobel, M.E. (1995) *Handbook of Statistical Modeling for the Social and Behavioral Sciences.* New York: Plenum Press.

Atkinson, A.C. (1985) *Plots, Transformations, and Regression.* Oxford: Clarendon Press.

Baltagi, B.H. (1995) *Econometric Analysis of Panel Data.* Chichester: Wiley.

Berkhof, J. and Snijders, T.A.B. (1998) 'Tests for a random coefficient in multilevel models'. *Submitted for Publication.*

Bishop, Y.M.M., Fienberg, S.E., and Holland, P.W. (1975) *Discrete Multivariate Analysis: Theory and Practice.* Cambridge, Mass.: The MIT Press.

Box, G.E.P., Hunter, W.G., and Hunter, J.S. (1978) *Statistics for Experimenters.* New York: Wiley.

Brekelmans, M. and Créton, H. (1993) 'Interpersonal teacher behavior throughout the career'. In: Wubbels, Th. and Levy, J. (eds.), *Do You Know What You Look Like? Interpersonal Relationships in Education.* London: The Falmer Press, pp. 46–55.

Breslow, N.E. and Clayton, D.G. (1993) 'Approximate inference in generalized linear mixed models'. *Journal of the American Statistical Association*, 88, 9–25.

Bryk, A.S. and Raudenbush, S.W. (1992) *Hierarchical Linear Models, Applications and Data Analysis Methods.* Newbury Park, CA: Sage Publications.

Bryk, A.S., Raudenbush, S.W., and Congdon, R.T. (1996) *HLM. Hierarchical Linear and Nonlinear Modeling with the HLM/2L and HLM/3L Programs.* Chicago: Scientific Software International.

Burstein, L. (1980) 'The analysis of multilevel data in educational research and evaluation'. *Review of Research in Education*, 8, 158–233.

Burstein, L., Linn, R.L., and Capell, F.J. (1978) 'Analyzing multilevel data in the presence of heterogeneous within-class regressions'. *Journal of Educational Statistics*, 3, 347–383.

Cameron, A.C., and Trivedi, P.K. (1998) *Regression Analysis of Count Data.* Cambridge: Cambridge University Press.

Chow, G.C. (1984) 'Random and changing coefficient models'. In: Z. Griliches and M.D. Intriligator (eds), *Handbook of Econometrics, Volume 2.* Amsterdam: North-Holland.

Cochran, W.G. (1977) *Sampling Techniques*, 3d edn. New York: Wiley.

Cohen, J. (1988) *Statistical Power Analysis for the Behavioral Sciences.* 2nd edn. Hillsdale, N.J.: Erlbaum.

Cohen, J. (1992) 'A power primer'. *Psychological Bulletin*, 112, 155–159.

Cohen, M. (1998) 'Determining sample sizes for surveys with data analyzed by hierarchical linear models'. *Journal of Official Statistics*, 14, 267–275.

Coleman, J.S. (1990) *Foundations of Social Theory.* Cambridge: Harvard University Press.

Commenges, D. and Jacqmin, H. (1994) 'The intraclass correlation coefficient: distribution-free definition and test'. *Biometrics*, 50, 517–526.

Commenges, D., Letenneur, L., Jacqmin, H., Moreau, Th., and Dartigues, J.-F. (1994) 'Test of homogeneity of binary data with explanatory variables'. *Biometrics*, 50, 613–620.

Cook, R.D. and Weisberg, S. (1982) *Residuals and Influence in Regression.* New York and London: Chapman & Hall.

Cook, R.D. and Weisberg, S. (1994) *An Introduction to Regression Graphics.* New York: Wiley.

Cook, T.D. and Campbell, D.T. (1979) *Quasi-experimentation. Design & Analysis Issues for Field Settings.* Boston: Houghton Mifflin Company.

Crowder, M.J. and Hand, D.J. (1990) *Monographs on Statistics and Applied Probability 41.* London: Chapman & Hall.

Davidian, M. and Giltinan, D.M. (1995) *Nonlinear Models for Repeated Measurement Data.* London: Chapman & Hall.

Davis, J.A., Spaeth, J.L., and Huson, C. (1961) 'A technique for analyzing the effects of group composition'. *American Sociological Review*, 26, 215–225.

De Leeuw, J. and Kreft, I. (1986) 'Random coefficient models for multilevel analysis'. *Journal of Educational Statistics*, 11 (1), 57–85.

De Weerth, C. (1998) *Emotion-related Behavior in Infants.* Ph.d. thesis, University of Groningen, The Netherlands.

Dekkers, H.P.J.M., Bosker, R.J., and Driessen, G.W.J.M. (1998) 'Complex inequalities of educational opportunities'. *Submitted for publication.*

Diggle, P.J., Liang, K.-Y., and Zeger, S.L. (1994) *Analysis of Longitudinal Data.* Oxford: Clarendon Press.

Dixon, W.J. (1992) *BMDP Statistical Software Manual.* Volume 2. Berkeley, Los Angeles: University of California Press.

Dixon, W.J., and Merdian, K. (1992) *ANOVA and Regression with BMDP.* Los Angeles: Dixon Statistical Associates.

Dogan, M. and Rokkan, S. (eds.) (1969) *Quantitative Ecological Analysis in the Social Sciences.* Cambridge, Mass.: The M.I.T. Press.

Donner, A. (1986) 'A review of inference procedures for the intraclass correlation coefficient in the one-way random effects model'. *International Statistical Review*, 54, 67–82.

Duncan, O.D., Curzort, R.P., and Duncan, R.P. (1961) *Statistical Geography: Problems in Analyzing Areal Data.* Glencoe, IL: Free Press.

Efron, B. and Morris, C.N. (1975) 'Data analysis using Stein's estimator and its generalizations'. *Journal of the American Statistical Association*, 74, 311–319.

Eisenhart, C. (1947) 'The assumptions underlying the analysis of variance'. *Biometrics*, 3, 1–21.

Engel, J. and Keen, A. (1994) 'A simple approach for the analysis of generalized linear mixed models'. *Statistica Neerlandica*, 48, 1–22.

Fisher, R.A. (1932) *Statistical Methods for Research Workers, 4th edn.* Edinburgh: Oliver & Boyd.

Gibbons, R.D. and Bock, R.D. (1987) 'Trends in correlated proportions'. *Psychometrika*, 52, 113–124.

Gibbons, R.D. and Hedeker, D. (1994) 'Application of random-effects probit regression models'. *Journal of Consulting and Clinical Psychology*, 62, 285–296.

Gibbons, R.D. and Hedeker, D. (1997) 'Random effects probit and logistic regression models for three-level data'. *Biometrics*, 53, 1527–1537.

Gilks, W.R., Richardson, S., and Spiegelhalter, D.J. (1996) *Markov Chain Monte Carlo in Practice.* London: Chapman & Hall.

Glass, G.V. and Stanley, J.C. (1970) *Statistical Methods in Education and Psychology.* Englewood Cliffs, NJ: Prentice-Hall.

Goldstein, H. (1986) 'Multilevel mixed linear model analysis using iterative generalized least squares'. *Biometrika*, 73, 43–56.

Goldstein, H. (1987) 'Multilevel covariance component models'. *Biometrika*, 74, 430–431.

Goldstein, H. (1991) 'Nonlinear multilevel models with an application to discrete response data'. *Biometrika*, 78, 45–51.

Goldstein, H. (1995) *Multilevel Statistical Models.* 2nd edn. London: Edward Arnold.
In electronic form at http://www.arnoldpublishers.com/support/goldstein.htm .

Goldstein, H. and Healy, M.J.R. (1995) 'The graphical presentation of a collection of means'. *Journal of the Royal Statistical Society*, Ser. A, 158, 175–177.

Goldstein, H., Healy, M.J.R., and Rasbash, J. (1994) 'Multilevel time series models with applications to repeated measures data'. *Statistics in Medicine*, 13, 1643–1655.

Goldstein, H. and Rasbash, J. (1996) 'Improved approximations for multilevel models with binary responses'. *Journal of the Royal Statistical Society*, Ser. A, 159, 505–513.

Goldstein, H., Rasbash, J., Plewis, I., Draper, D., Browne, W., Yang, M., Woodhouse, G., and Healy, M. (1998) *A user's guide to MLwiN.* London: Multilevel Models Project, Institute of Education, University of London.

Gouriéroux, C., and Montfort, A. (1996) *Simulation-based Econometric Methods.* Oxford: Oxford University Press.

Guldemond, H. (1994) *Van de Kikker en de Vijver* ('About the Frog and the Pond'). Ph.d. thesis, University of Amsterdam.

Hagle, T.M. and Mitchell II, G.E. (1992) 'Goodness-of-fit measures for probit and logit'. *American Journal of Political Science*, 36, 762–784.

Haldane, J.B.S. (1940) 'The mean and variance of χ^2, when used as a test of homogeneity, when expectations are small'. *Biometrika*, 31, 346–355.

Hamerle, A. and Ronning, G. (1995) 'Panel analysis for qualitative variables'. In: Arminger, Clogg, and Sobel (1995), pp. 401–451.

Hand, D. and Crowder, M. (1996) *Practical Longitudinal Data Analysis.* London: Chapman & Hall.

Hausman, J.A. and Taylor, W.E. (1981) 'Panel data and unobservable individual effects'. *Econometrica*, 49, 1377–1398.

Hays, W.L. (1988) *Statistics.* 4th edn. New York: Holt, Rinehart and Winston.

Hedeker, D. and Gibbons, R.D. (1994) 'A random effects ordinal regression model for multilevel analysis'. *Biometrics*, 50, 933–944.

Hedeker, D. and Gibbons, R.D. (1996a) 'MIXOR: A computer program for mixed-effects ordinal regression analysis'. *Computer Methods and Programs in Biomedicine*, 49, 157–176.

Hedeker, D. and Gibbons, R.D. (1996b) 'MIXREG: A computer program for mixed-effects regression analysis with autocorrelated errors'. *Computer Methods and Programs in Biomedicine*, 49, 229–252.

Hedeker, D., Gibbons, R.D., and Waternaux, C. (1999) 'Sample size estimation for longitudinal designs with attrition: Comparing time-related contrasts between two groups.' *Journal of Educational and Behavioral Statistics*, 24, 70–93.

Hedges, L.V. (1992) 'Meta-analysis'. *Journal of Educational Statistics*, 17, 279–296.

Hedges, L.V. and Olkin, I. (1985) *Statistical Methods for Meta-analysis.* New York: Academic Press.

Heyl, E. (1996) *Het Docentennetwerk. Structuur en Invloed van Collegiale Contacten Binnen Scholen.* Ph.d. thesis, University of Twente, The Netherlands.

Hilden-Minton, J.A. (1995) *Multilevel Diagnostics for Mixed and Hierarchical Linear Models.* Ph.d. dissertation, Department of Mathematics, University of California, Los Angeles.

Hill, P.W. and Goldstein, H. (1998) 'Multilevel modeling of educational data with cross-classification and missing identification for units'. *Journal of Educational and Behavioral Statistics*, 23, 117–128.

Hodges, J.S. (1998) 'Some algebra and geometry for hierarchical linear models, applied to diagnostics'. *Journal of the Royal Statistical Society, Ser. B*, 60, 497–536.

Holland, P.W. and Leinhardt, S. (1981) 'An exponential family of probability distributions for directed graphs' (with discussion). *Journal of the American Statistical Association*, 76, 33–65.

Hosmer, D.W. and Lemeshow, S. (1989) *Applied Logistic Regression.* New York: Wiley.

Hox, J.J. (1994) *Applied Multilevel Analysis.* Amsterdam: TT-Publikaties. Available in electronic form at http://www.ioe.ac.uk/multilevel/amaboek.pdf .

Hsiao, C. (1995) 'Panel analysis for metric data'. In: Arminger, Clogg, and Sobel (1995), pp. 361–400.

Hüttner, H.J.M. (1981) 'Contextuele analyse' (Contextual analysis). In: Albinski, M. (ed.), *Onderzoekstypen in de Sociologie*, p. 262–288. Assen: Van Gorcum.

Hüttner, H.J.M. and van den Eeden, P. (1995) *The Multilevel Design. A Guide with an Annotated Bibliography, 1980–1993.* Westport, Conn.: Greenwood Press.

International Social Survey Programme (1994) *Family and Changing Gender Roles II [computer-file].* Cologne: Zentral Archiv (Z.A. number 2620).

Kaplan, D. and Elliott, P.R. (1997) 'A didactic example of multilevel structural equation modeling applicable to the study of organizations'. *Structural Equation Modeling*, 4, 1–24.

Kasim, R.M. and Raudenbush, S.W. (1998) 'Application of Gibbs sampling to nested variance components models with heterogeneous within-group variance'. *Journal of Educational and Behavioral Statistics*, 23, 93–116.

Kelley, T.L. (1927) *The Interpretation of Educational Measurements.* New York: World Books.

Knapp, T.R. (1977) 'The unit-of-analysis problem in applications of simple correlation analysis to educational research'. *Journal of Educational Statistics*, 2, 171–186.

Kreft, I.G.G., De Leeuw, J., and Aiken, L. (1995) 'The effect of different forms of centering in hierarchical linear models'. *Multivariate Behavioral Research*, 30, 1–22.

Kreft, I.G.G. and De Leeuw, J. (1998) *Introducing Multilevel Modeling.* London: Sage Publications.

Kuk, A.Y.C. (1995) 'Asymptotically unbiased estimation in generalized linear models with random effects'. *Journal of the Royal Statistical Society*, Ser. B, 57, 395–407.

Laird, N.M. and Ware, J.H. (1982) 'Random-effects models for longitudinal data'. *Biometrics*, 38, 963–974.

Langford, I.H. and Lewis, T. (1998) 'Outliers in multilevel data'. *Journal of the Royal Statistical Society, Ser. A*, 161, 121–160.

Lesaffre, E. and Verbeke, G. (1998) 'Local influence in linear mixed models'. *Biometrics*, 54, 570–582.

Lin, X. (1999) 'Variance component testing in generalised linear models with random effects'. *Biometrika*, 84, 309–326.

Littell, R.C., Milliken, G.A., Stroup, W.W., and Wolfinger, R.D. (1996) *SAS System for Mixed Models* Cary, NC: SAS Institute.

Little, R.J.A. and Rubin, D.B. (1987) *Statistical Analysis with Missing Data.* New York: Wiley.

Long, J.S. (1997) *Regression Models for Categorical and Limited Dependent Variables.* Thousand Oaks, CA: Sage Publications.

Longford, N.T. (1987) 'A fast scoring algorithm for maximum likelihood estimation in unbalanced mixed models with nested random effects'. *Biometrika*, 74 (4), 812–827.

Longford, N.T. (1993a) *Random Coefficient Models.* New York: Oxford University Press.

Longford, N. T. (1993b) *VARCL: Software for Variance Component Analysis of Data with Nested Random Effects (maximum likelihood). Manual.* Groningen: ProGAMMA.

Longford, N.T. (1994) 'Logistic regression with random coefficients'. *Computational Statistics and Data Analysis*, 17, 1–15.

Longford, N.L. (1995) 'Random coefficient models'. In: Arminger, Clogg, and Sobel (1995), pp. 519–577.

Lord, F.M. and Novick, M.R. (1968) *Statistical Theory of Mental Test Scores.* Reading, Mass.: Addison-Wesley.

Maas, C. and Snijders, T.A.B. (1999) 'The multilevel approach to repeated measures with missing data'. Submitted for publication.

Maddala, G.S. (1971) 'The use of variance component models in pooling cross section and time series data'. *Ecnometrica*, 39, 341–358.

Mason, W.M. and Fienberg, S.E. (eds.) (1985) *Cohort Analysis in Social Research. Beyond the Identification Problem.* New York: Springer.

Mason, W.M., Wong, G.M., and Entwisle, B. (1983) 'Contextual analysis through the multilevel linear model'. In: Leinhardt, S. (ed.), *Sociological Methodology - 1983–1984* pp. 72–103. San Francisco: Jossey-Bass.

Maxwell, S.E. and Delaney, H.D. (1990) *Designing Experiments and Analyzing Data.* Belmont: Wadsworth, Inc.

McCullagh, P. and Nelder, J.A. (1989) *Generalized Linear Models.* 2nd edn. London: Chapman & Hall.

McCulloch, C.E. (1994) 'Maximum likelihood variance components estimation for binary data'. *Journal of the American Statistical Association*, 89, 330–335.

McCulloch, C.E. (1997) 'Maximum likelihood algorithms for generalized linear mixed models'. *Journal of the American Statistical Association*, 92, 162–170.

McKelvey, R.D. and Zavoina, W. (1975) 'A statistical model for the analysis of ordinal level dependent variables'. *Journal of Mathematical Sociology*, 4, 103–120.

Mealli, F. and Rampichini, C. (1999) 'Estimating binary multilevel models through indirect inference'. *Computational Statistics and Data Analysis*, 29, 313–324.

Meijer, E., Van Der Leeden, R., and Busing, F.M.T.A. (1995) 'Implementing the bootstrap for multilevel models'. *Multilevel Modeling Newsletter*, 7 (2), 7–11.

Miller, J.J. (1977) 'Asymptotic properties of maximum likelihood estimates in the mixed model of the analysis of variance'. *The Annals of Statistics*, 5, 746–762.

Moerbeek, M., van Breukelen, G.J.P., and Berger, M.P.F. (1997) 'A comparison of the mixed effect, the fixed effect, and the data aggregation model for multilevel data'. *Submitted for publication.*

Mok, M. (1995) 'Sample size requirements for 2-level designs in educational research'. *Multilevel Modeling Newsletter*, 7 (2), 11–15.

Mosteller, F. and Tukey, J.W. (1977) *Data Analysis and Regression.* Reading, Mass.: Addison-Wesley.

Muthén, B.O. (1994) 'Multilevel covariance structure analysis'. *Sociological Methods & Research*, 22, 376–398.

Muthén, L.K. and Muthén, B.O. (1998) *Mplus. The Comprehensive Modeling Program for Applied Researchers. User's Guide.* Los Angeles: Muthén and Muthén.

Neuhaus, J.M., Kalbfleisch, J.D., and Hauck, W.W. (1991) 'A comparison of cluster-specific and population-averaged approaches for analyzing correlated binary data'. *International Statistical Review*, 59, 25–35.

Neuhaus, J.M. (1992) 'Statistical methods for longitudinal and cluster designs with binary responses'. *Statistical Methods in Medical Research*, 1, 249–273.

Ochi, Y. and Prentice, R.L. (1984) 'Likelihood inference in a correlated probit regression model'. *Biometrika*, 71, 531–543.

Opdenakker, M.C. and Van Damme, J. (1997) 'Centreren in multi-level analyse: implicaties van twee centreringsmethoden voor het bestuderen van schooleffectiviteit'. *Tijdschrift voor Onderwijsresearch*, 22, 264–290.

Pedhazur, E.J. (1982) *Multiple Regression in Behavioral Research.* 2nd edn. New York: Holt, Rinehart and Winston.

Pregibon, D. (1981) 'Logistic Regression Diagnostics'. *The Annals of Statistics*, 9, 705–724.

Press, S.J. (1989) *Bayesian Statistics.* New York: Wiley.

Rasbash, J. and Goldstein, H. (1994) 'Efficient analysis of mixed hierarchical and cross-classified random structures using a multilevel model.' *Journal of Educational and Behavioral Statistics*, 19, 337–350.

Rasbash, J. and Woodhouse, G. (1995) *MLn: Command Reference.* London: Multilevel Models Project, Institute of Education, University of London.

Raudenbush, S.W. (1993) 'A crossed random effects model for unbalanced data with applications in cross sectional and longitudinal research'. *Journal of Educational Statistics*, 18, 321–349.

Raudenbush, S.W. (1995) 'Hierarchical linear models to study the effects of social context on development'. In: Gottman, J.M. (ed.), *The Analysis of Change*, pp. 165–201. Mahwah, N.J.: Lawrence Erlbaum Ass.

Raudenbush, S.W. (1997) 'Statistical analysis and optimal design for cluster randomized trials'. *Psychological Methods*, 2, 173–185.

Raudenbush, S.W. and Bryk, A.S. (1986) 'A hierarchical model for studying school effects'. *Sociology of Education*, 59, 1–17.

Raudenbush, S.W. and Bryk, A.S. (1987) 'Examining correlates of diversity'. *Journal of Educational Statistics*, 12, 241–269.

Raudenbush, S.W. and Chan, W.-S. (1994) 'Application of a hierarchical linear model to the study of adolescent deviance in an overlapping cohort design'. *Journal of Consulting and Clinical Psychology*, 61, 941–951.

Raudenbush, S.W. and Sampson, R. (1999) 'Assessing direct and indirect effects in multilevel designs with latent variables'. *Sociological Methods and Research*, to appear.

Raudenbush, S.W., Yang, M.-L., and Yosef, M. (1999) 'Maximum likelihood for generalized linear models with nested random effects via high-order, multivariate Laplace approximation'. Manuscript under review.

Rekers-Mombarg, L.T.M., Cole, L.T., Massa, G.G., and Wit, J.M. (1997) 'Longitudinal analysis of growth in children with idiopathic short stature.' *Annals of Human Biology*, 24, 569–583.

Richardson, A.M. (1997) 'Bounded influence estimation in the mixed linear model'. *Journal of the American Statistical Association*, 92, 154–161.

Robinson, W.S. (1950) 'Ecological correlations and the behavior of individuals'. *American Sociological Review*, 15, 351–357.

Rodríguez, G. and Goldman, N. (1995) 'An assessment of estimation procedures for multilevel models with binary responses'. *Journal of the Royal Statistical Society*, Ser. A, 158, 73–89.

Rosenthal, R. (1991) *Meta-analytic Procedures for Social Research (rev. edn.).* Newbury Park, CA: Sage Publications.

Rubin, D.R. (1996) Multiple imputation after 18+ years. *Journal of the American Statistical Association*, 91, 473–489.

Ryan, T.P. (1997) *Modern Regression Methods.* New York: Wiley.

Särndal, C.-E., Swensson, B., and Wretman, J. (1991) *Model-assisted Survey Sampling.* New York: Springer.

Searle, S.R. (1956) 'Matrix methods in components of variance and covariance analysis'. *Annals of Mathematical Statistics*, 29, 167–178.

Searle, S.R., Casella, G., and McCulloch, C.E. (1992) *Variance Components*. New York: Wiley.

Seber, G.A.F. and Wild, C.J. (1989) *Nonlinear Regression*. New York: Wiley.

Self, G.S. and Liang, K.-Y. (1987) 'Asymptotic properties of maximum likelihood estimators and likelihood ratio tests under nonstandard conditions'. *Journal of the American Statistical Association*, 82, 605–610.

Seltzer, M.H. (1993) 'Sensitivity analysis for fixed effects in the hierarchical model: A Gibbs sampling approach'. *Journal of Educational Statistics*, 18, 207–235.

Seltzer, M.H., Wong, W.H.H., and Bryk A.S. (1996) 'Bayesian analysis in applications of hierarchical models: Issues and Methods'. *Journal of Educational Statistics*, 21, 131–167.

Shavelson, R.J. and Webb, N.M. (1991) *Generalizability Theory. A Primer*. Newbury Park, CA: Sage Publications.

Singer, J.D. (1998a) 'Fitting multilevel models using SAS PROC MIXED.' *Multilevel Modeling Newsletter*, 10 (2), 5–8.

Singer, J.D. (1998b) 'Using SAS PROC MIXED to fit multilevel models, hierarchical models, and individual growth models.' *Journal of Educational and Behavioral Statistics*, 23, 323–355.

Snijders, T.A.B. (1996) 'Analysis of longitudinal data using the hierarchical linear model'. *Quality & Quantity*, 30, 405–426.

Snijders, T.A.B. and Bosker, R.J. (1993) 'Standard errors and sample sizes for two-level research'. *Journal of Educational Statistics*, 18, 237–259.

Snijders, T.A.B. and Bosker, R.J. (1994) 'Modeled variance in two-level models'. *Sociological Methods & Research*, 22, 342–363.

Snijders, T.A.B. and Kenny, D.A. (1999) 'Multilevel models for relational data'. *Personal Relationships* (in press).

Snijders, T.A.B. and Maas, C.J.M. (1996) 'Using *MLn* for repeated measures with missing data'. *Multilevel Modelling Newsletter*, 8 (2), 7–10.

Snijders, T.A.B., Spreen, M., and Zwaagstra, R. (1994) 'The use of multilevel modeling for analysing personal networks: networks of cocaine users in an urban area'. *Journal of Quantitative Anthropology*, 4, 85–105.

SPSS Inc. (1997) *SPSS Advanced Statistics 7.5*. Chicago, Ill: SPPS Inc.

StataCorp (1997) *Stata Statistical Software: Release 5.0*. College Station, TX: Stata Corporation.

Stevens, J. (1996) *Applied Multivariate Statistics for the Social Sciences*. Mahwah, N.J.: Lawrence Erlbaum Associates.

Stiratelli, R., Laird, N.M., and Ware, J.H. (1984) 'Random-effects models for serial observations with binary response'. *Biometrics*, 40, 961–971.

Swamy, P.A.V.B. (1971) *Statistical Inference in Random Coefficient Regression Models*. New York: Springer.

Swamy, P.A.V.B. (1973) 'Criteria, constraints, and multi-collinearity in random coefficient regression models'. *Annals of Economic and Social Measurement*, 2, 429–450.

Tacq, J. (1986) *Van Multiniveau Probleem naar Multiveau Analyse*. Rotterdam: Department of Research Methods and Techniques, Erasmus University.

Van Der Werf, G.Th., Smith, R.J.A., Stewart, R.E., and Meyboom - de Jong, B. (1998) *Spiegel op de Huisarts. Over Registratie van Ziekte, Medicatie en Verwijzingen in de Geautomatiseerde Huisartspraktijk*. Groningen: Department of General Medical Practice, University of Groningen.

Van Duijn, M.A.J., Van Busschbach, J.T., and Snijders, T.A.B. (1999) 'Multilevel analysis of personal networks as dependent variables'. *Social Networks*, in press.

Veall, M.R. and Zimmermann, K.F. (1992) 'Pseudo-R^2's in the ordinal probit model'. *Journal of Mathematical Sociology*, 16, 333–342.

Verbeke, G. and Lesaffre, E. (1997) 'The effect of misspecifying the random-effects distribution in linear mixed models for longitudinal data'. *Computational Statistics and Data Analysis*, 23, 541–556.

Verbeke, G. and Molenberghs, G. (1997) *Linear Mixed Models in Practice. A SAS-oriented Approach.* Lecture Notes in Statistics, 126. New York: Springer.

Vermeulen, C.J. and Bosker, R.J. (1992) *De Omvang en Gevolgen van Deeltijdarbeid en Volledige Inzetbaarheid in het Basisonderwijs.* Enschede: University of Twente.

Völker, B.G.M. (1995) *Should Auld Acquaintance Be Forgot? Institutions of Communism, the Transition to Capitalism and Personal Networks: The Case of East Germany.* Amsterdam: Thesis Publishers.

Völker, B. and Flap, H. (1997) 'The comrades' belief: Intended and unintended consequences of communism for neighbourhood relations in the former GDR'. *European Sociological Review*, 13, 241–265.

Wasserman, S. and Faust, K. (1994) *Social Network Analysis: Methods and Applications.* New York and Cambridge: Cambridge University Press.

Waternaux, C., Laird, N.M., and Ware, J.H. (1989) 'Methods for analysis of longitudinal data: Blood lead concentrations and cognitive development.' *Journal of the American Statistical Association*, 84, 33–41.

Windmeijer, F.A.G. (1995) 'Goodness-of-fit measures in binary choice models'. *Econometric Reviews*, 14, 101–116.

Wittek, R. and Wielers, R. (1998) 'Gossip in organizations'. *Computational and Mathematical Organization Theory*, 4, 189–204.

Wong, G.Y. and Mason, W.M. (1985) 'The hierarchical logistic regression model for multilevel analysis'. *Journal of the American Statistical Association*, 80, 513–524.

Yang, M., Rasbash, J., and Goldstein, H. (1998) *MLwiN Macros for Advanced Multilevel modelling.* London: Multilevel Models Project, Institute of Education, University of London.

Zeger, S.L. and Karim, M.R. (1991) 'Generalized linear models with random effects: a Gibbs sampling approach'. *Journal of the American Statistical Association*, 86, 79–86.

Index